LIGAND FIELD THEORY AND ITS APPLICATIONS

LIGAND FIELD THEORY AND ITS APPLICATIONS

BRIAN N. FIGGIS
University of Western Australia

MICHAEL A. HITCHMAN
University of Tasmania

WILEY-VCH
New York / Chichester / Weinheim / Brisbane / Singapore / Toronto

This book is printed on acid-free paper. ⊗

Copyright © 2000 by Wiley-VCH. All rights reserved.

Published simultaneously in Canada.

No part of this publication may be reproduced, stored in a retrieval system or transmitted in any form or by any means, electronic, mechanical, photocopying, recording, scanning or otherwise, except as permitted under Sections 107 or 108 of the 1976 United States Copyright Act, without either the prior written permission of the Publisher, or authorization through payment of the appropriate per-copy fee to the Copyright Clearance Center, 222 Rosewood Drive, Danvers, MA 01923, (978) 750-8400, fax (978) 750-4744. Requests to the Publisher for permission should be addressed to the Permissions Department, John Wiley & Sons, Inc., 605 Third Avenue, New York, NY 10158-0012, (212) 850-6011, fax (212) 850-6008, E-Mail: PERMREQ @ WILEY.COM.

For ordering and customer service, call 1-800-CALL-WILEY.

Library of Congress Cataloguing in Publication Data:

Figgis, Brian N.
    Ligand field theory and its applications   /   Brian N. Figgis,
Michael A. Hitchman.
        p.   cm.
    Includes index.
    ISBN 0-471-31776-4 (cloth: alk. paper)
    1. Ligand field theory.   I. Hitchman, Michael A.   II. Title.
QD475.F54 2000   99-28986
541.2'242–dc21

Printed in the United States of America.

10 9 8 7 6 5 4 3 2 1

To our wives, Jane and Annette, and children, Benjamin, Honor, Martin, Peter, and Ruth, who said it would never get done!

SPECIAL TOPICS IN INORGANIC CHEMISTRY
INTRODUCTION TO THE SERIES

This text represents the first in a series of one-volume introductions to major areas of inorganic chemistry written by leaders in the field. Inorganic chemistry covers a variety of diverse substances including molecular, coordination, organometallic, and nonmolecular compounds as well as special materials such as metallobiomolecules, semiconductors, ceramics, and minerals. The great structural diversity of inorganic compounds makes them vitally important as industrial feedstocks, fine chemicals, catalysts, and advanced materials. Inorganic compounds such as metalloenzymes also play a key role in life processes. This series will provide valuable, concise graduate texts for use in survey courses covering diverse areas of inorganic chemistry.

Department of Chemistry　　　　　　　　　　　　R. BRUCE KING, Series Editor
University of Georgia
Athens, Georgia

CONTENTS

3   THE ANGULAR OVERLAP MODEL   53

PREFACE

In the 30 years since the publication of *Introduction to Ligand Fields* the subject has undergone considerable change in its conceptual basis and in some aspects of its applications, while retaining its unrivaled usefulness despite advances in theoretical chemistry as applied to transition metal complexes. The origins of ligand field theory's basic premise, i.e., the "splitting" of the d orbitals, is now recognized to be largely in chemical bonding effects, including covalency, rather than the purely electrostatic perturbations produced by the charges of surrounding ligands. A new parameterization scheme that is chemically more transparent has been made available through the introduction of the angular overlap model. Although rigorous ab-initio and density functional calculations can now be carried out successfully on transition metal complexes, in practice they have not significantly reduced the need for ligand field treatments for interpreting, for example, experimental spectral and magnetic properties.

The subject has also matured. At the time of *Introduction to Ligand Fields*, ligand field theory was a subject in its own right, with the major emphasis being on the nature of the metal–ligand interaction and how it influenced experimental observations. Those investigations are now largely complete: It is understood how, why, and under what conditions ligand field theory can be successfully applied. Ligand field theory has become a valuable tool in the armories of inorganic chemistry and chemical physics and is widely used to interpret the properties of transition-metal ions in biologically important molecules. As such, there is a place for an updated version of *Introduction to Ligand Fields*, but one that emphasizes a different relationship between theory and application. This volume, under an appropriately modified title, addresses that need.

The chapters "Group Theory" and "Molecular Orbital Theory" have been deleted because these topics are now adequately covered in standard undergraduate textbooks. Two new chapters—"The Angular Overlap Model" and "The Origin and Calculation of Δ"—have been introduced. The chapters on the influence of weak, medium, and strong crystal fields have been merged and that on complexes of non-cubic stereochemistry has been subsumed into other parts of the book. In response to the changes in emphasis that have occurred, the chapter "Thermodynamic Aspects of Crystal Fields" has been replaced with "Influence of the d Configuration on the Geometry and Stability of Complexes." Finally, the chapters on the electronic spectra, magnetochemistry, and electronic paramagnetic resonance of transition metal complexes have been considerably expanded and updated, so that the theory and the applications of ligand field theory receive approximately equal weighting.

A number of people have read through all or part of the book, and we are grateful for their very valuable comments and suggestions. Particular thanks are owed to David Goodgame, David Johnson, Frank Mabbs, Dirk Reinen, and Claus Schäffer as well as to the editor of this series, Bruce King. Thanks are also due to Tim Astley and Horst Stratemeier for help in preparing the figures.

University of Western Australia   BRIAN N. FIGGIS

University of Tasmania   MICHAEL A. HITCHMAN

LIGAND FIELD THEORY AND ITS APPLICATIONS

CHAPTER 1

INTRODUCTION

1.1   THE CONCEPT OF A LIGAND FIELD

The basic notion of a ligand field was first developed by Bethe (1) in 1929 as an application of the newly developed quantum mechanics. In the following few years, the implications of his work for the spectral and magnetic properties of transition-metal complexes were recognized by Penney and Schlapp (2) and van Vleck (3), among others. Bethe undertook the study of the effect of the surrounding ions on the electron distribution within any one of the ions in a NaCl-type lattice. He supposed that the ions were undeformable spheres and the interactions that took place between them were solely the result of the electrostatic potentials set up by their charges. The charges were taken to be located at the centers of the ions. In the specific example of NaCl, the sodium ion was considered to be surrounded by six negative point charges located at the vertices of a regular octahedron. Bethe also took into account ions that were not immediate neighbors of the ion under consideration. It has transpired, however, that little is to be gained by dealing with the influence exerted by any but the immediate neighbors. Each of the six negative charges creates an electrostatic potential:

$$\mathbf{v}_{(i;x,y,z)} = e/r_{(i;x,y,z)} \tag{1.1}$$

at the point (x,y,z), where $\mathbf{v}_i$ is the potential caused by the ith of the six ions, and $r_{(i;x,y,z)}$ is the distance from the ith ion to the point (x,y,z).

  Let the origin of the set of Cartesian axes be located on the cation under consideration. The problem resolves itself into summing the potentials from the individual surrounding ions to give the total potential at any point near the central ion and then finding the effect of such a potential on the electrons of that ion. We have:

$$\mathbf{V}_{(x,y,z)} = \sum_{i=1}^{6} \mathbf{V}_{(i;\,x,y,z)} \qquad (1.2)$$

The process of performing this summation and examining the effect of $\mathbf{V}_{(x,y,z)}$ on the electrons of various central ions is a matter of some mathematical complexity, and its outline occupies much of the first chapters of this book. Before launching into the details of the process, we make some general remarks about the model employed as the starting point.

The facts that the original considerations were made on the potential developed near an ion that was part of a lattice and that such a lattice could occur only in a crystal led to the term *crystal field theory* to cover the subject. It has been found, however, that few of the results of crystal field theory depend on the existence of a lattice. We can therefore carry most of them over to the model that is at the basis of coordination chemistry: that of the *coordination cluster*. By a coordination cluster, of course, we mean a central metal ion associated with a number of attached ligands, the whole forming a distinguishable entity and possibly bearing a net electrical charge. The coordination clusters themselves pack into a lattice in the solid state. This packing is not considered of primary importance, because the principal contribution to the potential at the central metal ion comes from the atoms closest to it—the donor atoms.

In the more general model of a coordination compound, the electrons of the central ion are subject to a potential, not necessarily of simple electrostatic origin, from the ligand atoms. The term *ligand field theory* has been employed to cover all aspects of the manner in which the physical properties an ion or atom is influenced by its nearest neighbors. Ligand field theory, then, contains crystal field theory as a limiting special case.

Defined in the above manner, ligand field theory seems to include all theories of chemical bonding between an atom and its neighbors. Indeed, some authors take such an approach. They include in ligand field considerations subjects such as the strength of the metal–ligand bond and coordination number. Such efforts, however, meet with comparatively little success. We restrict our treatment of the ligand field effects to the results caused by the interactions of ligand atoms with d and f electrons. We do not inquire how the ligands got there, how firmly they are held, nor, too deeply, how the bonds came to possess their particular character.

A great many of the results of ligand field theory depend largely on the approximate symmetry of the ligand distribution around the central metal ion rather than on the particular ligands or the details of their locations. The term *approximate symmetry* needs further clarification. We use the nomenclature *octahedral* and *tetrahedral*, for example, to refer to a coordination cluster in which there are six ligands located near the vertices of a regular octahedron or four ligands near those of a regular tetrahedron, respectively, whether or not all the ligands are identical. For our purposes, $[Co(NH_3)_4Br_2]^+$ is an octahedral complex ion. Results of the type under consideration do not depend critically on the model chosen to describe the bonding between the central ion and the ligand atoms. Consequently, they may be obtained in the first place from crystal field theory and then generalized to other models of the bonding. Once the mathematical formalism of the crystal field approach has been developed, calculations using it are quite straightforward.

It is convenient, then, first to treat as many effects of ligand fields as possible by the crystal field method; later in the book we consider the effects that require the more general ligand field. It rapidly becomes evident that the success of ligand field theory in accounting for the principal features of the spectral, magnetic, and some thermodynamic and structural properties of coordination compounds is considerable. At first sight this seems remarkable because the effect of the electrostatic field produced by the ligands on the relative energies of the various d orbitals is too small, by at least an order of magnitude, to explain the observations. Moreover, as far as the energetics of the interaction are concerned, the refinement of the crystal field model to more realistic descriptions of the nature of the metal–ligand bond does little to improve on the main features of the results until high-quality *ab initio* theoretical calculations are reached. These last two points are elaborated in Chapter 4. It must be emphasized, however, that ligand field theory does not provide a method of calculating from first principles the energy levels in coordination complexes in absolute terms. The power of this model lies rather in its ability to yield the energies of the d orbitals in a complex, relative to an arbitrary zero, in terms of simple parameters, the values of which are always derived from experiment. The nature of the interaction between the ligands and the d orbitals is, therefore, relatively unimportant as far as the form of the energy splittings is concerned. As was recognized at the inception of crystal field theory, the relative disposition of the ligands about the metal ion—the symmetry of the complex—is the crucial factor that decides the pattern of the d orbital splitting that occurs upon the formation of a complex. The electrostatic crystal field theory as set out in the first part of this section provides a simple and historically important method of illustrating the way in which the d orbitals are affected by a set of ligands. The refinements following from the recognition that the nature of the metal ligand interaction is largely covalent (rather than electrostatic) are absolutely essential if some of the secondary features are to be accounted for and are considered later.

The progression from treating the ligand atoms as point charges through point dipoles to a simple molecular-orbital model still leaves the quantitative calculation of the ligand field effect on an unsatisfactory basis. Nevertheless, the last mentioned model makes it possible to rationalize the decrease of the central ion d electron repulsion and spin–orbit coupling parameters below the values known for the free ion, as well as phenomena related to the delocalization of electrons over both the central ion and the ligand atoms. A sophisticated molecular orbital treatment can give a quantitative account of all the aspects of the ligand field phenomenon. Its success, however, requires consideration of essentially all the perturbations acting on the electrons involved in the chemical bonding. Such a treatment lies outside the scope of this book except as mentioned in Chapter 4.

1.2 THE SCOPE OF LIGAND FIELD THEORY

Within the limitations stated in Section 1.1, the subject of ligand field theory consists of the correlation of the physical properties of coordination compounds with the nature and the positions of the ligand atoms that surround the central ion. So that ligand field theory may be used to interpret some particular physical property, it is necessary that a change in the ligand arrangement causes an alteration in the property that is large compared with changes caused by other factors that influence the property. Similarly, for any particular central metal ion to be treated by ligand field theory, it must show a marked dependence of at least one of its physical properties on the nature of its coordination environment; i.e., a dependence that is large compared to the dependence of the property on other changes.

The number of common physical properties that reflect changes in the ligand environment is somewhat limited. We somewhat arbitrarily divide them into three categories: thermochemical and structural, spectral, and magnetic properties. In the thermochemical–structural group are included subjects like the relative bond energies, stability-constant data, and distortions from a regular geometry, properties that depend directly on the ground state energetics of the coordination cluster system. As pointed out earlier, we neglect the absolute values of bond strengths and most stereochemistry considerations that, logically, could also come into this category. Properly, the spectral group of properties should include a treatment of vibrational motions of the coordination cluster and all electronic transitions that are not solely concerned with internal rearrangement of electrons on the ligands. We confine ourselves, however, largely to the so-called *ligand field* spectra that are concerned with electronic transitions within orbitals located essentially on the central ion. This restriction is imposed not because the remaining transitions of the *charge transfer* type are unimportant but rather because they are as yet comparatively ill understood. Similarly, it may be argued that the study of magnetic properties could be extended to include magnetically concentrated systems in which there are cooperative effects, such as ferrites, garnets, etc. Attention here is restricted to magnetically dilute systems, but includes metal cluster compounds in which there is magnetic exchange, because there are considerable complexities involved in going beyond these.

The elements that give rise to the central ions of the coordination clusters that are the concern of ligand field theory are essentially those of the three d-transition series. It is possible to detect small ligand field effects in the spectral and magnetic properties of the lanthanide element ions. However, there it is a case of rather minor perturbations from ligand field effects overwhelmed by the much more important effects of electron repulsions and spin–orbit coupling. Ligand fields can by no means be said to play an important role in the physical properties of that series of ions. The fact that ligand field perturbations are two orders of magnitude greater for the d electrons of the transition series ions than for the f electrons of the lanthanide ions is associated with the shielding produced by the s and p electrons that lie outside the f and go far to protect them from the ligand environment. No such protection is available for the d electrons of the transition-series ions. In the actinide series ions, it appears that ligand field effects are of an importance intermediate between those for transition-element ions and those for lanthanide element ions, although the position is not entirely clear. A brief survey of some applications of ligand field theory to actinide element chemistry is included in Chapter 11. Coordination compounds of ions with empty or filled d or f shells do not show discernible ligand field effects.

It is useful to recognize the relative importance of the terms in the Hamiltonian for different systems:

$$\text{First- and second-transition series}: \quad \mathbf{H}_{LF} \approx \mathbf{H}_{ER} > \mathbf{H}_{LS} \tag{1.3}$$
$$\text{Third transition series}: \quad \mathbf{H}_{LF} \approx \mathbf{H}_{ER} \approx \mathbf{H}_{LS} \tag{1.4}$$
$$\text{Lanthanides}: \quad \mathbf{H}_{ER} \gtrsim \mathbf{H}_{LS} > \mathbf{H}_{LF} \tag{1.5}$$
$$\text{Actinides}: \quad \mathbf{H}_{ER} \gtrsim \mathbf{H}_{LS} \approx \mathbf{H}_{LF} \tag{1.6}$$

$\mathbf{H}_{LF}$ takes into account the presence of the surrounding ligand molecules, whether treated as point charges, as in the crystal field model, or as at least partly covalently bound entities, as in ligand field methods. It will be discussed first in some detail, because it forms the whole reason for the presentation of this chapter. $\mathbf{H}_{ER}$ takes into account the repulsions within the d or f electron set, including the application of the constraints of the Pauli exclusion principle and electron spin. It consists of two components — the Coulomb repulsion and the exchange terms. This subject will be introduced only to the extent necessary for the application of the $\mathbf{H}_{LF}$ term to relevant physical properties (Chapter 6). $\mathbf{H}_{LS}$ takes into account the formal coupling between the spin and the angular momenta of the d or f electrons. In fact, it arises from the same basis as $\mathbf{H}_{ER}$ and is treated along with that term for the same purposes (Chapter 5).

1.3 THE d AND OTHER ORBITALS

At the heart of crystal field theory, and of ligand field theory in general, lie the properties of the set of five d orbitals. The following discussion summarizes

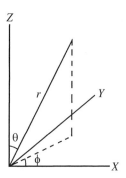

Figure 1.1. Spherical polar coordinates in relationship to Cartesian coordinates.

these properties and the properties of the hydrogen-like wavefunctions in general (4).

The hydrogen-like wavefunctions of a free ion (by *a free ion* is meant an ion not subject to any external influences) are eigenfunctions of the Hamiltonian operator:

$$\mathbf{H}_0 = -(h^2/8\pi^2 m)\nabla^2 - Z_{eff}\, e^2/r \tag{1.7}$$

where Z_{eff} is the effective nuclear charge. Referred to a set of spherical polar coordinates (Fig. 1.1), the wavefunctions are given by:

$$\Psi_{nlm} = \mathbf{R}_{nl}\mathbf{Y}_l^m \tag{1.8}$$

where $\mathbf{R}_{nl}$ is the radial and $\mathbf{Y}_l^m$ the angular component of the wavefunction. Here, n is the principal quantum number, l the quantum number specifying the orbital angular momentum of the electron, and m the quantum number specifying the component of this angular momentum in the z direction. For the present, electron spin is not considered. n is a positive integer, l takes on integral values from 0 to $(n-1)$, and m takes on integral values from $-l$ to l. In Figure 1.1, it may help to recognize that in referrence to the surface of the earth, θ corresponds to latitude, with zero at the North Pole, and ϕ to longitude, with zero at Greenwich.

The element of volume $d\tau$ is given by:

$$d\tau = r^2 \cdot \sin\theta \cdot d\theta \cdot d\phi \tag{1.9}$$

The function $r^2 \mathbf{R}_{nl}^2$, with $R_{nl,}$ taken to be real (as is always the case) describes the time-averaged probability of finding an electron in the specified orbital at a distance r from the nucleus. The factor r^2 occurs here simply because the volume of a shell of thickness dr is proportional to r^2. For reasons that are given later, it is of little use to employ radial distribution functions taken from free ion data in actual ligand field calculations. Thus we do not consider the details of these functions further than to note that, by definition, the relationships in Eq. 1.10 hold; i.e., the entire set of hydrogen-like wavefunctions form an orthonormal set:

$$\int R_{nl}R_{n'l'}Y_l^m Y_{l'}^{m'} d\tau = \delta(n, n'; l, l'; m, m') \tag{1.10}$$

This equality means that the integral is zero unless $n = n'; l = l'$ and $m = m'$. Some properties of the spherical harmonics Y_l^m are outlined in Appendix 1, and the functions up to order six are listed in Appendix 2.

Rather than talking about electrons described by wavefunctions, it is conventional to say that electrons occupy orbitals. These orbitals are specified by the foregoing angular parts of the wavefunctions and can be represented pictorially. Such diagrams are given in Figure 1.2 as electron densities corresponding to the s, p, d, and f orbitals. The plots show regions of space in which there is a high probability of finding the electron and have axial symmetry about the z axis. Considering the wavelike properties of the electron, the amplitude may have a positive ($+$) or negative ($-$) phase. The functions Y_l^0 are essentially concentrated along the z axis, the pairs $Y_l^{\pm m}$ form rings around the z axis, which look identical. They differ only in that one member of the pair may be regarded as corresponding to clockwise rotation of the electron cloud about the z axis and the other member to couter-clockwise rotation. The pair with $|m| = l$ is always located with maxima in the xy plane.

Actual calculations are usually performed with these solutions to the Schrödinger equation. On the other hand, it is often more convenient, for the purposes of visualizing the relationships among the different orbitals, to express them in alternative forms that are explicit expressions of the Cartesian coordinates. In these alternative forms, the wavefunctions for the orbitals do not contain $e^{im\phi}$ and consequently may be said to be the real orbitals. The real orbitals are simple linear combinations of those we gave above for cylindrical polar coordinates; the combinations are taken to relocate i. To combine the imaginary forms of the orbitals into the real forms we note that (de Moivre's theorem):

$$e^{\pm im\phi} = \cos(m\phi) \pm i\sin(m\phi) \tag{1.11}$$

and that

$$x = r \cdot \sin\theta \cdot \cos\phi, \ y = r \cdot \sin\theta \cdot \sin\phi, \ z = r \cdot \cos\theta \tag{1.12}$$

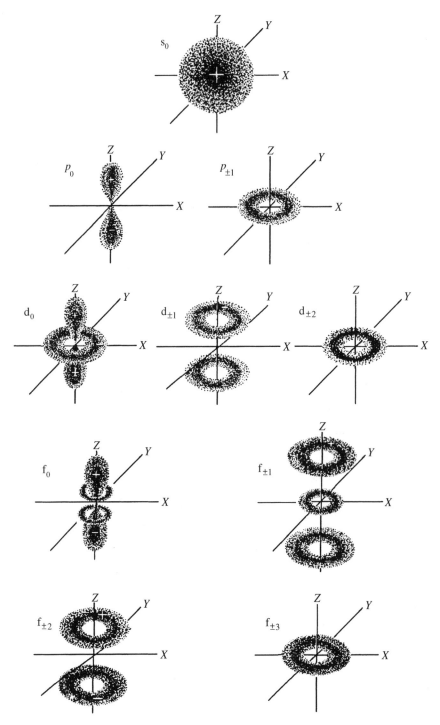

Figure 1.2. The angular dependence of the electron density distribution corresponding to some of the s, p, d, and f functions that are solutions of the Schrödinger equation.

We take, for example, the combination

$$(2) + (-2) \tag{1.13}$$

where (m) is an abbreviation for $Y_{n,2}^m$, that is for a d wavefunction, so that (2) stands for the wavefunction $R_{n2}Y_2^2$.

$$(2) = R_{n2} \cdot (15/16)^{1/2} \cdot (2\pi)^{-1/2} \cdot \sin^2\theta \cdot (\cos 2\phi + i\sin 2\phi) \tag{1.14}$$

$$(-2) = R_{n2} \cdot (15/16)^{1/2} \cdot (2\pi)^{-1/2} \cdot \sin^2\theta \cdot (\cos 2\phi - i\sin 2\phi) \tag{1.15}$$

$$(2) + (-2) = 2R_{n2} \cdot (15/16)^{1/2} \cdot (2\pi)^{-1/2} \cdot \sin^2\theta \cdot \cos 2\phi$$

$$= 2R_{n2} \cdot (15/16)^{1/2} \cdot (2\pi)^{-1/2} \cdot \sin^2\theta \cdot (\cos^2\phi - \sin^2\phi)$$

$$= 2R_{n2} \cdot (15/16)^{1/2} \cdot (2\pi)^{-1/2} \cdot (x^2 - y^2)/r^2$$

$$\propto (x^2 - y^2) \tag{1.16}$$

In the last step, all the factors other than functions of the Cartesian coordinates are omitted. Finally, it is necessary to make sure that the new orbital is normalized. To normalize a wavefunction Ψ we form the new wavefunction:

$$\Psi / \left(\int \Psi^* \Psi \, d\tau \right)^{1/2} \tag{1.17}$$

Now,

$$\int [(2) + (-2)]^* [(2) + (-2)] d\tau$$

$$= \int [(2)^*(2) + (2)^*(-2) + (-2)^*(2) + (-2)^*(-2)] d\tau$$

$$= 1 + 0 + 0 + 1$$

$$= 2 \tag{1.18}$$

Hence the *real* wavefunction, which is called $d_{x^2-y^2}$, is:

$$d_{x^2-y^2} = [(2) + (-2)]/\left\{ \int [(2) + (-2)]^* [(2) + (-2)] d\tau \right\}^{1/2}$$

$$= 2^{-1/2}[(2) + (-2)] \tag{1.19}$$

By making other suitable combinations the five d wavefunctions in their *real* forms are obtained as:

$$d_{z^2} = (0)(d_{z^2} \text{ is really } d_{(z^2 - r^2/3)}) \tag{1.20}$$

$$d_{yz} = -i2^{-1/2}[(1) + (-1)] \tag{1.21}$$

$$d_{xz} = 2^{-1/2}[(1) - (-1)] \tag{1.22}$$

$$d_{xy} = -i2^{-1/2}[(2) - (-2)] \tag{1.23}$$

$$d_{x^2-y^2} = 2^{-1/2}[(2) + (-2)] \tag{1.24}$$

Exactly the same procedure may be adopted with p and f orbitals (the s orbital is real in form anyway). For example, the p orbitals, with p_m the p wavefunction specified by m, are:

$$p_z = p_0 \tag{1.25}$$

$$p_y = -i2^{-1/2}(p_1 - p_{-1}) \tag{1.26}$$

$$p_x = 2^{-1/2}(p_1 + p_{-1}) \tag{1.27}$$

In Figure 1.3 the *real* forms of some of the p, d, and f orbitals are illustrated. The signs on different lobes of an orbital in these diagrams indicate the relative phases of the wavefunctions in that part of space. Of course, the probability density, from the square of the wavefunctions, is always positive. The shapes of the f orbitals are discussed more fully by King (5).

The orthonormal properties of the real d wavefunctions are readily demonstrated. For example,

$$\int d_{yz}^* d_{yz} d\tau = \int 2^{-1}[(1) - (-1)] \cdot [(1) - (-1)] d\tau$$

$$= \frac{1}{2} \int [(1) \cdot (1) - (1) \cdot (-1) - (-1) \cdot (1) + (-1) \cdot (-1)] d\tau$$

$$= \frac{1}{2} [1 - 0 - 0 + 1]$$

$$= 1 \tag{1.28}$$

and

$$\int d_{xy}^* d_{x^2-y^2} d\tau = i \int 2^{-1}[(2) - (-2)] \cdot [(2) + (-2)] d\tau$$

$$= \frac{i}{2} \int [(2) \cdot (2) + (2) \cdot (-2) - (-2) \cdot (2) - (-2) \cdot (-2)] d\tau$$

$$= \frac{i}{2} [1 + 0 - 0 - 1]$$

$$= 0 \tag{1.29}$$

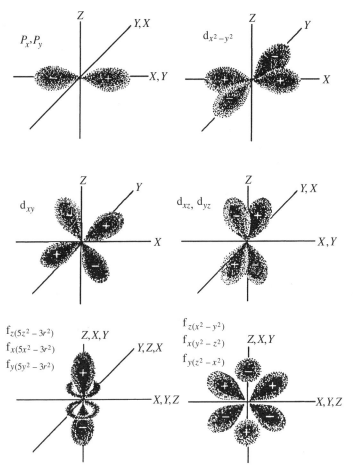

Figure 1.3. The angular dependence of some functions representing real forms of s, p, d, and f orbitals.

The difference in behavior under the operator for the z component of orbital angular momentum (l_z) exhibited by the d wavefunctions in the two forms is important and is a result that is required later. To find the observable value of orbital angular momentum that corresponds to a wavefunction ψ, we have to evaluate the quantity (6):

$$\int \psi^* l_z \psi \, d\tau \tag{1.30}$$

The actual form of l_z is:

$$l_z = (h/2\pi)(\partial/\partial\phi) \tag{1.31}$$

Thus

$$l_z(m) = m(\mathrm{h}/2\pi)(m) \tag{1.32}$$

For example:

$$l_z(2) = 2(\mathrm{h}/2\pi)(2) \tag{1.33}$$

We have, then:

$$
\begin{aligned}
\int (m)^* l_z(m) \mathrm{d}\tau &= \int (m)^*(\mathrm{h}/2\pi) m(m) \mathrm{d}\tau \\
&= (\mathrm{h}/2\pi) \int (m)^*(m) \mathrm{d}\tau \\
&= (\mathrm{h}/2\pi) m
\end{aligned} \tag{1.34}
$$

and

$$
\begin{aligned}
\int (m)^* l_z(m') \mathrm{d}\tau &= \int (m)^*(\mathrm{h}/2\pi) m'(m') \mathrm{d}\tau \\
&= 0
\end{aligned} \tag{1.35}
$$

Specifically,

$$\int (\pm 2)^* l_z(\pm 2) \mathrm{d}\tau = \pm 2(\mathrm{h}/2\pi) \tag{1.36}$$

$$\int (\pm 2)^* l_z(\mp 2) \mathrm{d}\tau = 0 \tag{1.37}$$

which show that the wavefunctions (2) and (-2) are associated, respectively, with 2 and $-2(\mathrm{h}/2\pi)$ units of orbital angular momentum in the z direction but that there is no cross-term ("off-diagonal element") of l_z between them.

On the other hand, when the real forms of the orbitals are used:

$$
\begin{aligned}
\int d_{xy}^* l_z d_{xy} \mathrm{d}\tau &= \int 2^{-1}[(2) - (-2)] \cdot l_z[(2) - (-2)] \mathrm{d}\tau \\
&= \frac{1}{2} \int [(2) \cdot l_z(2) - (2) \cdot l_z(-2) - (-2) \cdot l_z(2) \\
&\quad + (-2) \cdot l_z(-2)] \mathrm{d}\tau \\
&= \frac{1}{2} \int [2(2) \cdot (2) + 2(2) \cdot (-2) - 2(-2) \cdot (2) \\
&\quad + 2(-2) \cdot (-2)](\mathrm{h}/2\pi) \mathrm{d}\tau \\
&= (\mathrm{h}/2\pi)(2 + 0 - 0 - 2) \\
&= 0
\end{aligned} \tag{1.38}
$$

And

$$
\int d^*_{xy} l_z d_{x^2-y^2} d\tau = i \int 2^{-1}[(2) - (-2)] \cdot l_z[(2) + (-2)]d\tau
$$
$$
= \frac{i}{2} \int [(2) \cdot l_z(2) - (2) \cdot l_z(-2) - (-2) \cdot l_z(2)
$$
$$
+ (-2) \cdot l_z(-2)]d\tau
$$
$$
= \frac{i}{2} \int [2(2) \cdot (2) - 2(2) \cdot (-2) - 2(-2) \cdot (2)
$$
$$
+ 2(-2) \cdot (-2)](h/2\pi)d\tau
$$
$$
= \frac{i}{2} (h/2\pi)(2 - 0 - 0 + 2)
$$
$$
= 2i(h/2\pi) \tag{1.39}
$$

Similarly,

$$
\int d^*_{xz} l_z d_{xz} d\tau = \int d^*_{yz} l_z d_{yz} d\tau = 0 \tag{1.40}
$$

$$
\int d^*_{xz} l_z d_{yz} d\tau = h/2\pi \tag{1.41}
$$

In these *real* orbitals, all the orbital angular momentum in the z direction is associated with the cross-term between different members. Of course, for d_{z^2}, because $m = 0$

$$
\int d^*_{z^2} l_z d_{xy} d\tau(\text{etc.}) = \int d^*_{z^2} l_{z^2} d\tau = 0 \tag{1.42}
$$

The foregoing relationships have the following significance: The z component of orbital angular momentum, say, is associated with a torus of electron distribution about the z axis. It is zero if the electron distribution maximizes along that axis. In the orbitals specified by (m) such a distribution exists, whereas in those of the *real* type it does not. These latter orbitals have no z component of orbital angular momentum by themselves; however, because a rotation by $2\pi/8$ about the z axis transforms $d_{x^2-y^2}$ into d_{xy}, and a rotation by $2\pi/4$ transforms d_{xz} into d_{yz}, orbital angular momentum is to be associated with these two *pairs* of orbitals. The association of orbital angular momentum with such a rotation is conditional on the fact that the two orbitals that transform into each other under the rotation are degenerate. If the degeneracy is removed, the associated orbital angular momentum is lost, because the two orbitals cannot be rotated into each other without an energy change.

1.4   THE SYMMETRY PROPERTIES OF MOLECULES AND WAVEFUNCTIONS

The idea that a molecule is "highly symmetric" or of "low symmetry" is familiar in a qualitative sense to all chemists (7). At a more rigorous level, the symmetry of a molecule is described mathematically using group theory, which provides a powerful way of simplifying many chemical problems. It also forms the basis of a convenient notation system; the most widely used symbols are those of Schönflies. Indeed, group theory is now so important that it is included in many general textbooks; thus, the way in which symmetry properties are described using group theory is not covered here. A brief outline of the terminology is, however, appropriate.

1.4.1   The Molecular Point Groups

The symmetry properties of a complex are introduced formally through the *symmetry elements* it exhibits. These symmetry elements include mirror planes, an inversion center, and rotation axes. To identify these, a Cartesian coordinate system may be defined at an appropriate point in the molecule, and the effect of the *symmetry operations* associated with each symmetry element—reflection, inversion, and rotation—may be explored. The resulting set of symmetry elements allows the molecule to be assigned a *point group*, so called because the operations are applied with respect to a fixed point in the molecule (the symmetry elements relating the atoms in a crystal unit cell include also translations and define a *space group*).

Most coordination complexes exhibit a symmetry that, to at least a first approximation, conforms to that of one of the so-called Platonic solids. The commonest geometry involves six equivalent metal–ligand bonds at right angles, for instance, as in the CrF_6^{3-} ion. The Cartesian axes are drawn to coincide with the metal–ligand bonds, and the complex possesses a high degree of symmetry, having not only an inversion center but also many mirror planes and twofold, threefold, and fourfold rotation axes. The solid formed by joining the ligand atoms is the eight-faced *octahedron* shown in Figure 1.4a, and the complex belongs to the point group O_h. The most common geometry involving four metal–ligand bonds conforms to the symmetry properties of the four-faced tetrahedron; and a complex of this kind, e.g., the $CoCl_4^{2-}$ ion, belongs to the point group T_d. In this case, the Cartesian axes bisect the metal–ligand bond angles (Fig. 1.4b), and the complex does not have an inversion center.

The axis system is chosen in the manner that allows the symmetry of the complex to be described most easily. By convention, z is the axis associated with the largest number of rotational symmetry elements. The point groups O_h and T_d both have *cubic* symmetry in that, like a regular cube, the directions x, y, and z are equivalent. This is not true for a square planar complex involving four equivalent metal–ligand bond lengths at right angles; that belongs to the point group D_{4h}, where the unique symmetry axis z is normal to the plane of the complex, with the equivalent x and y axes passing along the metal–ligand bond directions.

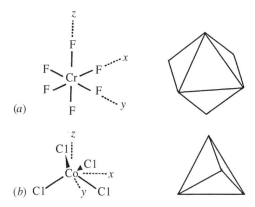

Figure 1.4. The relationship between the geometry of an octahedral (**a**) and a tetrahedral (**b**) complex with those of the Platonic solids.

Chemists often use the terminology describing symmetry quite loosely. For instance, the complex $Cr(H_2O)_6^{3+}$ is generally referred to as an octahedral complex of O_h symmetry, even though this is strictly true only if the hydrogen atoms are ignored. A complex such as $Co(NH_3)_5Cl^{2+}$ will also be described as octahedral, unless group theory is being used to interpret some property where the use of the true point group (neglecting the hydrogen atoms) C_{4v} is necessary, for instance the absorption of polarized light as discussed in Section 8.1.1.4.

1.4.2   The Representations of Wavefunctions

Features like the Cartesian axes and, most important, wavefunctions are conveniently classified by considering their behavior with respect to the various symmetry operations of the point group of the molecule. The traces, or *characters* (sums of diagonal elements), of the matrices used to represent the symmetry operations of the point group mathematically (*irreducible representations*) are used to generate labels that summarize the possible ways in which the features of interest conform to the symmetry elements of the point group. The symmetry properties of the irreducible representations are conveniently presented in the *character table* of the point group. Even when the symmetry is not directly relevant, it is common to refer to wavefunctions by their irreducible representations. This notation is useful because it incorporates considerable information.

For first-row transition ions, a good approximation is made by considering the spin and orbital parts of the wavefunctions separately (this aspect is discussed in some detail in chapter 5). The symmetry then refers just to the orbital part of each wavefunction. For common point groups such as O_h, T_d, and D_{4h}, the wavefunctions all transform as A, B, E, or T, where the symbol denotes not only the symmetry of the wavefunction with respect to rotations but also the number of orbital wavefunctions of that type. A and B refer to a single orbital wavefunction (a nondegenerate set); E, to a pair of orbital wavefunctions (a doubly degenerate set) and T to three wavefunctions of equal energy (a triply degenerate set). When appropriate, the behavior with respect to rotations or reflections is distinguished by a numerical subscript. When the point group is centrosymmetric, each wavefunction falls into one of two subgroups, depending on whether it changes sign on inversion through the center. The representations that are "even" carry the subscript g, whereas those that are "uneven" carry the subscript u (from the German *gerade* and *ungerade*). Generally, capital letters are used when referring to an energy state, and small letters are used for orbitals, or sets of orbitals. Often, Greek letters are used for vibrational, as opposed to electronic, wavefunctions.

As an illustration of this terminology, we note that the d orbitals of the complex $Cr(H_2O)_6^{3+}$ are split into two sets: a lower energy t_{2g} set and a higher energy e_g set. When describing the energy levels resulting from the interaction of the three valence d electrons with one another and with the water ligands, we say the lowest level is an A_{2g} state and the first excited state is a T_{2g} state. This terminology refers just to the orbital component of the states; the designation of the spin component is discussed in chapter 5. The *totally symmetric* vibration of the complex has α_{1g} symmetry.

1.4.3 Typical Applications in Ligand Field Theory

The power of group theory, as applied to the symmetry properties of transition metal complexes, lies in its ability to answer certain chemically relevant questions very simply. For instance, by simply inspecting the representations of the d orbitals in the character table for the point group O_h, it may be seen that the d_{xy}, d_{xz}, and d_{yz} orbitals transform as t_{2g}, while the $d_{x^2-y^2}$, d_{z^2} pair of orbitals transform as e_g. The fact that the d orbitals split into two sets in a molecule with O_h symmetry may be derived just from their symmetry labels alone, without the need for any calculation. It must be stressed, however, that group theory by itself gives *no* indication of the magnitude, or even the sign of the energy splitting between the sets. It provides a qualitative, but not a quantitative deduction. Like the octahedron, the cube belongs to the group O_h, but we shall see it produces a ligand field splitting of opposite sign (Section 2.5).

Group theory also provides a powerful means of telling whether certain integrals can be nonzero. This will be the case only if the *direct product* of the relevant wavefunctions and operator gives the totally symmetric representation of the point group of the complex. Thus the probability that an octahedral complex can undergo a transition in which an electron is excited from the t_{2g} orbitals to the e_g orbitals by absorbing a photon of light may be investigated by inspecting the result of the direct product:

$$t_{2g} \times t_{1u} \times e_g \qquad (1.43)$$

Here, the symmetry properties of the radiation correspond to its transformation as t_{1u} in O_h. Because two of the representations are even and one is odd, the products must be odd and cannot contain the totally symmetric representation a_{1g}, so the relevant integral must be zero. The transition is, therefore, forbidden to occur. In fact similar arguments apply to the electronic transitions between orbitals or states of the same parity for all centrosymmetric molecules. This restriction does not apply to noncentrosymmetric molecules, such as those of tetrahedral symmetry, so that for them most electronic transitions are allowed to occur. Thus group theory provides a simple explanation of the fact that octahedral complexes absorb visible light some 10–100 times less effectively than do tetrahedral complexes.

Finally, symmetry properties give a useful guide as to which orbitals participate in particular wavefunctions—only orbitals belonging to the same representations may mix in this way. For a complex of O_h symmetry, the metal s, p, and d orbitals all belong to different representations and so do not mix (neglecting the effects of vibrations). However, if the metal–ligand interaction along z differs from that along x and y, so that the point group is D_{4h}, the s and d_{z^2} orbitals belong to the same representation, and mixing between them becomes allowed by symmetry. Several lines of evidence suggest that such a mixing does indeed occur when the metal–ligand interaction departs drastically from cubic symmetry, as is the case for a planar complex (see Section 3.2.2).

1.5   QUALITATIVE DEMONSTRATION OF THE LIGAND FIELD EFFECT

With the availability of the d wavefunctions in their real forms it is possible to examine qualitatively the effect of various distributions of ligand atoms around the central ion on its d orbitals. The discussion is based on the interactions inherent in the electrostatic crystal field model, but the argument is not altered if the metal–ligand perturbation includes the effects of covalency in the bonding. Suppose that there are six negative point charges disposed at the vertices of a regular octahedron. Let these point charges represent the ligand atoms. They are placed on the axes of a Cartesian coordinate system with origin at the central ion at the points $(\pm a,0,0)$, $(0, \pm a,0)$ and $(0,0, \pm a)$ (Fig. 1.5). The negative charges repel the electrons of the central ion. The repulsion is greater the closer the electrons approach the charges. We choose the z axis used to define the wavefunctions to coincide with the z axis of the ligand distribution.

Consider the degree to which the electrons in the different d orbitals must venture into the vicinity of the charges. The d_{z^2} and $d_{x^2-y^2}$ orbitals have their electron density maxima directed along the z, axis and the x and y axes, respectively. The d_{xy}, d_{yz}, and d_{xz} orbitals have their maxima located in the regions between the Cartesian axes (Fig. 1.3). Consequently, the d electrons of the central ion tend to avoid the d_{z^2} and $d_{x^2-y^2}$ orbitals. They restrict themselves, as far as possible, to the remaining three orbitals. We may say that the d orbitals of the central ion in this octahedral crystal field are all affected by the presence of

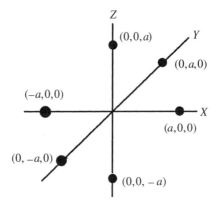

Figure 1.5. An octahedral array of point charges.

the ligand atoms, but the d_{z^2} and $d_{x^2-y^2}$ orbitals are more affected than the d_{xy}, d_{yz}, and d_{xz} orbitals. Figure 1.6 shows the effect of the crystal field on the energies of the five d orbitals. Because of their symmetry properties in the O_h point group, the d_{xy}, d_{yz}, and d_{xz} orbitals are degenerate and are labeled as the t_{2g} set.

It has become possible to use the diffraction of polarized neutrons to study experimentally the way in which the unpaired electrons are distributed in paramagnetic complexes. A contour plot of the unpaired spin density in the xy plane is shown for the complex ion CrF_6^{3-} in Figure 1.7. The metal ion has three unpaired electrons, one in each of the d orbitals of the t_{2g} set. The shape and orientation of the unpaired electron density indeed conforms to that calculated for the d_{xy} orbital. The regions marked with a negative sign correspond to

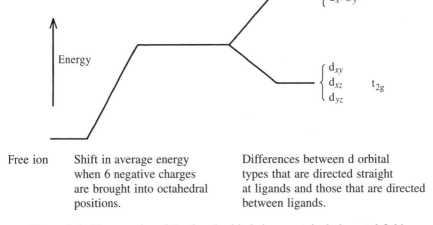

| Free ion | Shift in average energy when 6 negative charges are brought into octahedral positions. | Differences between d orbital types that are directed straight at ligands and those that are directed between ligands. |

Figure 1.6. The energies of the five d orbitals in an octahedral crystal field.

Figure 1.7. Unpaired spin density map in the xy plane derived for K_2NaCrF_6 from the diffraction of polarized neutrons (8).

hollows in the contour map, and the minor regions of unpaired spin density associated with the four fluoride ions indicate some small degree of covalency.

It is quite easy to see that the members of the t_{2g} set are equally affected by the presence of the ligand atoms because they are of identical shape and are symmetrically disposed relative to the ligands. The two members of the e_g set, d_{z^2}, and $d_{x^2-y^2}$, however, are also given as degenerate, and it is seen from Figure 1.3 that these two orbitals are not identical in shape. It is not obvious from inspection that they should be equally affected by the presence of the ligand atoms. It is possible to prove the degeneracy of the members of the e_g set by means of an argument based on the expressions for the two orbitals and the relationship of these to the two hypothetical orbitals $d_{y^2-z^2}$ and $d_{z^2-x^2}$. Because the proof is undertaken in the quantitative treatment given in Chapter 2, the argument is not followed through here. The equivalence may also be demonstrated by symmetry arguments, as discussed in the preceding section. Note that, because each of the three p orbitals has its electron density maxima directed at two ligand atoms, the qualitative arguments in use here predict that the p orbitals are *not* split by an octahedral crystal field.

In Figure 1.6, the mean energy of the five d orbitals is shown to be raised by the presence of the ligands. Because like charges repel one another, this is an obvious result for the electrostatic crystal field model. Of course, in Figure 1.6, the energy of only the d orbitals is considered. If the *total* energy of the system is plotted it may well be found that the energy gained from the attraction between the positive charge on the central ion and the ligands is larger than the destabilization energies of the d and other electrons of the central ion. Consequently, there may be a net stabilization energy for the coordination cluster. Similar arguments apply if the bonding is predominantly covalent rather than electrostatic. Although the d orbitals are antibonding and hence destabilized on complex formation, this is more than counterbalanced by the stabilization undergone by the ligand orbitals resulting from their interaction with the empty metal s and p orbitals.

The other arrangement of ligand atoms about the central ion with which it is necessary to deal frequently is the regular tetrahedron. In finding how the d orbitals are influenced by a tetrahedral disposition of ligands, it is convenient to remember that a regular tetrahedron can be obtained by taking the alternate corners of a cube (Fig. 1.8). The axes of the Cartesian system are chosen to pass through the cube face centers, as shown in the figure. The orbitals of the t_2 set are directed through the middles of the cube edges, whereas, of course, the e set is directed through the cube face centers. (Note that the subscript g has been dropped here, because the tetrahedron does not possess a center of symmetry.) An inspection shows that the t_2 orbitals are directed more toward the positions of the ligand atoms than are the members of the e set. Consequently, in a tetrahedral crystal field it is that set which is the less affected. *The relative energies of the e and t_2 sets* are *reversed with respect to the splitting in the octahedral crystal field.* For the cube stereochemistry see Section 2.5. Another cubic stereo-chemistry, that of the icosahedron, is met occasionally (e.g., in hydrated lanthanide double salts with magnesium). It has the peculiarity that it leads to a

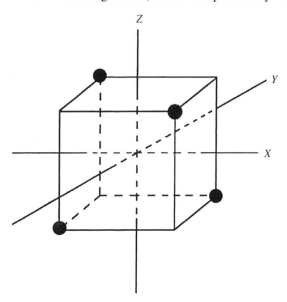

Figure 1.8. The relationship of the Cartesian axes, and hence of the d orbitals, to a tetrahedral arrangement of ligands.

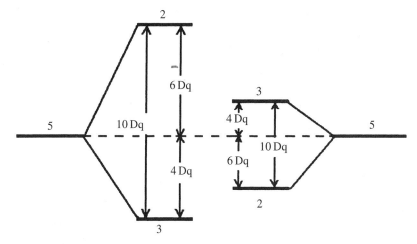

Figure 1.9. The splitting of the energies of the d orbitals by octahedral and tetrahedral crystal fields.

$t_{2g} - e_g$ orbital splitting of zero, because it essentially combines both octahedral and cube components.

For many purposes it is possible to disregard the shifts in the *average* energies of the d orbitals on the construction of the coordination clusters. It is customary to give the effect of the crystal field as simply the *splitting* of the various d orbitals relative to each other, leaving the mean of their energies unchanged. In Figure 1.9 the splitting diagrams on this basis for octahedral and tetrahedral crystal fields are set out. The separation between the $t_{2[g]}$ and $e_{[g]}$ sets in each stereochemistry is given as 10 Dq. The reason for this terminology will become clear in Chapter 2. The neglect of the shift in the center of gravity of the d orbitals is in line with the statement made earlier—that absolute binding energies of complex ions are not considered here. In the ligand field formalism, 10Dq is normally referred to simply as Δ.

1.6 THE PHYSICAL PROPERTIES AFFECTED BY LIGAND FIELDS

It has already been stated that it is convenient to treat the applications of ligand field theory under three headings: (1) thermochemical properties and geometric distortions, (2) spectral properties, and (3) magnetic properties. It is sufficient for the moment just to indicate why such a division is made and to note the general features of the manner in which these properties respond to the presence of ligand fields.

1.6.1 Thermochemical Properties and Geometric Distortions

In this category we include the physical properties that depend on the energy of only the ground levels of the coordination cluster. It was demonstrated in the previous section that there is a differentiation in energy between two types of d orbitals. When we deal with a central ion possessing a number of d electrons, *other things being equal*, the average energy of the coordination cluster is expected to correlate with the manner in which the d electrons are distributed among the $t_{2[g]}$ and $e_{[g]}$ orbitals. The d electrons tend to be found in the lower of the two sets; but when there are more than sufficient d electrons to provide one for each of the lower-lying orbitals, occupation of both sets can occur. Relatively speaking, the energy of the complex is destabilized by Δ whenever an electron occupies the higher energy set. In an octahedral ligand field where the $t_{2[g]}$ set is the lower, the first three electrons tend to go into these orbitals only, but additional electrons will generally occupy the higher energy e_g set.

Suppose there is some physical property that, ignoring the ligand field splitting of the d orbitals, is expected to vary smoothly with the number of d electrons and which reflects the energy of the coordination cluster. In fact, owing to the unequal energies of the d orbitals caused by the ligand field splitting, irregularities in the variation of the property with number of d electrons may be found. The most obvious property from this point of view is the bonding energy. In addition, whenever an electron occupies one of the less stable e_g orbitals, a lengthening of the metal–ligand distances will occur in the directions of the occupied d orbital lobes. The geometric distortions observed for metal ions with certain d configurations may be explained in this manner. In a more general sense than that covered by the crystal field approach, it may be expected that, by whatever mechanism the ligands are held to the central metal ion, the net energy of the coordination cluster reflects the relative occupation of the $t_{2[g]}$ and $e_{[g]}$ orbitals in the field from the ligand atoms.

1.6.2 Spectral Properties

Under the heading of spectral properties are included the electronic transitions that take place between the ground levels of coordination clusters and their excited levels. Transitions of this type correspond to frequencies from the far infrared to the far ultraviolet (uv). Because it is difficult to differentiate between the transitions of interest and skeletal vibrations of the coordination cluster or electronic transitions not essentially within the d shell of the central ion, the range is usually restricted to the optical and near infrared (ir) and uv regions of the spectrum, from about 5,000 to 40,000 cm^{-1}. It is obvious that to deal with the spectra of coordination complexes, it is necessary to understand the excited levels of the system as well as the ground levels. Thus the treatment of spectral properties takes place after that of the thermochemical aspects of ligand field theory.

The spectra of coordination compounds may be classified into *ligand field* bands and *charge transfer* bands. The ligand field bands are essentially concerned with transitions among the central metal ion d orbitals, which are differentiated as a result of the application of the ligand field. Thus, for the case of one d electron and an octahedral ligand field, it is expected that the excitation of the electron from the t_{2g} orbital set to the e_g set should correspond to a ligand field band. Indeed, in Ti^{3+} (d^1) in aqueous solution (Ti(H$_2$O)$_6^{3+}$) it is possible to observe a weak, broad absorption band at about 20,000 cm^{-1} which can be shown to correspond to the transition in question. When more than one d electron is present, the number of energy levels that arise from the configuration is larger. The investigation of the effect of the ligand field on these requires a knowledge of the theory of the spectra of free ions.

The charge transfer bands in the spectra of complexes are so named because they are the result of transitions between levels that correspond to differing electron distributions between the metal and the ligand atoms in the ground and the excited states. Thus a transition of this type corresponds to the transfer of charge from the central ion to the ligand atoms, or vice versa. Although it is possible to treat charge transfer spectra using the pure crystal field model, it is generally better to consider such spectra within the framework of a covalent ligand field model. The mixing of ligand and central ion orbitals introduced in the ligand field approach, as opposed to the crystal field approach, provides a facile mechanism for transitions of the charge transfer type. Indeed, in a simple molecular approach for rather covalent complexes, when the orbital mixing is taken into account it is sometimes difficult to distinguish sharply between ligand field and charge transfer spectra. In the process of transferring charge to or from the central metal ion, electrons are gained or lost. This changes the effective oxidation number of the metal, so that charge transfer spectra have also been called *redox* spectra.

1.6.3   Magnetic Properties

The magnetic properties of coordination compounds arise from the spin and orbital angular momenta of electrons in unfilled shells. Along with the paramagnetism from this source, a smaller diamagnetic effect from the precession of electron orbits in the applied magnetic field is always present. Diamagnetism is not treated here except insofar as it is assumed to be necessary to correct for its presence in the ligand atoms.

To understand the magnetic properties of transition metal ions in ligand fields it is necessary to have available a complete account of the effects of *spin–orbit coupling* on the ground levels of the ion in the ligand field. The spin and orbital angular momenta of an electron both confer on it some of the attributes of a magnetic dipole. These two magnetic dipoles are not independent. The influence that the orbital angular momentum of an electron exerts on its spin angular momentum, and vice versa, through the magnetic dipoles they induce means that the two angular momenta are coupled together. The interaction phenomenon is given the name *spin–orbit coupling*. The energies associated with this coupling range from about 100 to 5,000 cm^{-1} in the transition elements.

It is convenient to base the account of the effects of spin–orbit coupling on the information that is used in the interpretation of the spectral properties. Consequently, the section on magnetic properties follows the one on spectra. In dealing with magnetism in coordination complexes, we find that the sensitivity of the results to covalency effects and the details of the arrangement of the ground-level energies is greater than that for thermochemical or spectral properties. This sensitivity makes it necessary to consider the effects of departures from the electrostatic crystal field model, and of the ligand field from cubic symmetry, more closely than before.

The most general way of studying magnetic properties is by measuring the magnetic susceptibility. In certain cases, however, the orbital angular momentum of the ground state may be measured more directly by observing the transition between the components of the ground state when these are split by an applied magnetic field. The method of achieving this, electron paramagnetic (or spin) resonance, probably provides the most sensitive probe of the nature of the ground-state wavefunction of a complex. It, however, can be applied to only a limited range of complexes and is, therefore, considered after the magnetic properties have been developed.

1.7 CRYSTAL FIELDS AND LIGAND FIELDS

The previous sections show how crystal field theory provides a useful framework for the interpretation of at least some of the features met in various areas of coordination chemistry. It is now known, however, that the electrostatic basis of crystal field theory is a rather poor way of describing the effects of the ligands on the behavior of the electrons of the central ion. The modification of the description of the binding in a coordination cluster that proves to be the most useful in accounting for its spectra and magnetism is in terms of molecular orbital theory. In the molecular orbital treatment, the electrons of the bonding shell of the central ion, as well as those of the ligand atoms, cannot be entirely associated either with the central ion or with the ligand nuclei. They are associated, to some extent, with each of them.

The extent to which an electron is associated with the central ion can vary continuously from zero to unity. If the electrons of a free metal ion remain completely attached to it while a set of ligand atoms are brought up to the equilibrium distance in the coordination cluster, one has the pure crystal field model. Likewise, the electrons of the ligand remain completely attached to those atoms during the operation. On the other hand, if the electrons of the free metal ion and the ligand atoms mix during the assembly so that they are found with comparable likelihood on any atom of the system, whether the central ion or the ligand, then one has a completely covalent account of the bonding in the coordination cluster.

The two possibilities outlined in the previous paragraph represent the extreme types of bonding that have been considered in coordination complexes. Both are special cases of the molecular orbital treatment. A powerful aspect of this model is that all intermediate degrees of electron sharing can also be described. In principle, the term *crystal field theory* has often been reserved for the extreme case in which there is no mixing of the central ion and the ligand electrons and the name *ligand field theory* for all nonzero degrees of mixing. Here, as found elsewhere, the term *ligand field* is usually used generically to include all degrees of mixing, zero and nonzero, provided no confusion results.

A particularly powerful form of ligand field theory, derived from the molecular orbital approach, is the so-called *angular overlap model* (AOM) of the bonding in a coordination complex. This interprets the d orbital splittings in terms of weak covalent σ- and π-bonding interactions with ligand orbitals, which has the big advantage that it correlates directly with the language used in descriptive chemistry. The AOM is considered in Chapter 3.

The flexibility of molecular orbital–based methods is of great advantage in dealing with the magnetic and spectral properties of complexes. The advantage arises because an intensive study of these properties presents information that can be correlated in considerable detail with the aspects that depend on the delocalization of the d electrons onto the ligands. The information available on the thermochemical and structural properties of coordination complexes is generally not sufficiently precise to enable more than simple correlations to be drawn with the bonding. A treatment in the language of crystal or ligand field theory is, therefore, usually sufficient for these aspects.

REFERENCES

1. Bethe, H., *Ann. Phys.*, 1929, **3**, 133.
2. Penney, W., and Schlapp, R., *Phys. Rev.*, 1932, **41**, 194; 1932, **42**, 666; 1933, **43**, 486.
3. van Vleck, J. H., *J. Chem. Phys.*, 1935, **3**, 803, 807.
4. Eyring, H., Walter, J., and Kimball, G. E., *Quantum Chemistry*, Wiley, New York, 1944, chap. 6.
5. King, R. B., *Polyhedron*, 1994, **13**, 2005.
6. Ref. 4, chap. 3.
7. Cotton, F. A., *Chemical Applications of Group Theory*, 3rd ed., Wiley, New York, 1990.
8. Mason, R., Smith, A. R. P., Varghese, J. W., Chandler, G. S., Figgis, B. N., Reynolds, P. A., and Williams, G. A., *J. Amer. Chem. Soc.*, 1981, **103**, 1300.

General Reading

1. Ballhausen, C. J., *Introduction to Ligand Field Theory*, McGraw-Hill, New York, 1962.
2. Bersuker, I. B., *Electronic Structure and Properties of Transition Metal Complexes. Introduction to the Theory*. Wiley, New York, 1996.
3. Figgis, B. N., *Introduction to Ligand Fields*, Wiley-Interscience, New York, 1966.
4. Figgis, B. N., *Comprehensive Coordination Chemistry*, Vol. 1, Pergammon, New York, 1987, chap. 6.

5. Gerloch, M., and Slade, R. C., *Ligand Field Parameters*, Cambridge University Press, Cambridge, UK 1973.

6. Jorgensen, C. K., *Absorption Spectra and Chemical Bonding in Complexes*, Pergammon, London, 1962.

7. Jorgensen, C. K., *Modern Aspects of Ligand Field Theory*, North Holland, Amsterdam, 1971.

CHAPTER 2

QUANTITATIVE BASIS OF CRYSTAL FIELDS

2.1 CRYSTAL FIELD THEORY

It is now time to put the principles of crystal field theory, which were outlined at the beginning of Chapter 1, on a quantitative basis. We show that the crystal field model leads to a differentiation of the d orbitals into the t_{2g} and e_g subshells, as argued on qualitative grounds in Chapter 1, and we estimate the size of this splitting. The model used for the calculations is given in Section 1.5. First, the calculations are performed for a regular octahedral distribution of charges about the central ion, of magnitude ze, at a distance a; i.e., we use the system set out in Figure 1.5.

To find at any point in space the potential that arises from the octahedron of charges, a summation of the potentials from the individual charges must be made. The result is given by Eq. 2.1.

$$V_{(x,y,z)} = \sum_{i=1}^{6} v_{(i;\,x,y,z)}$$

$$= \sum_{i=1}^{6} e\,z_i/r_{ij} \tag{2.1}$$

Here, r_{ij} is used to denote the distance from the ith charge to the point (x, y, z), as a matter of convenience. The investigation of the effect of this potential on the d orbitals takes the form known as *zero*-th order perturbation theory, viz that for degenerate systems (1). The process involved is outlined here.

Suppose that the energy E_0 of a system corresponding to the Hamiltonian H_0 is n-fold degenerate, with eigenfunctions ψ_i (e.g., for the d orbitals $n = 5$, or 10, including spin), satisfying the equation:

$$\mathbf{H}_0\,\psi_i = E_0\,\psi_i \tag{2.2}$$

where $i = 1, 2 \ldots n$. If the Hamiltonian is changed slightly, the eigenfunctions ψ_i are, in general, no longer eigenfunctions of the new Hamiltonian $(\mathbf{H}_0 + \mathbf{H}')$, nor is E_0 an eigenvalue. The eigenvalues of $(\mathbf{H}_0 + \mathbf{H}')$ are, say, E'_j $(j \to 1, 2, \ldots n)$ and the corresponding eigenfunctions are ψ'_j.

$$(\mathbf{H}_0 + \mathbf{H}')\,\psi'_j = E'_j\,\psi'_j \tag{2.3}$$

The wavefunctions ψ'_j can be expressed as *linear combinations* of the wavefunctions ψ_i. In other words, it is possible to rearrange the set of original eigenfunctions to become eigenfunctions of $(\mathbf{H}_0 + \mathbf{H}')$:

$$\psi'_j = c_{1j}\psi_1 + c_{2j}\psi_2 + \cdots c_{nj}\psi_n$$
$$= \sum_{i=1}^{n} c_{ij}\psi_i \tag{2.4}$$

with, for normalization,

$$\sum_{i=1}^{n} c^*_{ij} c_{ij} = 1 \tag{2.5}$$

The process of finding the eigenvalues and eigenfunctions of the perturbed system is a matter of setting up and solving the *secular determinant* for values of E'_j. That determinant may be obtained by, for example, differentiating the ψ_i wavefunctions to minimize the energy (1). The determinant is:

$$
\begin{array}{c|cccc}
 & \psi_1 & \psi_2 & \cdots & \psi_n \\
\psi_1 & H'_{11} - E' & H'_{12} & \cdots & H'_{1n} \\
\psi_2 & H'_{21} & H'_{22} - E' & \cdots & H'_{2n} \\
\cdots & \cdots & \cdots & \cdots & \cdots \\
\psi_n & H'_{n1} & H'_{n2} & \cdots & H'_{nn} - E'
\end{array} = 0 \tag{2.6}
$$

where

$$H'_{ij} = \int \psi^*_i\, \mathbf{H}'\, \psi_j\, d\tau \tag{2.7}$$

There are, in general, n solutions, E_j, $j = 1$ to n. The determinant can be diagonalized by standard methods to derive the eigenvalues and eigenvectors (2).

The eigenvectors give the ψ'_j as a series of linear combinations of ψ_i. ψ'_j is then the eigenfunction corresponding to E'_j. For example, with c_{ij} a coefficient in the eigenvector corresponding to the eigenvalue E'_i

$$(H'_{11} - E'_i)\, c_{11} + H'_{12}\, c_{12} + H'_{13}\, c_{13} + \cdots = 0 \tag{2.8}$$

$$H_{21}\, c_{21} + (H'_{22} - E'_i)\, c_{22} + H'_{23}\, c_{23} + \cdots = 0 \tag{2.9}$$

etc. In the secular determinant as given above, the zero of energy is taken as the energy of the unperturbed system E_0. It makes no difference if some other zero is chosen.

In this particular instance, H_0 is the Hamiltonian for the hydrogen-like atom, and $H' = V_{(x,y,z)}$. The set of original eigenfunctions, ψ'_i, is the manifold of the five d wavefunctions. The secular determinant obtained is:

	(2)	(1)	(0)	(-1)	(-2)	
(2)	$H'_{2,2} - E$	$H'_{2,1}$	$H'_{2,0}$	$H'_{2,-1}$	$H'_{2,-2}$	
(1)	$H'_{1,2}$	$H'_{1,1} - E$	$H'_{1,0}$	$H'_{1,-1}$	$H'_{1,-2}$	
(0)	$H'_{0,2}$	$H'_{0,1}$	$H'_{0,0} - E$	$H'_{0,-1}$	$H'_{0,-2}$	
(-1)	$H'_{-1,2}$	$H'_{-1,1}$	$H'_{-1,0}$	$H'_{-1,-1} - E$	$H'_{-1,-2}$	
(-2)	$H'_{-2,2}$	$H'_{-2,1}$	$H'_{-2,0}$	$H'_{-2,-1}$	$H'_{-2,-2} - E$	(2.10)

where the wavefunctions (m) are defined in Section 1.3. The matrix elements $H'_{m,m'}$ are:

$$H'_{m,m'} = e \int (m)^* V_{(x,y,z)}\, (m')\, d\tau \tag{2.11}$$

m' may or may not be the same as m. This secular determinant is diagonalized to give the energies of the system of d orbitals (eigenvalues) under the potential. Finally, the energies are fed back into the secular equations to yield wavefunctions appropriate for the presence of the potential; i.e., the eigenvectors (also available directly and more readily as part of the diagonalization procedure). The potential $V_{(x,y,z)}$ contains no reference to the coordinates of electron spin, so it is not necessary to amplify the d orbital wavefunctions to include spin at this stage. $V_{(x,y,z)}$ acts directly only on the part of the wavefunction containing the coordinates, x, y, and z. It is shown later that the ligand field may have an indirect influence on electron spin.

It was stated in Section 1.6, that crystal field theory concerns itself only with the splitting of the d orbitals, and not with the relationship of their mean energy to their energy in the free ion. Thus it is customary to adjust $V_{(x,y,z)}$ so that it does not change the center of gravity of the d orbitals. In the form that only *splits* the d orbitals, we refer to the potential as V_{oct}.

2.2   THE OCTAHEDRAL CRYSTAL FIELD POTENTIAL

To evaluate the matrix elements of Eq. 2.11 it is desirable to have the octahedral crystal field potential $\mathbf{V}_{oct}$ in a form that facilitates the integrations involved. An expansion of the inverse of the distance r_{ij} between any two points i and j in space is available. The expression is in terms of spherical harmonics centered on the origin of the coordinate system (3).

$$r_{ij}^{-1} = \sum_{n=0}^{\infty} \sum_{m=-n}^{n} \frac{4\pi}{2n+1} \cdot \frac{r_<^n}{r_>^{n+1}} \cdot Y_{nj}^m \cdot Y_{ni}^{m*} \tag{2.12}$$

Here, $r_<$ is the lesser of the distances from the origin to the points i and j. Y_{nj}^m and Y_{ni}^m are spherical harmonics in which the angles θ_i and ϕ_i and θ_j and ϕ_j are involved, respectively. θ and ϕ specify the angles to the points i and j (Fig. 2.1).

For the crystal field model, it is the potential within the octahedron of charges that is of interest. For convenience, the distance from the origin at the central ion $r_<$ can be written simply as r, and $r_>$ can be defined as the distance a. The formal contributions from $r_< = a$ and $r_> = r$ are set out e.g. by Eyring *et al.* (3). Eq. 2.12 may now be put in the form:

$$r_{ij}^{-1} = \sum_{n=0}^{\infty} \sum_{m=-n}^{n} \frac{4\pi}{2n+1} \cdot \frac{r^n}{a^{n+1}} \cdot Y_{nj}^m \cdot Y_{ni*}^m \tag{2.13}$$

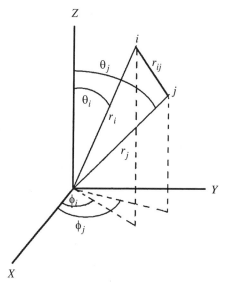

Figure 2.1. The angles θ and ϕ involved in the expansion of $1/r_{ij}$.

Because the six charges are located on the coordinate axes, the angles θ_i and ϕ_i, which specify them, are all 0, $\pm\pi/2$, or π; and the functions Y_{nj}^m reduce to simple numbers. The summation of r_{ij}^{-1} over all six charges considerably simplifies the resultant expression for $V_{(x,y,z)}$, because many terms cancel. The contributions from the terms with an odd n are not required if, as is usual, a plane of symmetry containing the z direction is present. Then, because Ξ_n^m is an odd function,

$$\int_0^\pi \Xi_l^{m'*} \cdot \Xi_n^m \cdot \Xi_l^{m''} \cdot \sin\theta \cdot d\theta = 0 \tag{2.14}$$

Consider the summation of Eq. 2.1 term by term. From each charge there is the term containing Y_0^0:

$$(4\pi ze/a) Y_0^0 \cdot Y_0^0 \tag{2.15}$$

which, because $Y_0^0 = 2^{-1/2}(2\pi)^{-1/2}$, is:

$$2^{1/2} \cdot (2\pi)^{1/2} \cdot (ze/a) Y_0^0 = ze/a \tag{2.16}$$

Thus the total contribution to $V_{(x,y,z)}$ from $n = 0$ is:

$$6 \cdot 2^{1/2} \cdot (2\pi)^{1/2} \cdot (ze/a) Y_0^0 = 6ze/a \tag{2.17}$$

For $n = 2$, $m = 0$, consider the expression Y_2^{0*}:

$$Y_{2i}^{0*} = Y_{2i}^0 = (5/8)^{1/2} \cdot (2\pi)^{-1/2} \cdot (3\cos^2\theta - 1) \tag{2.18}$$

For the points $(\pm a, 0, 0)$ and $(0, \pm a, 0)$:

$$\theta_i = \pi/2, \quad (3\cos^2\theta_i - 1) = -1 \tag{2.19}$$

For the point at $(0, 0, a)$:

$$\theta_i = 0, \quad (3\cos^2\theta_i - 1) = 2 \tag{2.20}$$

For the point at $(0, 0, -a)$:

$$\theta_i = \pi, \quad (3\cos^2\theta_i - 1) = 2 \tag{2.21}$$

The summation of $(3\cos^2\theta_i - 1)$ over the six points, then, yields zero. Consequently, the contribution to $V_{(x,y,z)}$ from $n = 2$, $m = 0$ is zero.

For $n = 2$, $m = \pm 1$

$$Y_{2i}^{\pm 1^*} = Y_{2i}^{\mp 1} = (15/4)^{1/2} \cdot (2\pi)^{-1/2} \cdot \cos \theta_i \cdot \sin \theta_i \cdot e^{\mp i\phi_i} \qquad (2.22)$$

For each of the points $\cos \theta_i \cdot \sin \theta_i = 0$, so that the contribution from $m = \pm 1$ is zero.

For $n = 2$, $m = \pm 2$

$$Y_{2i}^{\pm 2^*} = Y_{2i}^{\mp 2} = (5/16)^{1/2} \cdot (2\pi)^{-1/2} \cdot \sin^2 \theta_i \cdot e^{\mp 2i\phi_i} \qquad (2.23)$$

Now, $e^{\pm i\pi} = -1$, so that the contributions from the points are as follows:

Point	θ_i	ϕ_i	$\sin^2 \theta_i \cdot e^{\pm 2i\phi_i}$
$(0, 0, \pm a)$	$0, \pi$	—	0
$(0, \pm a, 0)$	$\pi/2$	$\pm \pi/2$	-1
$(\pm a, 0, 0)$	$\pi/2$	$0, \pi$	1
Total			$\underline{0}$

The contribution from $n = 2$, $m = \pm 2$ is zero. Therefore, the total contribution from $n = 2$ is zero

Consider now $n = 4$, $m = 0$

$$Y_{4i}^{0^*} = Y_{4i}^{0} = (9/128)^{1/2} \cdot (2\pi)^{-1/2} \cdot (35 \cos^4 \theta_i - 30 \cos^2 \theta_i + 3)$$
$$= (9/128)^{1/2} \cdot (2\pi)^{-1/2} \cdot P_4^0(\cos \theta_i) \qquad (2.24)$$

with

$$P_4^0(\cos \theta_i) = 35 \cos^4 \theta_i - 30 \cos^2 \theta_i + 3 \qquad (2.25)$$

Point	θ_i	$P_4^0(\cos \theta_i)$
$(0, 0, \pm a)$	0	8
$(0, \pm a, 0)$	$\pi/2$	3
$(\pm a, 0, 0)$	$\pi/2$	3

So that the contribution from $n = 4$, $m = 0$ is:

$$(4\pi/9) \cdot (2\pi)^{-1/2} \cdot (ze \, \overline{r^4}/a^5) \cdot (9/128)^{1/2} \cdot (2 \cdot 8 + 4 \cdot 3) \cdot Y_4^0 \qquad (2.26)$$

Consider now $m = \pm 1, \pm 2, \pm 3$;

The contributions from these values of m are found to be zero. For $m = \pm 4$

$$Y_{4i}^{\pm 4^*} = Y_{4i}^{\mp 4} = (315/256)^{1/2} (2\pi)^{-1/2} \cdot \sin^4 \theta_i \cdot e^{\mp 4i\phi_i} \qquad (2.27)$$

Point i	θ_i	ϕ_i	$\sin^4 \theta_i \, e^{\pm 4\phi_i}$
$(0, 0, \pm a)$	$0, \pi$	—	0
$(0, \pm a, 0)$	$\pi/2$	$\pm \pi/2$	1
$(\pm a, 0, 0)$	$\pi/2$	$0, \pi$	1

So that the contribution from $n = 4$, $m = \pm 4$ is:

$$(4\pi/9) \cdot (2\pi)^{-1/2} \cdot (ze\,\overline{r^4}/a^5) \cdot (315/256)^{1/2} \cdot 4Y_4^{\mp 4} \qquad (2.28)$$

The total contribution to $\mathbf{V}_{(x,y,z)}$ from $n = 4$ is, then,

$$(49/18)^{1/2} \cdot (2\pi)^{-1/2} \cdot (ze\,\overline{r^4}/a^5)\,[Y_4^0 + (5/14)^{1/2}(Y_4^4 + Y_4^{-4})] \qquad (2.29)$$

There is, for the present purposes, no need to carry the expansion of r_{ij}^{-1} to higher terms in the spherical harmonics. This is not because the series converges, but because the integrals of spherical harmonics with $l'' > 2l$ taken with the product of two spherical harmonics of order l (e.g., $l = 2$ the d wavefunctions) must vanish:

$$\int_0^\pi \int_0^{2\pi} Y_l^{m*} \, Y_{l'}^{m'} \, Y_l^m \sin\theta \cdot d\theta \, d\phi = 0, \qquad l' > 2l \qquad (2.30)$$

This matter is discussed in further detail in Appendix 1.

When dealing with the f wavefunctions, which involve $l = 3$, it is necessary to consider the expansion of r_{ij}^{-1} to the terms in Y_6. This subject is postponed to Section 2.10 and to Chapter 11 which deals with the effect of ligand fields on the f electrons.

The potential required for the d wavefunctions is, for the present purposes:

$$\mathbf{V}_{(x,y,z)} = \text{Eq. } 2.17 + \text{Eq. } 2.29 \qquad (2.31)$$

Because the d wavefunctions are also expressed as functions of the spherical harmonics, it is obvious that this form of the potential for the crystal field is suitable for performing the integrations suggested by Eq. 2.11.

2.3   THE EFFECT OF V$_{oct}$ ON THE d WAVEFUNCTIONS

To evaluate the matrix elements of Eq. 2.11 the terms in $\mathbf{V}_{(x,y,z)}$ are taken in turn. First, consider the portion of the potential that arises from Eq. 2.17.

$$\int (m)^* (6\,ze/a)\,(m')\,d\tau = (6\,ze/a) \int (m)^* (m')\,d\tau$$

$$= 6\,ze/a \quad (\text{if } m = m', \text{ otherwise zero.}) \qquad (2.32)$$

This term is independent of the (m) wavefunction chosen, and consequently it corresponds to an identical change in the energy of all the five d orbitals. As stated earlier, only the changes that take place in the relative energies of the different d orbitals, not their center of gravity, are of special interest. The term is not considered further here, except to say that it corresponds to the first energy change illustrated in Figure 1.6.

The second term in $\mathbf{V}_{(x,z,y)}$, that arising from Eq. 2.29, is responsible for the splitting of the d orbitals. It is convenient to give a special symbol to Eq. 2.29. $\mathbf{V}_{oct}$ is defined here as the part of $\mathbf{V}_{(x,y,z)}$ for an octahedron of charges that leads to the splitting of the d orbitals.

$$\mathbf{V}_{oct} = (49/18)^{1/2} \cdot (2\pi)^{1/2} \, (\mathrm{ze}\,\overline{r^4}/a^5) \, [\, Y_4^0 + (5/14)^{1/2} \, (\, Y_4^4 + Y_4^{-4})] \quad (2.33)$$

In evaluating the integrals with the second term of Eq. 2.31, which are:

$$\int (m')^* \, \mathbf{V}_{oct} \, (m') \, d\tau \qquad (2.34)$$

r may be integrated out immediately by the use of the relationship Eq. 1.10, giving:

$$\int_0^\infty R_{n,2}^* \cdot r^4 \cdot R_{n,2} \cdot r^2 \cdot dr = \overline{r^4} \qquad (2.35)$$

The subscripts for the quantum numbers n and l have been dropped in this result and the following material, because it is seldom necessary to specify the set of electrons that is being dealt with. The parameter $\overline{r^4}$ is, of course, most commonly the mean fourth power radius of the d electrons of the central ion. Because it is known (Section 1.3) that:

$$(m) = R_{n,2} \, Y_2^m$$

the expression for the matrix elements of the operator in Eq. 2.33 are given by:

$$\int (m)^* \mathbf{V}_{oct} \, (m') \, d\tau = (49/18)^{1/2} (2\pi)^{-1/2} \cdot \overline{r^4} \cdot (\mathrm{ze}/a^5) \cdot$$

$$\int_0^\pi \int_0^{2\pi} [Y_2^{m^*} \cdot Y_4^0 \cdot Y_2^{m'} \cdot \sin\theta \cdot d\theta d\phi + (5/14)^{1/2}$$

$$(Y_2^{m^*} \cdot Y_4^4 \cdot Y_2^{m'} \cdot \sin\theta \cdot d\theta \, d\phi + Y_2^m \cdot Y_4^{-4} \cdot Y_2^{m'} \cdot$$

$$\sin\theta \cdot d\theta \, d\phi)] \qquad (2.37)$$

The integrations involved in Eq. 2.37 are considerably simplified by noting that for

$$\int_0^{2\pi} \mathbf{Y}_{l_1}^{m_1} \cdot \mathbf{Y}_{l_2}^{m_2} \cdot \mathbf{Y}_{l_3}^{m_3} \, d\phi \neq 0 \tag{2.38}$$

it is necessary that, irrespective of the values l_1, l_2 and l_3, $m_1 + m_2 + m_3 = 0$ (Appendix 1). Consequently, the integrals in Eq. 2.37 that involve $\mathbf{Y}_4^0$ are zero, unless $(m') = (-m^\star) = (m)$. That is, the matrix element is zero unless it is between a d wavefunction and itself, not one of the other d wavefunctions. Such a matrix element is said to be *diagonal*, because it lies in the main diagonal of the secular determinant. The matrix elements of $\mathbf{Y}_0^0$, arising from the term from Eq. 2.17, are also diagonal.

Using the same argument, the integrals involving $\mathbf{Y}_4^4$ are zero unless $m' - m = -4$, which is possible only if $m' = -m = -2$. Those involving $\mathbf{Y}_4^{-4}$ are zero, unless $m' = -m = 2$. Eq. 2.37 may now be reduced, after integrating over ϕ, which yields 0 or 2π, to

$$\int_0^{2\pi} (m)^\star \mathbf{V}_{oct}(m) \, d\tau = (29/18)^{1/2} (ze \, \overline{r^4}/a^5) \int \Xi_2^{m^\star} \cdot \Xi_4^0 \cdot \Xi_2^m \cdot \sin \theta \cdot d\theta \tag{2.39}$$

and

$$\int_0^{\pi} (\pm 2)^\star \mathbf{V}_{oct}(\mp 2) \, d\tau = (35/36)^{1/2} (ze \, \overline{r^4}/a^5) \int \Xi_2^{\pm 2^\star} \Xi_4^{\pm 4} \Xi_2^{\mp 2} \cdot \sin \theta \cdot d\theta \tag{2.40}$$

All the relevant Ξ functions are given explicitly in Section 1.3 and Appendix 2, and the integration of their products is set out in Appendix 1. There results:

$$\int_0^{\pi} \Xi_2^{0^\star} \cdot \Xi_4^0 \cdot \Xi_2^0 \cdot \sin \theta \cdot d\theta = 18^{1/2}/7 \tag{2.41}$$

$$\int_0^{\pi} \Xi_2^{\pm 1^\star} \cdot \Xi_4^0 \cdot \Xi_2^{\pm 1} \cdot \sin \theta \cdot d\theta = -8^{1/2}/7 \tag{2.42}$$

$$\int_0^{\pi} \Xi_2^{\pm 2^\star} \cdot \Xi_4^0 \cdot \Xi_2^{\pm 2} \cdot \sin \theta \cdot d\theta = 2^{1/2}/14 \tag{2.43}$$

$$\int_0^{\pi} \Xi_2^{\pm 2^\star} \cdot \Xi_4^{\pm 4} \cdot \Xi_2^{\mp 2} \cdot \sin \theta \cdot d\theta = 35^{1/2}/7 \tag{2.44}$$

and hence

$$\int (0)^* \mathbf{V}_{oct}(0)\, d\tau = (ze\,\overline{r^4}/a^5) \tag{2.45}$$

$$\int (\pm 1)^* \mathbf{V}_{oct}(\pm 1)\, d\tau = -(2/3)(ze\,\overline{r^4}/a^5) \tag{2.46}$$

$$\int (\pm 2)^* \mathbf{V}_{oct}(\pm 2)\, d\tau = (1/6)(ze\,\overline{r^4}/a^5) \tag{2.47}$$

$$\int (\pm 2)^* \mathbf{V}_{oct}(\mp 2)\, d\tau = (5/6)(ze\,\overline{r^4}/a^5) \tag{2.48}$$

With this information, it is possible to construct the secular determinant for the effect of $\mathbf{V}_{oct}$ on the d wavefunctions. Writing, for reasons to be made clear presently,

$$Dq = (1/6)(ze^2\,\overline{r^4}/a) \tag{2.49}$$

the secular determinant is:

$$
\begin{array}{ccccc}
(2) & (1) & (0) & (-1) & (-2) \\
\end{array}
$$

$$
\begin{vmatrix}
Dq - E & & & & 5\,Dq \\
& -4\,Dq - E & & & \\
& & 6\,Dq - E & & \\
& & & -4\,Dq - E & \\
5\,Dq & & & & Dq - E
\end{vmatrix} \tag{2.50}
$$

This determinant may immediately be reduced to:

$$
\begin{array}{lll}
(1) \text{ and } (-1) & \text{at} \quad E = -4\,Dq & \tag{2.51} \\
(0) & \text{at} \quad E = 6\,Dq & \tag{2.52}
\end{array}
$$

and the determinant:

$$
\begin{array}{cc}
(2) & (-2) \\
\end{array}
$$

$$
\begin{vmatrix}
Dq - E & 5\,Dq \\
5\,Dq & Dq - E
\end{vmatrix} \tag{2.53}
$$

This last determinant is readily diagonalized to yield the eigenvalue energies: $-4\,Dq$ and $6\,Dq$. From this 2×2 determinant, the secular equations:

$$(Dq - E)c_2 + 5\,Dqc_{-2} = 0 \tag{2.54}$$

$$5\,Dqc_2 + (Dq - E)c_{-2} = 0 \tag{2.55}$$

may be obtained. Substituting $E = -4\,Dq$, we can derive the eigenvectors:

$$c_{-2} = -c_2 \tag{2.56}$$

The wavefunction corresponding to the substituted energy is:

$$\psi = c_2\,(2) + c_{-2}\,(-2) \tag{2.57}$$

with, for normalization:

$$c_2^{\star} c_2 + c_{-2}^{\star} c_{-2} = 1 \tag{2.58}$$

The requirements on c_2 and c_{-2} are met, e.g., by: $c_2 = 2^{-1/2}$ or $c_2 = -i \cdot 2^{-1/2}$. Then the energy $-4Dq$ would correspond to the wavefunction:

$$\psi = 2^{-1/2}[(2) - (-2)] \quad \text{or} \quad -i \cdot 2^{-1/2}[(2) - (-2)] \tag{2.59}$$

The latter is, of course, the wavefunction d_{xy} as defined in Section 1.3. Similarly, substituting the energy $6\,Dq$ into the same 2×2 determinant, we find it to correspond to:

$$\psi = 2^{-1/2}\,[(2) + (-2)] \tag{2.60}$$

which is the $d_{x^2-y^2}$ wavefunction.

Collecting the results of the last few paragraphs, we find at $E = -4\,Dq$ the $d_{t_{2g}}$ set of wavefunctions:

(1) d_{xz}

$-i\,2^{-1/2}\,[(2) - (-2)]$ d_{xy}

(-1) d_{yz}

and at $E = 6\,Dq$ the d_{e_g} set of wavefunctions:

(0) d_{z^2}

$2^{1/2}\,[(2) + (-2)]$ $d_{x^2-y^2}$

The combination of (1) and (-1) into d_{xz} and d_{yz} can proceed as outlined in Section 1.3, because they are still degenerate in the presence of the crystal field, and any linear combination of them is permissable. The combination of (2) and (-2) into $d_{x^2-y^2}$ and d_{xy}, however, is *required* by the presence of the crystal field. The resultant two orbitals are *not* degenerate, and it is *not* permitted to take linear combinations of them and, e.g., turn them back into (2) and (-2).

The effort of the last few pages brings us to the same conclusion, on a quantitative basis, that was reached by the qualitative reasoning of Section 1.5. The d orbitals are split as shown in Figure 1.6. The most obvious points of progress have been the demonstration that d_{z^2} and $d_{x^2-y^2}$ are indeed degenerate and a quantitative expression for Dq. There is also the overall shift in the energy of the center of gravity of the d orbitals (Fig. 1.6) owing to the term from Eq. 2.17.

The potential V_{oct} may, by employing the expressions relating spherical polar to Cartesian coordinates, be converted into $V_{oct} = D(x^4 + y^4 + z^4 - 3r^4/5)$, where $D = 35ze/4a^5$. This form of the potential from the octahedron of charges is convenient to remember and to visualize, but it is not generally useful when performing the type of calculations that were carried out above, although it does lead to the approach taken in Section 2.8. The quantity q is, then,

$$q = 2e\,\overline{r^4}/105 \tag{2.61}$$

Dq is defined here so that the separation between the d_{e_g} and $d_{t_{2g}}$ orbitals is 10 Dq.

Because the center of gravity of the d orbitals must be unchanged by V_{oct}:

$$4\,E_{d_{t_{2g}}} + 6\,E_{d_{e_g}} = 0 \tag{2.62}$$

But

$$E_{d_{e_g}} - E_{d_{t_{2g}}} = 10\,Dq \tag{2.63}$$

The two equations are satisfied by:

$$E_{d_{e_g}} = 6\,Dq \tag{2.64}$$

$$E_{d_{t_{2g}}} = -4\,Dq \tag{2.65}$$

Therefore, given the definition of 10 Dq as the separation between the two types of d orbitals, and accepting the reasoning of Section 1.5 about the qualitative effect of the presence of the ligand atoms and the degeneracy of d_{z^2} and $d_{x^2-y^2}$, the splitting pattern may be obtained with little labor.

The quantity of 10 Dq, is often abbreviated to Δ, here specifically Δ_{oct}, particularly for use in the ligand as opposed to the crystal field sense, and that symbol for the d orbital splitting is used whenever convenient from now on.

2.4   THE EVALUATION OF Δ

If the value of $\overline{r^4}$ is known, it is possible to use the above electrostatic crystal field model to calculate the magnitude of the splitting of the d orbitals. Values of

$\overline{r^4}$ can be calculated readily for free transition metal ions. Using, say, $5.0\,a_0^4$ for $\overline{r^4}$, and taking a as, e.g., $4.0\,a_0$ (240 pm), one obtains Δ_{oct} of an order of magnitude too low. The figures just quoted lead, for $z = 1$, to:

$$10\,\mathrm{Dq}_{\text{oct}} = \sim (10/6)\,(20/5^5)$$
$$= \sim 10^{-2}\,R$$
$$= \sim 1,000\,\text{cm}^{-1} \tag{2.66}$$

In fact, Δ_{oct} is usually observed to be on the order of $10{,}000\,\text{cm}^{-1}$ for the first transition series octahedral complexes.

The lack of success in the calculation of the magnitude of Δ using the preceding model, with any reasonable value for $\overline{r^4}$, led to the development of a model in which the point charges representing the ligand atoms were replaced by point dipoles. It was thought that the dipoles might provide a more realistic representation of ligands such as water or ammonia. The point dipole assembly produces a potential at the point j, which is the difference between the potentials produced by a set of point charges at distance a and a set of opposite but equal charges in similar positions but at distance $(a + \mathrm{d}a)$. The point dipole moment μ is then $\mu = ze\,\mathrm{d}a$. Ignoring the shift in the center of gravity of the d orbitals, the model gives:

$$\mathbf{V}_{\text{oct}} = (49/18)^{1/2}\,(ze\,\overline{r^4})[(1/a^5) - (1/(a + \mathrm{d}a)^5)][Y_4^0 + (5/14)^{1/2}(Y_4^4 + Y_4^{-4})]$$
$$= (35/18^{1/2})(\mu\,\overline{r^4}/a^6)[Y_4^0 + (5/14)^{1/2}(Y_4^4 + Y_4^{-4})] \tag{2.67}$$

Note that all the matrix elements of $\mathbf{V}_{\text{oct}}$ with the d wavefunctions are of the same form in the point dipole model as they were for point charges. The only difference lies in Dq_{oct}, which is now

$$\mathrm{Dq}_{\text{oct}} = 5\mu\,\overline{r^4}/6a^6 \tag{2.68}$$

The success of the point dipole approximation in estimating practical values of Δ is little if any better than that of the point charge model (4). If μ is taken to be about 1 Debye, Δ_{oct} is again $\sim 1{,}000\,\text{cm}^{-1}$.

Much more sophisticated calculations show that the magnitude of Δ can be obtained with fair accuracy from first principles, but that any attempt to do so on either of the above simple models must fail. The more sophisticated calculations are mentioned in Chapter 4: For the present, Δ is treated as an empirical quantity. The fact that the crystal field model fails to yield Δ with reasonable accuracy does not invalidate its use as far as it is an expression of the effect of the symmetry of the ligand environment. Thus we pursue the study of the influence of $\mathbf{V}_{\text{oct}}$ on transition element ions, without, for the time being, further attempts to correlate it with the bonding in the complex.

2.5   THE TETRAHEDRAL AND CUBIC POTENTIALS

The arrangement of four point charges at the vertices of a regular tetrahedron, as shown in Section 1.5, results in a splitting of the d orbitals of the same form, but opposite in sign, to that produced by the octahedral arrangement. The quantitative treatment of the tetrahedral problem is based on Figure 1.8. There are charges of magnitude ze at the points $(\pm a/3^{1/2}, \pm a/3^{1/2}, a/3^{1/2})$ and $(\pm a/3^{1/2}, \pm a/3^{1/2}, -a/3^{1/2})$; and the summation over the contributions to the potential at any point in space, owing to these charges, proceeds as for the octahedron; there results:

$$\mathbf{V}_{(x,y,z)} = \sum_{i=1}^{4} ze/r_{ij} \tag{2.69}$$

Dropping the term in Y_0^0, which affects only the d orbital center of gravity, and labeling the remainder of the potential $\mathbf{V}_{tet}$, we have:

$$\begin{aligned}\mathbf{V}_{tet} &= -(392/729)^{1/2} (ze\,\overline{r^4}\,a^5)[Y_4^0 + (5/14)^{1/2}\,(Y_4^4 + Y_4^{-4})] \\ &= -(4/9)\,\mathbf{V}_{oct}\end{aligned} \tag{2.70}$$

So that the matrix elements

$$\int (m)\,{}^*\mathbf{V}_{tet}\,(m')\,d\tau \tag{2.71}$$

may be obtained from those for $\mathbf{V}_{oct}$ by multiplying by $-4/9$. The energies of the d orbitals in the tetrahedral crystal field are d_{t_2} at $-4\,Dq$ and d_e at $6\,Dq$, with

$$Dq_{tet} = -(4/9)\,Dq_{oct} \tag{2.72}$$

and

$$\Delta_{tet} = -(4/9)\,\Delta_{oct} \tag{2.73}$$

The d_e set is the lower lying in the tetrahedral crystal field.

Again, the elementary considerations of Chapter 1 are substantiated by the detailed calculations. The sign of the splitting of the d orbitals is reversed on passing from octahedral to tetrahedral stereochemistry. The factor $-4/9$, relating the tetrahedral to the octahedral crystal field magnitude, is based on the same point charges at the same distance from the central ion. If it is desired to compare Δ_{tet} for the coordination cluster ML_4 with Δ_{oct} of ML_6, it must be remembered that the ratio $-4/9$ holds only if the metal–ligand separation does not change between the two species.

In dealing with tetrahedral stereochemistry it is worth mentioning that a cube is composed of two interlocking tetrahedra (Fig. 1.8). The sets of d_{t_2} and d_e orbitals of one of the tetrahedra are the same as for the other tetrahedron. Consequently, the potential for the crystal field from eight ligands at the vertices of a cube may be obtained by adding together the potentials from the two tetrahedra. The result is, of course,

$$Dq_{cube} = -(8/9)\,Dq_{oct} \qquad (2.74)$$

The effect of the crystal field from a cube of ligands is to split the d orbitals into $d_{t_{2g}}$ and d_{e_g} sets at $-4Dq_{cube}$ and $6Dq_{cube}$ respectively, with the d_{e_g} set the lower lying. Note the restoration of the g subscript, because the cube has inversion symmetry.

2.6 NAMING THE NEW d ORBITALS

To conform with common practice, from now on we refer to the two sets of orbitals as split by the crystal field by just their symmetry labels: $d_{t_{2g}}$ as t_{2g}, d_{t_2} t_2, d_{e_g} as e_g, and d_e as e.

2.7 POTENTIALS FOR LOWER SYMMETRIES

The procedure for describing the potential for any other group of ligand point charges follows directly from that set out for the examples of cubic symmetry, above. However, fewer cancellations and equivalences of terms arise, so that the resultant expressions are more complicated. In particular, the difficulty arises that the averages of different powers of r are involved—specifically $\overline{r^2}$ and $\overline{r^4}$. It is obvious from the proceeding sections that the effective values of r^4 required to produce observed d orbital splittings are not closely related to the R_{3d} functions. Presumably, the same situation applies to the effective values of r^2. Thus the low-symmetry potentials involve (at least) two purely empirical parameters, which normally must be obtained from experiment.

Two low-symmetry stereochemistries are considered: the tetragonally distorted octahedron, which in the limit leads to the square-planar or linear geometry, and the trigonally distorted octahedron, with the hexagonal plane or linear system being the extreme limits.

2.7.1 Tetragonally Distorted Octahedron

The stereochemistry involved here may be derived from that of an octahedron by changing the bond lengths (Fig. 2.2a). The net effect is to change the positions of the ligand point charges along z relative to those along x and y. The distances may be denoted b and a, respectively.

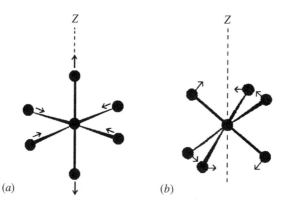

Figure 2.2 Change in the ligand positions associated with a tetragonal (**a**) and a trigonal (**b**) distortion from a regular octahedral geometry.

Ignoring the constant term in Y_0^0, the summation over the indices in Eq. 2.13 yields nonvanishing components as follows: For $n = 2$, $m = 0$, points $(\pm a, 0, 0)$ and $(0, \pm a, 0)$ each give:

$$(5/8)^{1/2} (4\pi/5) (2\pi)^{-1/2} (ze\,\overline{r^2}/a^3)(-1) \tag{2.74}$$

points $(0, 0, \pm b)$ each give:

$$(5/8)^{1/2} (4\pi/5) (2\pi)^{-1/2} (ze\,\overline{r^2}/b^3) \cdot 2 \tag{2.75}$$

Total:

$$(2\pi)^{1/2} (8/5)^{1/2} (ze\,\overline{r^2})(b^{-3} - a^{-3)} \tag{2.76}$$

For $n = 4$, $m = 0$, points $(\pm a, 0, 0)$ and $(0, \pm a, 0)$ each give:

$$(9/128)^{1/2} (2\pi)^{-1/2} (4\pi/9) (ze\,\overline{r^4}/a^5) \tag{2.77}$$

points $(0, 0, \pm b)$ each give:

$$(9/128)^{1/2} (2\pi)^{-1/2} (4\pi/9) (ze\,\overline{r^4}/a^5) \tag{2.78}$$

Total:

$$(2\pi)^{1/2} 18^{-1/2} (ze\,\overline{r^4}) (4b^{-5} + 3a^{-5}) \tag{2.79}$$

For $n = 4$, $m = 4$, points $(\pm a, 0, 0)$ and $(0, \pm a, 0)$ each give:

$$(315/256)^{1/2} (4\pi/9) (ze\,\overline{r^4}/a^5) \tag{2.80}$$

points $(0, 0, \pm b)$ each give zero. Total:

$$(2\pi)^{1/2} (35/36)^{1/2} (\mathrm{ze}\, \overline{r^4}/a^5) \tag{2.81}$$

The crystal field potential for the tetragonally distorted octahedron is, then:

$$\mathbf{V}_{\mathrm{tetrag}} = (2\pi)^{1/2} 2\mathrm{ze} \left[-(2/5)^{1/2} (b^{-3} - a^{-3}) \overline{r^2} \, \mathrm{Y}_2^0 + 72^{-1/2} (4b^{-5} + 3a^{-5}) \overline{r^4} \right.$$
$$\left. (\mathrm{Y}_4^0 + (35/2)^{1/2} a^{-5} (\mathrm{Y}_4^4 + \mathrm{Y}_4^{-4}) \right] \tag{2.82}$$

The result for the stereochemistry of a square plane is obtained by setting $b = \infty$ in this expression and is, then:

$$\mathbf{V}_{\mathrm{sq-pl}} = (2\pi)^{1/2} 2\mathrm{ze} [-(2/5)^{1/2} a^{-3} \overline{r^2} \, \mathrm{Y}_2^0$$
$$+ 8^{-1/2} \overline{r^4} \, \mathrm{Y}_4^0 + (35/2)^{1/2} (\mathrm{Y}_4^4 + \mathrm{Y}_4^{-4})] \tag{2.83}$$

Just as it is convenient to introduce the quantity Dq to paramatrize the cubic-symmetry crystal field energies, so we define an equivalent quantity in connection with the lower symmetry components that contain the term in $\overline{r^2}$. We set (5):

$$\mathrm{Cp} = (2/7) (\mathrm{ze}\, \overline{r^2}/a^3) \tag{2.84}$$

Now the crystal field energies are paramatrized by Cp and Dq. As might be expected from the failure to calculate Δ above, one cannot obtain the practical relationship between Cp and Dq from calculations with free ion radial wavefunctions. However, any simple consideration indicates that $\mathrm{Cp} > \mathrm{Dq}$, and this is borne out by experiment. Generally, it seems that for square-planar complexes

$$\mathrm{Cp} >\sim 3\mathrm{Dq} \tag{2.85}$$

The integrals involving the various d orbitals and the potential $\mathbf{V}_{\mathrm{sq-pl}}$ yield:

$$\int (0)\, {}^\star \mathbf{V}_{\mathrm{sq-pl}} (0) \cdot d\tau = -2\,\mathrm{Cp} + 18\,\mathrm{Dq}/7 \tag{2.86}$$

$$\int (\pm 1)\, {}^\star \mathbf{V}_{\mathrm{sq-pl}} (\pm 1) \cdot d\tau = -\mathrm{Cp} - 12\,\mathrm{Dq}/7 \tag{2.87}$$

$$\int (\pm 2)\, {}^\star \mathbf{V}_{\mathrm{sq-pl}} (\pm 2)\, d\tau = 2\,\mathrm{Cp} + 3\,\mathrm{Dq}/7 \tag{2.88}$$

$$\int (\pm 2)\, {}^\star \mathbf{V}_{\mathrm{sq-pl}} (\mp 2)\, d\tau = -35\,\mathrm{Dq}/7 \tag{2.89}$$

Proceeding as before, the energies corresponding to the d orbitals are:

$$
\begin{array}{lll}
d_{z^2} & \text{at} & -2Cp + 18\,Dq/7 \\
d_{xz}, d_{yz} & \text{at} & -Cp - 12\,Dq/7 \\
d_{x^2-y^2} & \text{at} & 2Cp + 38\,Dq/7 \\
d_{xy} & \text{at} & 2Cp - 32\,Dq/7
\end{array}
$$

Note that, contrary perhaps to intuition, the d_{z^2} orbital does not necessarily lie lowest in energy in the crystal field treatment of the square plane. The relationship between the energies of the d orbitals and the ratio Cp/Dq are set out in Figure 2.3.

2.7.2 Trigonally Distorted Octahedron

Referred to a threefold axis z passing through a pair of opposite face centers, the vertices of an octahedron can be defined by a vector of length a, which makes an angle θ with z and whose projection on the $x-y$ plane makes an angle ϕ with x (Fig. 2.2b). For a regular octahedron, $\theta = \cos^{-1}(1/3) = 54.74°$ or $125.26°$. ϕ has the values π, $+2\pi/6$ and $-2\pi/6$ or 0, $2\pi/3$ and $-2\pi/3$ for the first, second, and third vertices, respectively.

If the octahedron is not regular but is distorted by "squeezing" or "pulling" the ligands along the threefold axis, the resultant stereochemistry is specified as

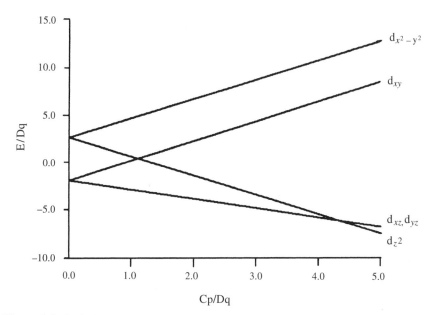

Figure 2.3. Variation of the energy of the d orbitals (units Dq) as a function of the ratio of the parameters relating to $\overline{r^2}$ and $\overline{r^4}$, viz Cp and Dq.

above, but θ is no longer 54.74° but rather some other angle. The summation of the components of the potential for the trigonally distorted octahedron proceed as before. Ignoring the term in Y_0^0, the nonzero contributions are as follows. For $n = 2$, $m = 0$, the six points each give:

$$(5/8)^{1/2} (4\pi/5) (2\pi)^{-1/2} (ze\,\overline{r^2}/a^3) (3\cos^2\theta - 1) \tag{2.90}$$

For $n = 4$, $m = 0$, the six points each give:

$$(9/128)^{1/2} (4\pi/5) (2\pi)^{-1/2} (ze\,\overline{r^4}/a^5)(35\cos^4\theta - 30\cos^2\theta + 3) \tag{2.91}$$

For $n = 4$, $m = \pm 3$, the points at θ, $\phi = 0$; $\pi - \theta$, $\phi = \pi$ each give:

$$\pm(315/32)^{1/2} (4\pi/9) (2\pi)^{-1/2} (ze\,\overline{r^4}/a^5) \cos\theta \cdot \sin^3\theta \tag{2.92}$$

the points at θ, $\phi = \pm 2\pi/3$ *together* give:

$$\pm 2(315/32)^{1/2} (4\pi/9) (2\pi)^{-1/2} (ze\,\overline{r^4}/a^5) \cos\theta \cdot \sin^3\phi \tag{2.93}$$

the points at θ, $\phi = \pm 2\pi/6$ give:

$$\pm 2(315/32)^{1/2} (4\pi/9) (2\pi)^{-1/2}(ze\,\overline{r^4}/a^5) \cos\theta \cdot \sin^3\phi$$

Total:

$$12(315/32)^{1/2} (2\pi)^{1/2} (ze\,\overline{r^4}/a^5) \cos\theta \cdot \sin^3\phi \tag{2.94}$$

Thus:

$$\mathbf{V}_{\text{trig}} = (18/5)^{1/2} (2\pi)^{1/2} (ze\,\overline{r^2}/a^3)(3\cos^2\theta - 1) - 8^{-1/2} (2\pi)^{1/2} (ze\,\overline{r^4}/a^5) \cdot$$
$$[35\cos^4\theta - 30\cos^2\theta + 3) Y_4^0 + 140^{1/2} \cos\theta \cdot \sin^3\phi\,(Y_4^3 - Y_4^{-3})] \tag{2.95}$$

A trigonal distortion splits the t_{2g} but not the e_g orbitals of an octahedral complex. The energies of the d orbitals obtained from the application of $\mathbf{V}_{\text{trig}}$ are set out in Appendix 3.

2.8   OTHER PARAMETERIZATION SCHEMES

The development of the preceding sections began with the simplest and most straightforward lines available, at, it must be admitted, the expense of some tedium. It is possible by using more sophisticated mathematical techniques to reduce the labor of deriving crystal field potentials and of evaluating the integrals that occur on their application. These techniques are based in various ways on operator methods (6). We do not intended to examine these techniques in any detail here, but it is pertinent to point out some of the features connected with them, because they receive frequent mention in the physics literature.

2.8.1 The A and B Coefficients

The general crystal field potential may be written in the succinct form (6):

$$\mathbf{V}_{cf} = \sum_{n=0}^{2l} \sum_{k=-n}^{n} \mathbf{B}_n^k \mathbf{O}_n^k \tag{2.96}$$

$$= \sum_{n=0}^{2l} \sum_{k=-n}^{n} \mathbf{A}_n^k \mathbf{Q}_n^k \tag{2.97}$$

where l is the quantum number for the type of electrons under consideration ($d \to 2$, $f \to 3$). In practical cases n is restricted to even integers and k to a limited number of the possible values, determined by the symmetry; furthermore:

$$\mathbf{B}_n^k = (-1)^{-k} \mathbf{B}_n^{-k*} \quad \text{and} \quad \mathbf{A}_n^k = \mathbf{A}_n^{-k} \tag{2.98}$$

$\mathbf{O}_n^k$ is an operator form of $r_l^n Y_l^k (\theta, \phi)$, and $\mathbf{Q}$ is an operator based on the expansion of these spherical harmonics as functions of the Cartesian coordinates x, y, and z.

Some relationships between the B and A parameters and those of the preceding sections are of the type:

$$\mathbf{B}_2^k = \mathbf{Cp} \, F_4^k \tag{2.99}$$

$$\mathbf{B}_4^k = \mathbf{Dq} \, F_4^k \tag{2.100}$$

where F_2 and F_4 are functions of θ and ϕ, appropriate to the symmetry involved, e.g.:

$$\mathbf{B}_2^0 = -ze \, r^2 \, (3\cos^2\theta - 1)/7a^3 = \mathbf{Cp} \, F_2^0 \tag{2.101}$$

Generally, however no attempt is made to relate the A's or B's to the details on the ligand distribution; they are used as empirical parameters to reproduce the data. Some relationships between the A and B parameters are given in Appendix 4 along with a list of the $\mathbf{Q}$ functions.

When dealing with a general distribution of ligand charges about the central metal atom, it is not even necessary to write down the crystal field potential. For the d orbitals, the matrix elements corresponding to *all* possible cases have been tabulated and arranged into a formalism that allows them to be obtained directly as expressions of the charges, distances, and angles involved. The procedure is set out in Appendix 5.

2.8.2   The Parameters Ds and Dt

The parametrization of the potential V_{sq-pl} in terms of Cp and Dq as outlined above is straightforward. If, however, the same procedure is applied to the potential V_{tetrag}, a complication arises: Two distances are involved, a and b. It is quite possible to redefine Cp as some average that includes both distances:

$$Cp = (2/7)\, ze\, \overline{r^2}\, (b^{-3} - a^{-3}) \tag{2.102}$$

The effect of V_{tetrag} on the d orbitals is then developed as before. Often an alternative scheme for tetragonal symmetry is used, and it has some advantages. Setting

$$Cp(a) = (2/7)\, ze\, \overline{r^2}/a^3 \tag{2.103}$$
$$Cp(b) = (2/7)\, ze\, \overline{r^2}/b^3 \tag{2.104}$$
$$Dq(a) = (1/6)\, ze\, \overline{r^4}/a^5 \tag{2.105}$$
$$Dq(b) = (1/6)\, ze\, \overline{r^4}/b^5 \tag{2.106}$$

we derive the quantities (7):

$$Ds = Cp(a) - Cp(b) \tag{2.107}$$
$$Dt = (4/7)\, [Dq(a) - Dq(b)] \tag{2.108}$$

Then, with $Dq = Dq(a)$, the energies of the d orbitals are:

$$d_{z^2} \text{ at } 6\,Dq - 2\,Ds - 6\,Dt$$
$$\text{Splitting} = 4\,Ds + 5\,Dt$$
$$d_{x^2-y^2} \text{ at } 6\,Dq + 2\,Ds - Dt$$

and

$$d_{xz,yz} \text{ at } -4\,Dq - Ds + 4\,Dt$$
$$\text{Splitting} = 3\,Ds - 5\,Dt$$
$$d_{xy} \text{ at } -4\,Dq + 2\,Ds - Dt$$

2.9   LIMITATIONS OF CRYSTAL FIELD THEORY: LIGAND FIELD THEORY

The artificial nature of the representation of a metal ion and its ligands as point charges in the original electrostatic form of the crystal field model was recognized almost from the time of its introduction. Although this model succeeded in rationalizing many facets of the spectral and magnetic properties of transition-metal complexes, it did this only because it was based on a most important feature of a coordination complex: the symmetry of the arrangement of the ligands about the metal ion. The nature of the interaction between the d electrons and the ligands has a number of implications.

The development of the crystal field potential in this and the previous chapter has been based on the assumption that only the immediate ligand neighbors of a metal ion are important. At first sight, the assumption might appear to be reasonable, because the potential, $\mathbf{V}_{oct}$ say, contains an a^{-5} term, and in many structures ligands not in the first coordination sphere are several times as far away from the metal ion as are those in it. A summation to include the charges on other metal and ligand ions outside the first coordination sphere is an obvious test of the assumption. Details of such a summation are not developed here: It precedes along the same lines as in the previous sections, but is quite tedious. The results, of course, differ substantially from lattice to lattice, depending on the packing arrangements of the more remote ions. For many lattices, the longer range symmetry about a metal ion is low, even though its *local* symmetry, that of the first-coordination sphere, is cubic (or close to this). This requires the introduction of components that fall off only as r^{-3} or even r^{-1}, which produces a term comparable to $\mathbf{V}_{oct}$ itself.

The discussion in the previous paragraph leads to the conclusion that the crystal field potential should depend critically on the crystal lattice in which the transition metal ion is housed. Experimentally, the opposite is observed. Crystal field effects are to a first approximation independent of the disposition of the ions outside the first-coordination sphere, and even of the transition from the solid to the liquid state. The spectra, magnetic, and other properties of complex ions in solution are much the same as in their crystalline form. Furthermore, experiments that follow Dq as a function of a, such as the observation of the spectrum of NiO under high pressure, give agreement with the relationship

$$\text{Dq} \propto a^{-5} \tag{2.109}$$

expected from nearest-neighbour-only participation. It should be noted that this relationship will not be followed if the atoms concerned are part of *different* ligand molecules. For example in chromium(III) and cobalt(III) complexes, the metal–carbon separations are 230 and 200 pm when a methyl group is bonded and 208 and 189 pm with cyanide, but the d orbital splitting is little different for methyl groups and for the cyanide ion.

Attempts have also been made to correlate the components of the crystal fields of symmetry lower than cubic—principally the term in Y_2^0 such as appears in $\mathbf{V}_{tetrag}$ — with the more distant ions in the lattice. Complicated scaling procedures have been invoked to try to make crystal field parameters transferable from crystal to crystal. It cannot be said, however, that the results have been worth the trouble involved.

As well as failing to reproduce d orbital splittings quantitatively or to account for the success of the first coordination sphere approximation, the electrostatic form of the crystal field model has other limitations. As will be set out in detail later, it is found that the parameters representing several other properties of the free ion are changed, sometimes substantially, when a metal ion is incorporated into the ligand environment of a complex. These properties are principally interelectronic repulsion energies, spin–orbit coupling, and orbital angular momentum.

An obvious deduction from all these factors is that the d orbital splitting phenomenon is a product not so much of classical electrostatic interactions between a metal cation and surrounding negative charges, but of the interactions involved in the covalent contributions to the chemical bonding of the metal atom with its ligands. The overriding importance of the first coordination sphere ligands and the variation of the constants representing metal atom properties are then, in principle at least, explained quite simply. Thus the qualitative success of the crystal field model is seen to arise only because it provides an account of effects owing to the *symmetry* of the ligand environment.

A development of crystal field theory that admits that the d orbital splitting arises from chemical bonding, rather than an electrostatic interaction, but that retains the formalism of that theory, is generally called *ligand field theory*. Ligand field theory often uses the same potentials — $\mathbf{V}_{oct}$, $\mathbf{V}_{tetrag}$, etc. — as does crystal field theory, but acknowledges that these lack a proper physical basis. The parameters and constants involved in the use of these potentials (Dq, Cp, Ds, Dt) remain, but their relationship to atomic properties is sought in the nature of the chemical bond rather than in classical electrostatics. Ligand field theory provides a flexibility not available in a purely ionic approach, one that is not constrained to employ unchanged the parameters defining the properties of unbonded ions.

Although the ligand field development removes some of the difficulties apparent from an entirely ionic model, an obvious question arises: How *are* the parameters that have been carried over from crystal field theory indeed related to the covalent bonding interactions between atoms? Often, no attempt is made to answer this question. The parameters are simply taken as experimental observations related to the chemical bonding only in quite general terms. One form of ligand field theory, the angular overlap model (Chapter 3), does go some way toward a chemical interpretation. It describes the d orbital splittings using parameters which represent the σ and π interactions between the metal ion and each ligand in the complex. Even here, however, the model is simply a method of parametrizing experimental observations, albeit in a way that relates directly to the framework generally used to discuss chemical behavior. The treatment of a metal–ligand cluster by the proper procedures of theoretical chemistry leads to results that cannot be directly compared with the parameters of ligand field theory. It seems then that the crystal field / ligand field model is not a satisfactory basis for a physically real account of the interaction of a metal ion with its ligands. Nevertheless, it provides a formalism within which such interactions may be parametrized. Because the rigorous treatment of a complex ion by the methods of theoretical chemistry is not likely to become generally available, at least in the foreseeable future, such parametrization is of value and constitutes the best that can be done in most circumstances.

2.10 f ORBITALS AND THE CRYSTAL FIELD POTENTIAL

It is straightforward, although tedious, to develop the crystal field potential for an octahedron to include the sixth-order spherical harmonies that are needed to deal with the f orbitals. The result referred to the axis as z is

$$\mathbf{V}_{oct} = (2\pi)^{1/2}((49/18)^{1/2} (ze^2 \overline{r^4}/a^5)[Y_4^0 + (5/14)^{1/2} (Y_4^4 + Y_4^{-4})]$$
$$+ (9/832)^{1/2} (ze^2 \overline{r^6}/a^7)[Y_6^0 - (7/2)^{1/2} (Y_6^4 + Y_6^{-4})]) \qquad (2.110)$$

The matrix elements involved in investigating the effect of this potential on the f orbital set evaluated with the help of the results for expressions of the type:

$$\int \Xi_3^m \cdot \Xi_3^{m'} \Xi_6^{m-m'} \cdot \sin\theta \cdot d\theta \qquad (2.111)$$

are given in Appendix 1. We write:

$$\langle f_m | \mathbf{V}_{oct} | f_{m'} \rangle = \int f_m^* \mathbf{V}_{oct} f_{m'} \, d\tau \qquad (2.112)$$

The results are:

$$\langle 0 | \mathbf{V}_{oct} | 0 \rangle = 6\,Dq' \quad + 20\,Fr \qquad (2.113)$$
$$\langle \pm 1 | \mathbf{V}_{oct} | \pm 1 \rangle = Dq' \quad - 15Fr \qquad (2.114)$$
$$\langle \pm 2 | \mathbf{V}_{oct} | \pm 2 \rangle = -7Dq' \quad + 6Fr \qquad (2.115)$$
$$\langle \pm 3 | \mathbf{V}_{oct} | \pm 3 \rangle = 3Dq' \quad - Fr \qquad (2.116)$$
$$\langle \pm 1 | \mathbf{V}_{oct} | \mp 3 \rangle = 15^{1/2}Dq' + 735^{1/2}Fr \qquad (2.117)$$
$$\langle \pm 2 | \mathbf{V}_{oct} | \mp 2 \rangle = 5Dq' \quad - 42Fr \qquad (2.118)$$

where D has the same significance as before with the d orbitals, while:

$$Dq' = (2/165) (e\,\overline{r^4}/a^5) \qquad (2.119)$$
$$Fr = (5/572)^{1/2} (ze\,\overline{r^6}/a^7) \qquad (2.120)$$

The notation Fr has been introduced to conform with the series Cp for the term in $\overline{r^2}$ and Dq for that in $\overline{r^4}$. F is the coefficient that would be used in connection with the operator $\mathbf{Q}_6^0$ (Appendix 4).

The f orbital energies are obtained by the same procedure as in the d orbital case, and the linear combinations corresponding to them are deduced. We obtain:

$$
t_{1u} \quad
\begin{aligned}
&f_{z(x^2-y^2)} \\
&f_{x(y^2-z^2)} \\
&f_{y(z^2-x^2)}
\end{aligned}
\qquad \text{at} \quad 6\,\mathrm{Dq}' + 20\,\mathrm{Fr}
$$

$$
t_{2u} \quad
\begin{aligned}
&f_{z(5z^2-r^2)} = f_0 \\
&f_{x(5x^2-r^2)} \\
&f_{y(5y^2-r^2)}
\end{aligned}
\qquad \text{at} \quad -2\,\mathrm{Dq}' - 36\,\mathrm{Fr}
$$

$$
a_{2u} \quad f_{xyz} \qquad \text{at} \quad -12\,\mathrm{Dq}' + 48\,\mathrm{Fr}
$$

The relationship between the two parameters determining the f orbital energies depends on the ratio $\overline{r^4}/\overline{r^6}$, which is at least as uncertain as $\overline{r^2}/\overline{r^4}$ for d orbitals, although one may expect $\mathrm{Dq}' < \mathrm{Fr}$, for the same reasons that $\mathrm{Cp} > \mathrm{Dq}$. Consequently, even the relative f orbital energy separations cannot be predicted quantitatively from first principles. The order $t_{1u} > t_{2u} > a_{2u}$, however, should be maintained. In the literature, the separation $t_{1u} - t_{2u}$ has been parameterized as $\theta (= 8(B_4^0 + 35B_6^0/13)/33)$, and the separation $t_{2u} - a_{2u}$ as $\Delta\,(= 2(5B_4^0 - 210B_6^0/13)/33)$ (8). Because octahedral (or other cubic) stereochemistry is rare among lanthanide element compounds, the f orbital splitting pattern observed there is more complicated still. No attempt is made, in general, to relate the f orbital splittings to anything more than the symmetry of the ligand environment. Most usually a set of A or B parameters for the operators $\mathbf{O}$ or $\mathbf{Q}$ is employed, and the parameters are treated entirely empirically. For example, when the symmetry of the ligand environment is trigonal, as is not uncommon, the set of parameters

$$
B_2^0, B_4^0, B_4^{\pm 3}, B_6^0 \quad \text{and} \quad B_6^{\pm 3} \tag{2.121}
$$

might be invoked.

For the actinide elements, the higher symmetry environments are often known, and the splittings involved are larger. Then a set of parameters such as Cp, Dq', Fr could be more useful. The f electron systems, particularly the actinides, are discussed at more length in Chapter 11.

REFERENCES

1. Eyring, H., Walter, J., and Kimball, G. E., *Quantum Chemistry*, Wiley, New York, 1944, chap. 7.

2. Margenau, H., and Murphy, G. M., *The Mathematics of Physics and Chemistry*, Van Nostrand, New York, 1956.

3. Ref. 1, append. 5.

4. Polder, D., *Physica*, 1942, **9**, 709.

5. Gerloch, M., and Slade, R. C., *Ligand Field Parameters*, Cambridge University press, Cambridge, UK, 1973, chap. 4.

6. Abragam, A., and Bleaney, B., *Electron Paramagnetic Resonance of Transition Ions*, Clarendon, Oxford, UK, 1970, chap. 16.

7. Ballhausen, C. J., *Introduction to Ligand Field Theory*, McGraw-Hill, New York, 1962, chap. 5.

8. Reisfield, M. J., and Crosby, I. C., *Inorg. Chem.*, 1965, **4**, 65.

CHAPTER 3

THE ANGULAR OVERLAP MODEL

3.1 BASIS OF THE ANGULAR OVERLAP MODEL

At the end of Chapter 2, the limitations of the electrostatic crystal field theory were outlined. As pointed out, some of these may be overcome by decreasing the values of the parameters that describe features such as interelectron repulsion and spin–orbit coupling from their free ion values to allow for covalency. Although, the resultant development, generally called ligand field theory, is more flexible and apparently more realistic, it still does not stand up to a rigorous scrutiny. The conceptual framework now most widely used to discuss the general chemistry of the transition metals is a simple molecular orbital (MO) approach. This describes the metal–ligand bonding in terms of covalent σ and π bonding interactions, rather than using the electrostatic perturbations of crystal field theory. The concepts inherent in the simple MO approach form the basis of a powerful method of describing metal–ligand interactions which retains and builds on the empiricism of that ligand field approach. It is known as the angular overlap model (AOM) (1–4).

It is important to recognize that the AOM involves severe approximations and was never meant to be a quantitative method of *calculating* physically real values of the energy levels in a transition-metal complex. Rather, it is a way of parametrizing the metal–ligand interactions. The resulting parameters are always derived from experiment, so that it is essentially a form of ligand field theory. It may be argued that the AOM is superior to crystal field theory because its basis—that weak covalent interactions between the ligand orbitals and metal d orbitals cause properties such as color—is closer to physical reality than is the electrostatic basis that underlies crystal field theory. In our view, however, the most important advantages of the AOM are not theoretical in origin; rather they stem from the fact that it correlates directly with the way most chemists interpret experimental observations. In addition, the bonding parameters refer to a particular metal–ligand interaction, not the complex as a whole, and it appears that, within limits, it may be possible to transfer the parameters from one compound to another. This means that it offers the potential that parameters derived from spectroscopic or magnetic measurements may be used to quantify the σ and π bonding interactions that have long been used to describe the behavior of ligands in transition-metal complexes qualitatively. Although it is still too early to be certain of the overall self-consistency of the model, this facet offers a distinct advantage over the traditional modified crystal field approach.

Because the simple MO picture of the bonding in transition-metal complexes is now included in most general chemistry text books, it is not presented here. It is however, pertinent to review briefly the concepts of the model that underlie the AOM in its most commonly used form.

3.1.1 Simple MO Picture of the Bonding in Transition-Metal Complexes

The MO approach assumes that when ligands bind to a metal ion their atomic orbitals overlap to yield a set of molecular orbitals, and the electronic structure of the complex is obtained by feeding the valence electrons into these MOs, obeying the Pauli exclusion principle and Hund's rule. Major features of the pattern of MOs may be obtained by noting that if a combination of ligand orbitals has an identical symmetry to a set of metal atomic orbitals these interact to produce a set of bonding and antibonding MOs. As a simple example, we consider a complex such as the $Cr(NH_3)_6^{3+}$ ion. The nitrogen atom of each ammonia ligand is assumed to use the lone electron pair occupying an sp^3 hybrid orbital to bond to the metal ion, and its remaining electrons are involved in bonding to the hydrogen atoms. Ignoring the hydrogen atoms, the complex belongs to the point group O_h, and combinations of ligand lone pair orbitals exist that form σ molecular orbitals with the metal 4s and 4p orbitals, as well as with the $e_g^\star$ set of d orbitals. The bonding combination involving the $d_{x^2-y^2}$ member of the $e_g^\star$ set is shown in Figure 3.1.

The energies of the MOs in a complex are conveniently represented using a schematic energy diagram. One for the $Cr(NH_3)_6^{3+}$ ion is shown in Figure 3.2a. Here, any interaction between the ligand lone pair orbitals is ignored. Note that, at the level considered, no combination of ligand lone pair orbitals has t_{2g} symmetry, so that this set of metal d orbitals does not form MOs and remains nonbonding. It must be stressed that this is not simply because the orientation of the metal t_{2g} orbitals is unfavorable for π bond formation. The symmetry of the overlap is such that any bonding overlap is exactly counterbalanced by a corresponding antibonding overlap, as illustrated in Figure 3.3a. Most ligands, however, have orbitals available that can interact with the metal t_{2g} orbitals to form π bonds. Usually, these are lower in energy than the metal orbitals and are

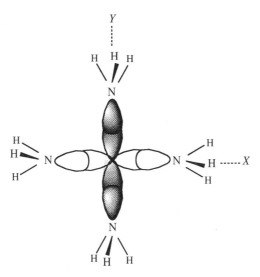

Figure 3.1. The way in which the bonding combination of the lone pair orbitals of ammonia ligands overlap with the $d_{x^2-y^2}$ orbital in a metal hexaammine complex. The unshaded and shaded lobes represent the positive and negative phases, respectively, of the wavefunctions.

filled with electrons. A typical example is the fluoride ion, and the π interaction between the d_{xz} and the fluoride p_z orbital is shown in Figure 3.3b. The schematic MO energy diagram for the complex CrF_6^{3-} is given in Figure 3.2b. Again, interactions between the ligand orbitals are neglected.

It may be seen that the main difference from $Cr(NH_3)_6^{3+}$ as far as the relative energies of the d orbitals are concerned is that the t_{2g} set of orbitals becomes weakly antibonding in the fluoro complex. For ligands that contain multiple bonds, such as the cyanide ion or α,α'-bipyridyl, the empty antibonding π^* orbitals are of the correct symmetry and orientation to interact with the metal t_{2g} orbitals (Fig. 3.3c). The effect of the π interaction is then to make the metal t_{2g} set weakly bonding, as shown for the $Cr(CN)_6^{3-}$ ion in Figure 3.2c.

The Cr^{3+} ion has three valence electrons, and in an octahedral complex these occupy the t_{2g} (or t_{2g}^*) set of orbitals. In the hexaammine complex, these orbitals are nonbonding so that the electrons remain localized in the d orbitals. In the fluoro complex, however, the interaction with the ligand π orbitals is accompanied by some delocalization and because the ligand orbitals are filled, whereas the metal orbitals are only half filled, the net result is a transfer of electron density from the ligand to the metal atom. Thus the fluoride ion is said to be a π *donor* ligand. When the interaction is with empty ligand π^* orbitals, electron density is effectively transferred from the partly full metal t_{2g} set to the ligand, so that ligands such as cyanide ion are deemed to be π *acceptors*.

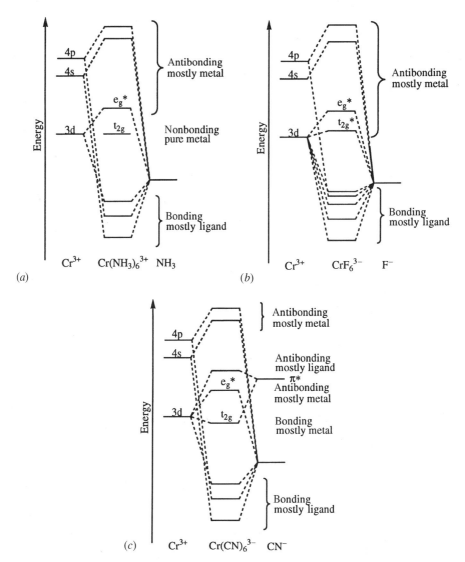

Figure 3.2. The relative energies of the most important MOs in a typical complex involving a non π bonding ligand such as NH_3 (a), a π donor such as F^- (b), and a π acceptor such as CN^- (c).

The important features of the simple MO picture of the bonding in coordination complexes relevant to the AOM may be summarized as follows:

1. Although the complexes are ionic in the sense that the metal and often the ligands are ions, the metal-ligand interaction is presumed to have a substantial covalent component.

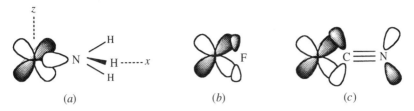

Figure 3.3. Interaction of the d_{xz} orbital with the lone pair orbital of an ammonia ligand (a), a filled p orbital of π symmetry of a fluoride ion (b), and the empty antibonding π^* orbital of the cyanide ion (c).

2. The bond energy is derived mainly from the interaction between the ligand orbitals and metal s and, to a lesser extent, p orbitals. Energetically, the interaction with the metal d orbitals is much less important.

3. Although the formal oxidation state of the transition metal decides its d configuration, the covalent interactions involving the metal s and p orbitals mean that the effective positive charge on the metal is considerably lower than its formal oxidation state. This causes the d orbitals to expand significantly compared to the wavefunctions of a free transition ion.

4. The e_g^* molecular orbitals consist of a linear combination of metal d orbitals and ligand orbitals, but with a much larger d orbital contribution. Similar arguments hold for the t_{2g} orbital set when π bonding is present, although here the ligand participation is even smaller because the bonding is weaker.

5. The magnitude of the interaction between the d orbitals and the ligand orbitals depends on the amount by which they overlap and the energy separation between the metal and ligand orbitals. The larger the energy separation, the weaker the interaction, in terms of both energy and orbital mixing.

6. Because the d orbitals are relatively compact, the overlap between these and the ligand orbitals increases upon their expansion. Other things being equal, overlap thus increases as the principal quantum number rises when going from the first- to the second- to the third-transition series. Overlap decreases as the effective nuclear charge rises on crossing each transition series and when the oxidation state of a metal is increased.

7. The energy gap between the metal and ligand orbitals depends on the oxidation state of the metal and the electronegativity of the ligand. For filled ligand orbitals, a high oxidation state and a low ligand electronegativity lead to a small energy separation. For empty ligand π^* orbitals, the lower the oxidation state of the metal, the smaller the energy separation from the metal d orbitals.

3.1.2   Derivation of AOM Parameters Using the Wolfsberg–Helmholtz Approximation

As discussed in the preceding section, molecular orbitals are generally expressed as linear combinations of atomic orbitals:

$$\phi = \sum_{jk} C_{jk}\psi_{jk} \tag{3.1}$$

where ψ_{jk} is the kth wavefunction of the jth atom of the molecule. The molecular wavefunctions are eigenfunctions of the Hamiltonian operator $\mathbf{H}$ so that for a complex with molecular orbitals ϕ_i, the energies E_i are obtained from Eq. 3.2:

$$\mathbf{H}\phi_i = E_i\phi_i \tag{3.2}$$

This Hamiltonian operator contains the kinetic energy operators for each electron of the molecule and various potential terms for the interactions of the charges on the nuclei and the electrons. The solution of the set of differential equations represented by Eq. 3.2 in the general (*ab initio*) case is difficult and exceedingly tedious. The problem is greatly simplified by limiting the number and quality of the atomic orbitals considered, by omitting or estimating some of the terms in the Hamiltonian operator (approximate methods), or by using experimental data to define them (empirical methods). Of course, such simplifications are obtained at the cost of losses in rigor, generality, accuracy, and predictive power.

The AOM makes use of each of these simplifications. The number of orbitals is reduced to the most elementary possible case: the bonding between one ligand orbital and a d orbital on the metal (it may be an advantage in some situations to extend the basis set to include the metal 4s orbital (Section 3.2.2). The molecular orbital then becomes:

$$\phi_i = C_{iM}\phi_M + C_{iL}\phi_L \qquad i = 1 \text{ or } 2 \tag{3.3}$$

and the energies are the roots of the determinant derived from Eq. 3.2.

$$
\begin{array}{cc}
\phi_M & \phi_L \\
\end{array}
$$
$$
\left| \begin{array}{cc}
H_M - E & H_{ML} - S_{ML}E \\
H_{ML} - S_{ML}E & H_L - E
\end{array} \right| = 0 \tag{3.4}
$$

Where, $H_M = \int \phi_M \mathbf{H} \phi_M d\tau$; $H_L = \int \phi_L \mathbf{H} \phi_L d\tau$; $H_{ML} = \int \phi_M \mathbf{H} \phi_L d\tau$; and $S_{ML} = \int \phi_M \phi_L d\tau$ (overlap integral). After multiplying out the determinant and introducing approximations based on the assumption that H_M and H_L are well separated, with $H_L < H_M$ and S_{ML} being small, we have:

$$E_a = H_M + (H_{ML}^2 - 2H_{ML}H_M S_{ML} + H_M^2 S_{ML}^2)/(H_M - H_L) \qquad (3.5a)$$
$$E_b = H_L - (H_{ML}^2 - 2H_{ML}H_L S_{ML} + H_L^2 S_{ML}^2)/(H_M - H_L) \qquad (3.5b)$$

Here, E_b is the energy corresponding to bonding overlap, E_a to antibonding overlap. The diagonal matrix elements H_M and H_L represent the valence state ionization energies of the metal and ligand orbitals, respectively, in the complex. The overlap S_{ML} may, in principle at least, be calculated numerically. The remaining quantity, the off-diagonal matrix element H_{ML}, is not directly available from experiment and is troublesome to calculate. The AOM obtains this by making use of the Wolfsberg–Helmholtz approximation, which is:

$$H_{ML} \cong S_{ML}(H_M + H_L)/2 \qquad (3.6)$$

Then

$$E_a \cong H_M + H_L^2 S_{ML}^2/(H_M - H_L) \qquad (3.7)$$
$$E_b \cong H_L - H_M^2 S_{ML}^2/(H_M - H_L) \qquad (3.8)$$

In this approximation, the energy of the antibonding molecular orbital E_a, which is largely metal in character, is raised by an amount proportional to the square of the overlap integral:

$$e \approx KS^2; \quad K \approx H_L^2/(H_M - H_L) \qquad (3.9)$$

The overlap between the two orbitals is generally classified as being of σ or π symmetry referred to the internuclear axis. Four σ interactions are shown in Figure 3.1 for the ammonia ligands coordinated along the x and y axes of an octahedral complex, and two different kinds of π interactions are shown in Figure 3.3. In principle, interactions of δ symmetry may also occur. These involve an overlap region concentrated off the internuclear axis, with a charge density having fourfold symmetry about that axis. In general, δ interactions could be important only for donor atoms such as phosphorus which have low-energy d orbitals. From simple overlap considerations they are expected to be much less important than the σ and π interactions, so they are almost always ignored. Just three interactions are, therefore, used to describe the energy changes occurring when one ligand atom bonds to a transition metal at this level of approximation.

The effect arising from a ligand orbital of σ symmetry is represented by the parameter e_σ, and those from orbitals of π symmetry are given by $e_{\pi x}$ and $e_{\pi y}$. Here x and y define orthogonal axes normal to the bond axis. For planar ligands such as pyridine, x is chosen to be parallel to the plane of the ligand. Because the d orbitals in a complex are almost always expected to be antibonding with respect to σ interactions, e_σ is positive. On the other hand, π interactions may produce MOs which are either raised or lowered relative to the metal d orbitals, depending on whether the ligand orbitals are lower or higher in energy than the metal orbitals (Fig. 3.2b and c). This affects the sign of K in Eq. 3.9; and the e_π parameters are positive for the former antibonding situation but negative for the latter bonding situation.

3.1.3   Derivation of the d Orbital Energies and Wavefunctions in a Complex Using the AOM

In the AOM, the above description is extended to the complex as a whole simply by summing the effects of the individual ligand atoms. That is, each metal–ligand interaction is described by a set of parameters—e_σ, $e_{\pi x}$, and $e_{\pi y}$—and the energy levels of the complex are obtained by summing these interaction terms, taking into account the geometry of the complex by means of the angular overlap matrix. The energies and wavefunctions are given by the eigenvalues and eigenvectors of this matrix, respectively. For this approximation to hold, the effects of ligand–ligand interactions must be small. An important consequence of this facet of the model is that the e_σ and e_π parameters are, in principle at least, characteristic of a particular metal ion and ligand and, within limits, may be transferable from one complex to another. The situation is often more complicated for many polydentate ligands.

Neglecting for the moment these last-mentioned complications (see Section 3.3), the d orbital energies in a complex are obtained by considering the geometric relationships between the x, y, z coordinate systems used to define the $e_{\pi x}, e_{\pi y}$, and e_σ parameters of each ligand and a global coordinate system X, Y, Z centered on the metal. The bond direction is given in terms of two angles: that between the Z axis and the M–L direction θ and that between the line defined by the intersection of the XY plane and the plane containing the M–L direction and the Z axis, and the X axis, ϕ. (Fig. 3.4). The orientation of the ligand x and y axes is given by a final rotation ψ about the metal–ligand bond axis. The angular overlap factors that indicate the effect on the metal–ligand overlap integrals of a rotation through angles θ, ϕ, ψ, are given in Table 3.1. The effects of a set of ligands is obtained by simple addition as has been set out in detail (1–4).

A great advantage of the AOM is that it is quite straightforward to treat a complex of low symmetry—it is a matter of simple trigonometry to use the ligand positions to define the appropriate angular overlap matrix. The metal–ligand interactions in such a low-symmetry complex, however, involve nonzero off-diagonal matrix elements of the angular overlap matrix, so that the energy levels are not given by simple equations involving e_σ, e_π, etc., and the wavefunction associated with each energy level is in general a complicated mixture of the d functions. Of course, many transition metal complexes belong to high-symmetry point groups. Then the eigenvectors of the angular overlap matrix generally correspond to separate sets of d functions, and the corresponding eigenvalues are simple arithmetic equations involving the AOM bonding parameters. Some typical examples are given below.

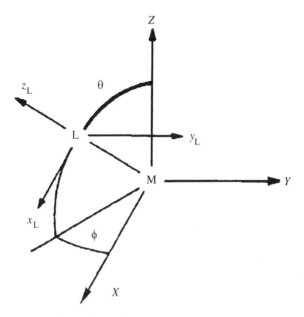

Figure 3.4. Angles used to relate a ligand-based coordinate system x, y, z to a global coordinate system X, Y, Z.

3.2 AOM EXPRESSIONS FOR COMPLEXES OF VARIOUS SYMMETRY

3.2.1 Octahedral Complexes

A typical octahedral complex is shown in Figure 3.5. Here it is assumed that each ligand is independent and has a local z, x, and y coordinate system oriented as

TABLE 3.1. Angular Overlap Factors Indicating the Effect on the d Orbitals of the Angular Rotations θ, ϕ, and ψ

	$F_{\sigma}[d, L(\theta, \phi, \psi)]$	$F_{\pi y}[d, L(\theta, \phi, \psi)]$	$F_{\pi x}[d, L(\theta, \phi, \psi)]$
$d_{x^2-y^2}$	$(\sqrt{3}/4)\cos 2\phi(1-\cos 2\theta)$	$-\sin 2\phi\sin\theta\cos\psi- \frac{1}{2}\cos 2\phi\sin 2\theta\sin\psi$	$-\sin 2\phi\sin\theta\sin\psi+ \frac{1}{2}\cos 2\phi\sin 2\theta\cos\psi$
d_{z^2}	$(1+3\cos 2\theta)/4$	$(\sqrt{3}/2)\sin 2\theta\sin\psi$	$(-\sqrt{3}/2)\sin 2\theta\cos\psi$
d_{xy}	$(\sqrt{3}/4)\sin 2\phi(1-\cos 2\theta)$	$\cos 2\phi\sin\theta\cos\psi- \frac{1}{2}\sin 2\phi\sin 2\theta\sin\psi$	$\cos 2\phi\sin\theta\sin\psi+\frac{1}{2}\sin 2\phi \sin 2\theta\cos\psi$
d_{xz}	$(\sqrt{3}/2)\cos\phi\sin 2\theta$	$-\sin\phi\cos\theta\cos\psi- \cos\phi\cos 2\theta\sin\psi$	$-\sin\phi\cos\theta\sin\psi+\cos\phi \cos 2\theta\cos\psi$
d_{yz}	$(\sqrt{3}/2)\sin\phi\sin 2\theta$	$\cos\phi\cos\theta\cos\psi- \sin\phi\cos 2\theta\sin\psi$	$\cos\phi\cos\theta\sin\psi+\sin\phi \cos 2\theta\cos\psi$

^{a} See figure 3.4 and the text.

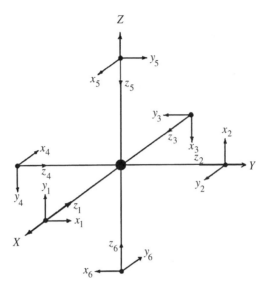

Figure 3.5. Global and ligand-based coordinate systems in a typical orthoaxial complex.

indicated. The metal-centred global X, Y, and Z axes lie along the metal–ligand bond directions, and the d orbital energy matrix obtained by adding the effects of the six ligands (making use of the expressions in Table 3.1) is shown in Table 3.2. Although the d_{xz}, d_{yz}, and d_{xy} orbital energies are given by simple expressions, the d_{z^2} and $d_{x^2-y^2}$ orbitals are connected by an off-diagonal element which depends upon the difference in the σ interaction along the X and Y axes. The

TABLE 3.2. Angular Overlap Matrix for the Complex Shown in Figure 3.5

	$d_{x^2-y^2}$	d_{z^2}	d_{xy}	d_{xz}	d_{yz}
$d_{x^2-y^2}$	$0.75[e_\sigma(1) + e_\sigma(2) + e_\sigma(3) + e_\sigma(4)]$	$0.25\{3^{1/2}[e_\sigma(1) + e_\sigma(3) - e_\sigma(2) - e_\sigma(4)]$	0	0	0
d_{z^2}	$0.25\{3^{1/2}[e_\sigma(1) + e_\sigma(3) - e_\sigma(2) - e_\sigma(4)]\}$	$e_\sigma(5) + e_\sigma(6) + 0.25[e_\sigma(1) + e_\sigma(2) + e_\sigma(3) + e_\sigma(4)]$	0	0	0
d_{xy}	0	0	$e_{\pi x}(1) + e_{\pi x}(4) + e_{\pi y}(2) + e_{\pi y}(3)$	0	0
d_{xz}	0	0	0	$e_{\pi y}(1) + e_{\pi y}(6) + e_{\pi x}(5) + e_{\pi x}(3)$	0
d_{yz}	0	0	0	0	$e_{\pi x}(2) + e_{\pi x}(6) + e_{\pi y}(5) + e_{\pi y}(4)$

wavefunctions involving these orbitals are, therefore, linear combinations of these d functions, and the energy levels cannot be given by simple arithmetic expressions.

Note that in this example the effects of the ligands along each of the metal-centred Cartesian axes X, Y, and Z add arithmetically. The energy of the interaction is the same whether a ligand lies on the positive or negative side of an axis. The same result occurs for all centrosymmetric complexes, and the *total* ligand interaction along each Cartesian axis is used to define the so-called *holohedrized symmetry* of such a complex.

If all the metal–ligand bond lengths are equal, the interactions may be described using just one set of bonding parameters and the d orbital energy matrix is diagonal, the energies being given by:

$$E(x^2 - y^2) = E(z^2) = 3e_\sigma \tag{3.10}$$

$$E(xz) = E(yz) = E(xy) = 2e_{\pi x} + 2e_{\pi y} \tag{3.11}$$

The d orbitals are then split into two sets, the energy separation being the familiar Δ, equivalent to the electrostatic crystal field splitting $10\,\mathrm{Dq}$. The three bonding parameters cannot be determined from this single experimental observable, a problem that often arises in the AOM formalism. Even when the π bonding has cylindrical symmetry about the metal–ligand bond axis, such as occurs for a hexahalide complex, so that $e_{\pi x} = e_{\pi y} = e_\pi$, the most that can be said is that the energy separation between the e_g and t_{2g} orbital sets depends on the difference in σ and π bonding parameters of the ligands according to the relationship:

$$\Delta = 3e_\sigma - 4e_\pi \tag{3.12}$$

Note that the σ and π bonding parameters act quite independently — no d orbital is influenced by both types of interaction. This situation always applies when all the ligand atoms lie on the metal-centred Cartesian coordinate axes, and such complexes are referred to as *orthoaxial*.

For mixed–ligand octahedral complexes involving ligands that form π bonds that are symmetrical about the metal–ligand bond axis, the AOM expressions for the d orbital energies are rather simple. For a complex of the form ML_5X (C_{4v} symmetry):

$$E(x^2 - y^2) = 3e_\sigma(L) \tag{3.13}$$

$$E(z^2) = 2e_\sigma(L) + e_\sigma(X) \tag{3.14}$$

$$E(xz) = E(yz) = 3e_\pi(L) + e_\pi(X) \tag{3.15}$$

$$E(xy) = 4e_\pi(L) \tag{3.16}$$

and for the corresponding complex *trans*-ML_4X_2 (D_{4h} symmetry) the expressions are:

$$E(x^2 - y^2) = 3e_\sigma(L) \tag{3.17}$$

$$E(z^2) = e_\sigma(L) + 2e_\sigma(X) \tag{3.18}$$

$$E(xz) = E(yz) = 2e_\pi(L) + 2e_\pi(X) \tag{3.19}$$

$$E(xy) = 4e_\pi(L) \tag{3.20}$$

The $e_g^\star$ and $t_{2g}^\star$ orbital sets of the parent octahedral complex ML_6 are thus split by amounts that depend on the difference in σ and π bonding between L and X, respectively (here it is assumed that the $t_{2g}^\star$ orbital set is weakly antibonding). Note that the splittings of the disubstituted complex are twice those of the monosubstituted complex. For the corresponding complex *cis*-ML_4X_2 (C_{2v} symmetry but with the metal centered X and Y axes defined to lie along the L-M-X bond directions), the expressions are:

$$E(x^2 - y^2) = 1.5e_\sigma(L) + 1.5e_\sigma(X) \tag{3.21}$$

$$E(z^2) = 2.5e_\sigma(L) + 0.5e_\sigma(X) \tag{3.22}$$

$$E(xz) = E(yz) = 3e_\pi(L) + e_\pi(X) \tag{3.23}$$

$$E(xy) = 2e_\pi(L) + 2e_\pi(X) \tag{3.24}$$

Note that the energy separations $E(z^2) - E(x^2 - y^2)$, and $E(xz, yz) - E(xy)$ are both exactly half those in the *trans*-complex, although the signs are reversed.

3.2.2 Tetragonally Distorted Octahedral, Planar, and Linear Complexes: The Importance of d-s Mixing

The AOM expressions for all orthoaxial complexes can in fact be related back to the general expressions in Table 3.2 by simply noting which ligands are equivalent or absent. Thus tetragonally distorted octahedral complexes involving one ligand type with π bonding that is symmetric about the metal–ligand bond axis have identical expressions to those for trans-disubstituted complexes (Eqs. 3.17–3.20) where L and X refer here to the parameters for this ligand at different bond distances. The expressions for planar complexes involving similar ligands are identical to these, but with the parameters representing ligand X now set to zero:

$$E(x^2 - y^2) = 3e_\sigma \tag{3.25}$$

$$E(z^2) = e_\sigma \tag{3.26}$$

$$E(xz) = E(yz) = 2e_\pi \tag{3.27}$$

$$E(xy) = 4e_\pi \tag{3.28}$$

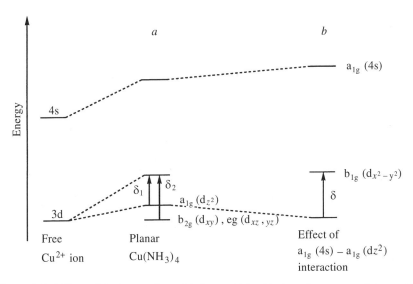

Figure 3.6. Effect of the interaction with the $a_{1g}(4s)$ orbital on the energy of the $a_{1g}(d_{z^2})$ orbital in a planar complex such as the $Cu(NH_3)_4^{2+}$ ion (one-electron orbital description).

It has been found, however, that the energy levels deduced from the electronic spectra of square planar complexes are not in good agreement with this set of equations. For example, for complexes of $3d^9$ metal ions such as Cu^{2+}, for which the d-d transitions correlate directly with the energy differences between the d orbitals, the electronic spectra are expected to consist of three bands of energy: $2e_\sigma$, $3e_\sigma - 4e_\pi$, and $3e_\sigma - 2e_\pi$. If the ligand is a saturated amine such as ammonia, however, the parameter e_π should be close to zero, because the ligand has no orbitals available to participate in π bonding (Section 3.1.1). Thus two bands are expected, at energies of $2e_\sigma$ and $3e_\sigma$, corresponding to the transitions illustrated as δ_1 and δ_2 in Figure 3.6a. In fact for complexes such as the $Cu(NH_3)_4^{2+}$ ion and similar diamagnetic planar Ni^{2+} complexes, just a *single* composite band is observed in the electronic spectrum corresponding to the transition δ in Figure 3.6b. This may be explained in terms of an interaction between the metal $3d_{z^2}$ and $4s$ orbitals, both of which transform as a_{1g} in the D_{4h} point group (5–7). The effect of this interaction is to depress the $a_{1g}(d_{z^2})$ orbital until it is effectively nonbonding. This means that in the planar tetraammine complexes it is approximately equal in energy to the $b_{2g}(d_{xy})$ and $e_g(d_{xz}, d_{yz})$ sets of orbitals (Fig 3.6b).

Such an interpretation is confirmed by the electronic spectrum of planar $CuCl_4^{2-}$ complexes, as discussed in Section 8.1.4.1. For these, the parameter e_π is no longer zero, and the three bands implied by Eq. 3.25–3.28 are indeed observed. The energy separation between the transitions from the $e_g(d_{xz}, d_{yz})$ and $b_{2g}(d_{xy})$ orbitals to the $b_{1g}(d_{x^2-y^2})$ orbital yields e_π, whereas the absolute value of these transitions provides e_σ. It is found that the transition from the d_{z^2} orbital occurs at a much higher energy than predicted by Eq 3.26, close to $3e_\sigma$ rather than $2e_\sigma$, so that here also this orbital is effectively nonbonding.

The AOM may be extended to include the effect of the $d_{z^2} - 4s$ interaction by adding an additional parameter e_{ds} to describe each metal–ligand interaction. For a square planar complex this has the effect of altering the expression for the energy of the d_{z^2} orbital to:

$$E(z^2) = e_\sigma - 4e_{ds} \qquad (3.29)$$

Because the d_{z^2} orbital is approximately nonbonding in these planar complexes, e_{ds} is about one-quarter e_σ in magnitude.

In addition to the shift in energy, the interaction also causes the orbitals to mix. If the matrix element connecting the two orbitals is H_{ds}, then perturbation theory predicts that the energy shift ΔE caused by the interaction is given by:

$$\Delta E \approx H_{ds}^2/(H_s - H_d) \qquad (3.30)$$

whereas the coefficient of admixture of the perturbing wavefunction c is given by:

$$c \approx H_{ds}/(H_s - H_d) \qquad (3.31)$$

These equations suggest that the square of the coefficient, which indicates the probability that an electron will be found in that component of the wavefunction, is given by $c^2 \approx \Delta E/(H_s - H_d)$. As the energy separation $(H_s - H_d)$ between the 3d and 4s orbitals in a transition-metal complex is on the order of magnitude 80,000 cm^{-1}, the energy shift of $\sim 6{,}000$ cm^{-1} inferred for the planar $CuCl_4^{2-}$ ion implies an admixture of $\sim 7.5\%$ 4s character in the $a_{1g}(3d_{z^2})$ orbital.

The addition of an extra bonding parameter to describe the metal–ligand interactions often has the unfortunate consequence that insufficient experimental data are available to determine all the parameters independently. In practice, d–s mixing only has a significant effect upon the d orbital energy levels of a complex when the geometry departs drastically from cubic symmetry. For mixed ligand and tetragonally distorted octahedral complexes, the AOM equations may be modified by adding a term E_{ds} to the expression for $E(z^2)$ in Eq. 3.18. It defines the energy by which the d_{z^2} orbital is lowered by the d-s interaction. This term differs from the AOM bonding parameters e_σ etc. in that it represents the effect of the departure from cubic symmetry of the σ-bonding interactions of *all* of the ligands; i.e., it is a *global* rather than a *local* ligand field parameter. In complexes such as the *trans*-$Co(NH_3)_4 X_2^+$ ion, in which the departure from cubic symmetry is caused only by the fact that two different ligands are present, d-s mixing can safely be ignored unless the ligands have different σ-bonding characteristics. d-s mixing, however, may be significant for complexes with large tetragonal distortions owing to Jahn–Teller coupling and should increase in proportion to the magnitude of the distortion.

Occasionally, Cu^{2+} complexes adopt a tetragonally compressed octahedral geometry. The unpaired electron then occupies the $a_{1g}(d_{z^2})$ orbital, and participation of the 4s orbital in the wavefunction may be estimated by measuring the contact term in the hyperfine interaction with the copper nuclear spin (Section 10.3.2). A contribution in the range of 3 to 9% is generally obtained, which agrees well with the indirect estimates from the optical spectra of planar complexes.

The limiting geometry resulting from tetragonal compression is a linear complex. Then, because of the drastic departure from cubic symmetry, the effect of d–s mixing is pronounced. The $d_{x^2-y^2}$ and d_{xy} orbitals are nonbonding in a linear complex, and the energies of the remaining orbitals are given by:

$$E(z^2) = 2e_{\sigma} - 4e_{ds} \tag{3.32}$$
$$E(xz, yz) = 4e_{\pi} \tag{3.33}$$

Not many complexes of this geometry are known for metal ions with partly filled d shells, although it has long been recognized that the propensity for linear coordination by d^{10} metal ions, such as Hg^{2+}, is readily explained if the metal uses $nd - (n+1)s$ hybrids orbitals (7). Analysis of the hyperfine splitting observed in the electron paramagnetic resonance (EPR) spectrum of the CuF_2 molecule implies a participation of ~25% by the 4s orbital into the $\Sigma(z^2)$ orbital (8). Detailed study of the fluorescence spectrum of the $CuCl_2$ molecule in the gaseous phase shows that here the interaction between the metal $3d_{z^2}$ and 4s orbitals is so large that the molecule actually has a $^2\Pi(xz, yz)$ ground state (9). In these linear complexes it therefore seems that it would be quite unrealistic to restrict ligand field treatments to the metal d orbitals alone.

3.2.3 Tetrahedral, Distorted Tetrahedral, and Square-Based Pyramidal Complexes

For a tetrahedral complex with symmetrical π-bonding ligands ($e_{\pi x} = e_{\pi y} = e_{\pi}$) the d orbital energies are given by the expressions:

$$E(xz, yz, x^2 - y^2) = (4/3)e_{\sigma} + (8/9)e_{\pi} \tag{3.34}$$
$$E(z^2, xy) = (8/3)e_{\pi} \tag{3.35}$$

Note that both σ- and π-bonding parameters contribute to the energy of the triply degenerate orbital set, as may occur for a complex that does not have orthoaxial symmetry. The Cartesian axes are defined here as shown in Figure 1.8. The energy separation between the t_2^* and e orbital sets is $(4/9)(3e_{\sigma} - 4e_{\pi})$, or 4/9 that between the e_g^* and t_{2g} sets in the corresponding octahedral complex, a result identical to that obtained using the electrostatic crystal field theory.

Tetrahedral complexes that would have orbital degeneracies in a regular situation are subject to a Jahn–Teller distortion, as discussed in Section 7.1.2.2. This generally takes the form of an angular distortion which produces a complex of D_{2d} symmetry, (Fig. 7.5). The distortion is conveniently described by the angle α between each metal–ligand bond and the Z axis; the X and Y axes lie in the planes containing two of the four M–L bonds (α is half the angle θ; it should be noted that this definition of the X and Y axes differs from that conventionally used for the D_{2d} point group). The AOM expressions for such a complex are:

$$E(x^2 - y^2) = 3(\sin^4 \alpha)e_\sigma + (\sin^2 2\alpha)e_\pi \tag{3.36}$$

$$E(z^2) = (3\cos^2 \alpha - 1)^2 e_\sigma + 3(\sin^2 2\alpha) e_\pi - 4(3\cos^2 \alpha - 1)^2 e_{ds} \tag{3.37}$$

$$E(xy) = 4(\sin^2 \alpha)e_\pi \tag{3.38}$$

$$E(xz) = E(yz) = (3/2)(\sin^2 2\alpha)e_\pi + 2(\cos^2 \alpha + \cos^2 2\alpha)e_\pi \tag{3.39}$$

In tetrahedral and near-tetrahedral complexes, the metal 4p orbitals are allowed by symmetry to mix with the metal orbitals of t_2 symmetry, and an additional parameter e_{dp} might be added to Eq. 3.39 to account for this effect. Experimental evidence suggests, however, that, unlike the 4s-3d$_{z^2}$ interaction, the influence of the metal 4p orbitals is negligible and may safely be ignored. This may well be because the metal 4p orbitals are considerably higher in energy than the 4s orbital and hence are quite far removed from the d orbitals.

Because the AOM treats the ligands in the ZX and ZY planes independently, the expressions for the metal–ligand interactions of the basal ligands in a square-based pyramidal ML$_4$X complex are identical to those given above for a distorted tetrahedral complex of C_{4v} symmetry. If α defines here the angle between the axial and basal ligands, the complete expressions:

$$E(x^2 - y^2) = 3(\sin^4 \alpha) e_\sigma(L) + (\sin^2 2\alpha)e_\pi(L) \tag{3.40}$$

$$E(z^2) = e_\sigma(X) + (3\cos^2 \alpha - 1)^2 e_\sigma(L) + 3(\sin^2 2\alpha)e_\pi(L) - E_{ds} \tag{3.41}$$

$$E(xy) = 4(\sin^2 \alpha)e_\pi(L) \tag{3.42}$$

$$E(xz) = E(yz) = (3/2)(\sin^2 2\alpha)e_\pi(L) + 2(\cos^2 \alpha$$
$$+ \cos^2 2\alpha)e_\pi(L) + e_\pi(X) \tag{3.43}$$

are obtained by adding the effect of the single axial ligand X. Here, the effect of d–s mixing is represented by the term E_{ds}, because it is influenced by the bonding parameters of both X and L.

3.2.4 Trigonal Bipyramidal Complexes

The d orbital energies in a regular trigonal bipyramidal complex of the form ML$_3$X$_2$, where X represents the axial and L the equatorial ligands, are given by the expressions:

$$E(z^2) = 2e_\sigma(X) + (3/4)e_\sigma(L) - E_{ds} \tag{3.44}$$

$$E(x^2 - y^2, xy) = (9/8)e_\sigma(L) + (3/2)e_\pi(L) \tag{3.45}$$

$$E(xz) = E(yz) = 2e_\pi(X) + (3/2)e_\pi(L) \tag{3.46}$$

Note that in this stereochemistry, the parameters of the equatorial and axial ligands are expected to differ even when identical ligands occur in these two coordination sites. This follows because these generally involve rather different metal–ligand bond distances. If included, the effect of d-s mixing should again be treated using a global parameter, because it results from the difference in the perturbation of the axial and equatorial ligands.

3.2.5 Variation of AOM Parameters with Bond Distance

An important advantage of the AOM is that it offers the possibility that the bonding parameters may be transferred from one situation to another; however, for this to be realized, corrections must often be made for differences in the metal–ligand bond length (10, 11). Simple crystal field theory predicts that the d orbital splitting in a complex should depend inversely on the fifth power of the metal–ligand bond distance. For a dipolar form of the theory an inverse sixth-power relationship would hold. Measurements of the pressure dependence of the electronic spectra of simple metal oxides show that the ligand field splitting Δ does indeed vary as r^{-n}, where r is the metal–ligand distance, and n lies between 5 and 6 (Sections 2.9 and 7.1.2). Initially, this was thought to be a rather surprising result, because it is known that the basic cause of the d orbital splitting is a covalent, rather than an electrostatic interaction. Calculations suggest, however, that the square of the diatomic overlap integral between the metal and ligand orbitals is also expected to vary as r^{-n}, where $n \sim 5(11)$. Thus the AOM also provides a satisfactory explanation of the pressure dependence of d-d transition energies.

   In principle, the variation in e_σ and e_π as a function of metal–ligand bond distance may be deduced by calculating the overlap integral S at the different bond lengths, but this involves a number of approximations (12). Not only must the form of the metal and ligand wavefunctions be represented correctly but the effective nuclear charges acting on these must be estimated. For this reason, most researchers have been content to assume that e_σ and e_π vary empirically as about r^{-n}, where n lies between 5 and 6. Indeed this range does provide a rough guide to the uncertainty in the transferability of the parameters as far as bond distance is concerned. It should be noted that e_π is expected to vary somewhat more steeply with bond distance than does e_σ, because the π overlap is more sensitive to the internuclear separation than is the σ overlap.

   The metal–ligand parameters e_σ derived from the optical spectra of a range of Ni^{2+} complexes of saturated amines are shown plotted against the Ni-N bond

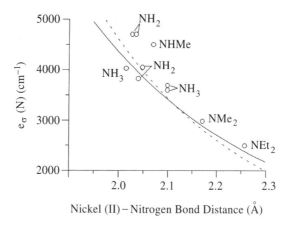

Figure 3.7. Variation of e_σ as a function of bond distance for saturated amines bonded to Ni^{2+}. The *solid* and *dashed lines* show the variation expected if the parameter varied inversely as the fifth and sixth power, respectively, of the metal–ligand bond distance. [From Lever, A. B. P., Walker, I. M., McCarthy, P. J., Mertes, K. B., Jircitano, A. and Sheldo, R., *Inorg. Chem.*, 1983, **22**, 2252, used with permission.]

distances in Figure 3.7. The bonding parameters follow the curves remarkably well, considering the approximations involved. In particular, the AOM parameters are expected to depend not only on the overlap integral but also on the difference in valence state ionization energies of the metal and ligand orbitals (Eq. 3.9). These are expected to vary as a function of the effective nuclear charge, which is strongly influenced by the metal–ligand bond distance. This effect, however, may be smaller than might be expected because the effective nuclear charge of the metal depends on the overall influence of all the ligands in the complex. Usually, when the metal–ligand bond lengths of different complexes of a particular metal ion are compared, a shortening of one metal–ligand bond is compensated for by the lengthening of another. Thus it is likely that the effective nuclear charge of the metal remains approximately constant. This concept gains support from the observation that although individual e_σ and e_π parameters may vary substantially, their sum is approximately constant. There are some exceptions, but for the divalent first-row transition-metal ions the sum for all i ligands: $\sum_i (e_\sigma + e_{\pi x} + e_{\pi y}) \approx 22,000\,cm^{-1}$, irrespective of the shape of the complex or the nature of the ligands (13).

3.2.6 Typical AOM Parameters

A major advantage of the AOM is its potential to provide chemical information. To illustrate the kind of information provided by the model, some typical bonding parameters are given in Table 3.3. It must be stressed that these are only representative. Most of them are the average of values obtained for several complexes, and in some cases considerable variation occurs on going from one complex to another. It is often not clear whether these changes represent true

TABLE 3.3. Representative AOM Parameters for Metal Ions in Various Coordination Geometries [a]

Metal Ion	Ligand	e_σ	e_π	e_{ds}	References
		Octahedral			
Ni^{2+}	NH_3	3,600	⋆		14
	en	4,000	⋆		14–16
	pyridine	4,500	900 [#]		16
Co^{2+}	pyridine	3,860	110 [#]		14
Fe^{2+}	pyridine	3,700	100 [#]		14
Co^{3+}	pyridine	6,100	−750 [#]		15
Cr^{3+}	NH_3	7,180	⋆		14,15,17
	en	7,260	⋆		14,15
	pyridine	6,150	−330 [#]		14
	F^-	8,200	2,000		15
	Cl^-	5,700	980		15
	Br^-	5,380	950		15
	I^-	4,100	670		15
	OH^-	8,600	2,150		15
	H_2O	7,550	1,850		15
	glycine(NH_2)	6,700	⋆		15
	glycine(O)	8,800	2,000		15
	CN^-	7,530	−930		17
		Tetrahedral			
Ni^{2+}	$P(Ph)_3$	5,000	−1,750		15
	Cl^-	3,900	1,500		15
	Br^-	3,600	1,000		15
	I^-	2,000	6,00		15
	quinoline	4,000	−500		15
Co^{2+}	$P(Ph)_3$	3,800	−1,000		15
	Cl^-	3,600	1,400		15
	Br^-	3,300	1,000		15
	quinoline	3,500	−500 [#]		15
		Planar			
Cu^{2+}	Cl^-	5,030	900	1320	18
Pd^{2+}	Cl^-	10,150	2,000	2,540	19
	Br^-	9,500	1,800	2,380	19
Pt^{2+}	Cl^-	12,400	2,800	3,100	19
	Br^-	10,900	2,200	2,725	19

[a] See text for a discussion of the limitations that apply to their interpretation. Unless designated otherwise, the π bonding is assumed to be isotropic about the metal–ligand bond axis.
⋆ π bonding assumed negligible; [#] π bonding in the plane of the amine assumed negligible.

chemical differences or are simply the result of experimental uncertainty. It must also be recognized that the chemical environment of a ligand often influences its capacity to bond to a metal. For instance, in an oxide lattice of the form AMO_n, where A is a cation, the nature of A and its disposition in the crystal lattice may have a significant effect on the bonding interactions experienced by the transition metal M. Furthermore, it is often the case that specific assumptions are made in the derivation of the bonding parameters. Thus it is usually assumed that the π bonding interactions of aliphatic amines and heteroaromatic ligands in the plane of the ring are negligible. For this reason, before making use of any set of parameters, it is important to check the particular conditions that applied in their derivation. Useful tabulations of parameters are given by Bencini and co-workers (14) and Lever (15).

Even allowing for the above uncertainties, several interesting trends may be inferred from Table 3.3. The e_σ parameters of octahedral and tetrahedral complexes of divalent first-row transition ions are broadly similar and much smaller than those of the corresponding octahedral complexes of trivalent metal ions. This is consistent with the fact the metal and ligand orbitals are closer in energy when the metal is in a higher oxidation state (Section 3.1.1) as well as with the shorter metal-ligand bond lengths that occur for trivalent metal ions. The e_σ parameters of planar Pd^{2+} and Pt^{2+} complexes are high, reflecting the fact that because of their larger size the 4d and 5d orbitals are able to overlap better with the ligand orbitals than is the case for the more compact 3d orbitals. Moreover, planar d^8 metal ions exhibit short metal–ligand bond lengths (Section 7.1.3.3). The e_σ and e_π parameters of the halide ions increase in the sequence $I^- < Br^- < Cl^- < F^-$. This presumably reflects an increase in overlap between the ligand orbitals and the d orbitals. The halide ion e_π parameters are always positive and the fluoride ion is a particularly strong π donor toward Cr^{3+}, a characteristic shared by oxygen ligands bonded to this metal ion. For octahedral and planar complexes, the halide ion e_σ parameters are about four or five times larger than the e_π parameters. For tetrahedral complexes, this ratio is lower, about three for the chloride and bromide ions. The data are, however, too sparse to decide if this difference is general.

The signs of the e_π parameters suggest that the ligand pyridine apparently acts as a weak π donor toward the divalent metal ions Ni^{2+}, Co^{2+}, and Fe^{2+} but as a weak π acceptor toward the trivalent metal ions Co^{3+} and Cr^{3+}. The similar ligand quinoline, on the other hand, is a weak π acceptor in tetrahedral Ni^{2+} and Co^{2+} complexes. Again, it is probably unwise to draw firm conclusions from the data at this stage; however, the net π-bonding capacity of an aromatic amine depends on the balance between the interactions involving the bonding and antibonding ligand π orbitals. This must depend on a range of factors, so it is quite possible that a ligand such as pyridine may act as a π donor in some complexes but as π acceptor in others. Both cyanide ion and triphenylphosphine have negative e_π parameters and are moderate π acceptors. In each case this accords with accepted chemical belief.

For planar complexes, the parameter e_{ds} describing the interaction between the metal s and d_{z^2} orbitals is conventionally included in the analysis. For planar tetrahalide complexes, e_{ds} is found to be somewhat larger than the π bonding parameter e_π so that here the experimental data can be interpreted satisfactorily only by including the d–s interaction.

All in all, it appears that the AOM has the capacity to set the qualitative framework long used by experimental chemists on a more quantitative basis. Definite confirmation of this conclusion, however, must await the gathering of a more detailed and broadly based set of data than is currently available.

3.3   EXTENSIONS OF THE AOM FOR SOME POLYATOMIC LIGANDS

The d orbital energies are generally dominated by interactions with the ligand atoms in the first-coordination sphere. It must be remembered, however, that when polyatomic ligands are involved, the bonding between the atoms within the ligand is considerably stronger than that between the ligand and the metal. There are potentially several ways in which intraligand bonding may complicate the AOM for polyatomic ligands. The effects are difficult to separate and appear to be generally relatively small, but it is worthwhile to consider them briefly.

3.3.1   Phase-Coupled Ligators

Even in early considerations of metal–ligand bonding, it was realized that for ligands with conjugated π systems the form of the conjugation might affect the metal–ligand π interaction (20, 21). In particular, the metal–ligand π interaction may be influenced by the relative phases of the ligand p orbitals of π symmetry. This relationship, in turn, depends on the number of atoms involved in the ligand π system. An example for which it appears that the d orbital energies are affected by the form of the intraligand π bonding is the planar complex Co(salen), shown in Figure 3.8a. In this complex, the metal–ligand bonds are almost exactly at right angles to one another. Thus, considering just the p orbitals of π symmetry on the four atoms bonded to the metal, the "holahedrized" symmetry is D_{4h} and the d_{xz} and d_{yz} orbitals are expected to be degenerate. In fact, the d_{yz} orbital is some 4000 cm^{-1} higher in energy than the d_{xz}, and the unpaired electron occupies this orbital in the complex, which is low spin. The delocalized π system in the ligand involves the overlap of the p orbitals of five atoms, and the relative phases of these in the highest occupied molecular orbital (HOMO), said to be of ψ symmetry, are of the correct symmetry to interact with the d_{yz} orbital. The phases of the lowest unoccupied molecular orbital (LUMO), designated χ, are of the correct symmetry to interact with the d_{xz} orbital. The phase-coupling of the ligand π orbitals thus splits the d_{yz}, d_{xz} orbitals (Fig. 3.9a). The effect may be represented by using the AOM expressions for a complex of D_{4h} symmetry and simply adding terms the $2E_\psi$ and $2E_\chi$ to the equations representing the energies of the d_{yz} and d_{xz} orbitals, respectively.

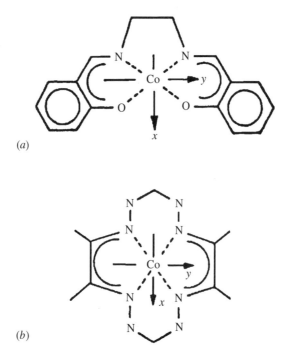

Figure 3.8. The geometries of two planar complexes: Co(salen) (*a*) and $Co(C_{10}H_{20}N_8)^{2+}$ (*b*).

If this explanation of the splitting of the d_{yz} and d_{xz} orbitals is correct, then their relative energies should reverse if the number of atoms involved in the ligand π system is reduced from five to four, because this switches the relative phases of the HOMO, and LUMO. This is indeed found to be the case. In the complex $Co(C_{10}H_{20}N_8)^{2+}$, the unpaired electron occupies the d_{xz} orbital, which is ~ 6000 cm^{-1} higher in energy than the d_{yz} (Fig. 3.8b). Here the symmetry of the phase-coupled π orbitals of π symmetry, denoted as χ, means that it is d_{xz} that is raised in energy with respect to d_{yz} (Fig 3.9b). The phase coupling of the ligand π orbitals thus appears to provide a self-consistent picture of the metal–ligand π bonding involving ligands of this type. Similar effects have been reported in complexes formed by ligands such as acetylacetonate and oxalate ions, although there the splitting of the d_{xz} and d_{yz} orbitals is smaller, possibly because the complexes may be more ionic in their bonding.

3.3.2   Off-Axis Bonds and Interactions with Nonbonding Electron Pairs

For polyatomic ligands involving light atoms such as oxygen and nitrogen, the valence electrons are conventionally described in terms of hybrid orbitals. Because the metal–ligand interaction is considerably weaker than the bonding

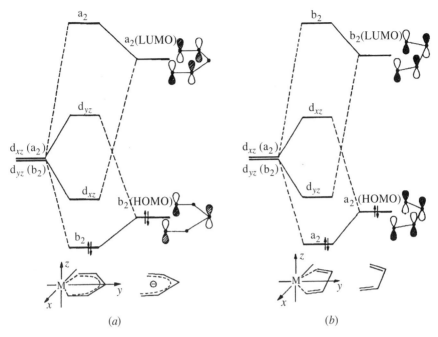

Figure 3.9. The phases of the atomic orbitals forming the HOMO and LUMO for the ligands shown in Figure 3.8. The way in which these interact with the $b_2(d_{yz})$ and $a_2(d_{xz})$ orbitals in the complexes is also set out. [From Ceulemans, A., Debugst. R., Dejehet, F., King, G.S.D., Vanhecke, M. and Vanquickenborne L. G. J. Phys. Chem., 1990, **94**, 105; used with permission].

between the atoms in the ligand, the orientation of the hybrid orbitals is largely decided by intraligand bonding interactions. It has been suggested that this may affect the AOM description of the metal–ligand bonding in two ways (22). In the first way, the lone pair electrons not directed at the metal may still influence its d orbital energies. The trigonal bipyramidal complex Co(picoline-N-oxide)$_5^{2+}$ is an example. Assuming that the lone pair electrons on the ligand oxygen atom occupy sp^2 orbitals, the formally nonbonding electron pair not directed at the metal may still affect the energy of the d orbitals of π symmetry (Fig. 3.10a). In this complex, the angle Co–O–N is close to 120°, as expected for a set of ligand sp^2 hybrid orbitals. The second way may be illustrated by considering the tetrahedral complex Co[OP(C$_6$H$_5$)$_3$]$_2$Cl$_2$. Here, if the electrons on the ligand oxygen atom are localized in sp^2 orbitals with an orientation decided by the neighboring phosphorus atom, the axis defining the lone pair directed at the metal is off-set from the Co-O vector by $\sim 30°$ (Fig. 3.10b). This gives rise to the concept of misdirected or off-axis bonding. The tilting of the ligand orbital away from the bond axis lowers the σ interaction and introduces an interaction with the d orbitals of π symmetry. Because d orbitals of both σ and π symmetry are simultaneously influenced by the effect, it is conventionally represented by

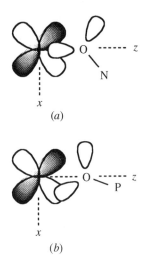

Figure 3.10. (*a*), The way in which a formally nonbonding lone pair of electrons may interact with a d orbital. (*b*), Off-axis bonding.

adding an additional parameter $e_{\pi\sigma}$ to the description of the metal–ligand bonding.

The parameter $e_{\pi\sigma}$ may be used to represent both nonbonding and off-axis interactions with the d orbitals of π symmetry. These interactions differ from a bond of true π symmetry in that only one of the d orbital lobes is affected by the interaction (Fig. 3.10). This means that the parameter may be positive or negative, depending on the relative orientations of the ligand lone pair orbitals in the complex. For $Co(picoline-N-oxide)_5^{2+}$, an optimal fit to the experimental data is obtained with $e_{\pi\sigma} = 500$ cm^{-1} for the equatorial ligands. The parameter here represents the influence of the ligand nonbonding lone pair electrons. For $Co[OP(C_6H_5)_3]_2Cl_2$ the corresponding value is $e_{\pi\sigma} = -100$ cm^{-1}, and it is thought that this is owing to a combination of a positive contribution from the nonbonding electron pairs, with a negative contribution from the effects of off-axis bonding.

3.4.   APPROXIMATIONS IN THE DERIVATION OF BONDING PARAMETERS

A problem in applying the AOM is that there is often insufficient experimental data to allow a unique set of bonding parameters to be determined independently. This situation frequently occurs when the set is extended to include effects such as those described in the preceding section. At most, four energy differences may be determined from five d orbital energies of a low-symmetry complex. If the conventional parameters e_σ, $e_{\pi x}$, and $e_{\pi y}$ are supplemented by an additional parameter e_{ds}, it is impossible to determine $e_{\pi\sigma}$ unless the other parameters are constrained in some way. This means that for systems of this kind the interpretation of data is often ambiguous. For instance, the energy levels of complexes such as Co(salen) discussed in Section 3.3.1 in terms of the π system of the ligand have also been explained by including interactions with nonbonding electron pairs (22). The situation is greatly improved if further information on the composition of the ground state wavefunction is obtained, for example by EPR spectroscopy.

As data on a wide range of compounds are accumulated, the confidence with which parameters can be transferred, with appropriate corrections, from one complex to another should improve. At present, however, it is prudent not to extend the AOM beyond its simple form, unless supported by strong experimental evidence. Such strong evidence appears to be available for d–s mixing, where the effect is suggested quite independently by EPR and optical spectroscopy. The parametrization schemes introduced for the other extensions of the AOM, however, should be approached with caution.

In mixed ligand complexes, in which two or more sets of bonding parameters are involved, it is generally impossible to determine all of these independently without imposing constraints. Various strategies have been developed to deal with this problem, and when comparing parameters derived from different studies, it is important to be aware of the approximations used in their derivation. A summary of the main approximations used to derive AOM parameters follows:

1. For saturated amines, such as ammonia, covalent metal–ligand π interactions are expected to be negligible, so that e_π may be set to zero. Studies of complexes for which sufficient data are available to determine the π-bonding parameters suggest that they are indeed small, and that this approximation is quite good.

2. For similar reasons, the in-plane π-bonding parameter of heterocyclic ligands, such as pyridine, is often set to zero. This also appears to be a good approximation. A similar approach has sometimes been made for planar oxygen-donor ligands, such as a chelating nitrate ion. The assumption here is that the lone electron pairs not involved in bonding to the metal are inactive. As yet, little experimental evidence is available to decide to what extent that assumption is justified.

3. For complexes involving the same ligand at different bond distances r, it is often assumed that the bonding parameters are proportional to r^{-5} or r^{-6}. Provided the bonds are not too dissimilar in length, this appears to be a good approximation (Section 3.2.5).

4. The ratio e_σ / e_π has sometimes been set equal to that calculated for S_σ^2 / S_π^2. This approximation is suspect because it is not known how well the overlap integrals can be estimated. Moreover, the effect of interactions with the ligand σ orbitals, which will affect e_σ but not e_π, is unclear.

5. The value of a parameter, or the ratio e_σ / e_π, may be set equal to that in another complex. This approximation appears to hold quite well for similar

complexes. Suitable corrections must be made when going from one metal ion to another and when the bond lengths differ. In addition, the bonding parameters of a ligand often change when it bridges to another atom. Moreover, whenever a complex departs significantly from cubic symmetry, the effect of d-s mixing must be included in the treatment. Tabulations of parameters are given in the books and review articles cited in this chapter.

3.5 ADVANTAGES OF THE AOM COMPARED TO THE ELECTROSTATIC CRYSTAL FIELD THEORY

It is important to stress again that neither the AOM nor the crystal field theory (CFT) should be thought of as a method of calculating real energy levels in a metal complex. Although it is true that sophisticated calculations suggest that the bonding in transition-metal complexes is predominantly covalent rather than ionic, it is misleading to think that the AOM is a better model than the CFM just for that reason. Both models are far too simplistic to provide a realistic picture of the bonding in a complex. Rather, they provide a way of parametrizing metal–ligand interactions; the magnitudes of the parameters are always defined by experiment. Their value, therefore, lies in providing support for the conceptual framework used to describe the chemical behavior of the transition elements, and the models should be compared on this basis.

As already mentioned, the AOM is, initially at least, a *local* rather than a *global* model—the parameters refer to individual metal–ligand bonds rather than to the complex as a whole. This contrasts with the CFT, for which the parameters describe the total ligand field experienced by the metal ion. For a complex of cubic symmetry, there is no real advantage in using the AOM, because only one parameter, the d orbital energy separation Δ, is available from experiment. It is, therefore, impossible to separate the AOM σ and π bonding parameters ($\Delta = 3e_\sigma - 4e_\pi$ for an octahedral complex with symmetrically π bonding ligands) unless the π bonding parameters can be assumed to be negligible. Such an assumption has been justified for saturated amines. For lower-symmetry complexes such as *trans*-ML_4X_2 (D_{4h} symmetry), however, the d orbital splittings provide more information. Three energy separations occur in this example, and they may be used to estimate the crystal field parameters Δ, Ds, and Dt (Chapter 2) or, using the AOM, the Δ value for L and the difference between the σ and π bonding parameters of X and L for the metal ion concerned.

The real advantage of the AOM as far as chemists are concerned arises from the fact that the parameters e_σ and e_π relate directly to the conceptual framework used by, for example, kineticists and preparative chemists. This advantage is less important for physicists, and it is probably for this reason that the parameters of the electrostatic CFT are widely used in their discipline.

Another potential advantage of the AOM is that the bonding parameters depend on two factors: the square of the metal–ligand overlap integral and the inverse of the difference in the valence state ionization energies of the metal and ligand orbitals (Eq. 3.9). This contrasts with the CFT, for which, as discussed in Chapter 2, the interaction depends simply on the distance between the d electrons and the ligands. In principle, the AOM has more flexibility when comparisons are made of the bonding parameters of a series of metal ions or ligands. For instance, if the effect on Δ of the oxidation state of the metal is considered (Section 8.4.1), it is clear that if the metal–ligand bond distance does not alter, the CFT predicts that the increase in effective nuclear charge associated with a rise in oxidation state could cause a decrease in Δ, because of the contraction of the d orbitals.

For the AOM, the situation is less clear cut. The metal–ligand orbital overlap will decrease, but the energy separation between the metal and ligand orbitals will also become smaller. In practice, an increase in oxidation state is almost always accompanied by an increase in Δ. Of course, the metal–ligand bond distance generally does not remain constant with a change of oxidation state. As the oxidation state increases, the metal–ligand bonds shorten, which tends to counterbalance the contraction of the d orbitals. When an alteration in the oxidation state is not accompanied by a change in spin state, however, the bond length difference may be quite small. Under these circumstances, the change in Δ is also quite modest and the AOM may be able to provide a more realistic interpretation of this behavior than does the CFT.

That being said, it must be stressed that as yet no general attempt has been made to interpret AOM bonding parameters in terms of the separate influences of the metal–ligand overlap and valence state ionization energies. It is, therefore, too early to say whether the AOM can provide a self-consistent and chemically useful picture of the bonding in metal complexes at this level. The same is true of another potential advantage of the model: the fact that it offers the possibility to estimate the extent to which the d orbitals become contaminated by ligand orbitals on complex formation. That is clearly impossible in the CFT, for which covalency is accounted for by simply applying a scaling factor to the free ion interelectron repulsion and spin–orbit coupling parameters.

In the original formulation of the AOM, perturbation theory was used to estimate the energy shift produced in a d orbital upon overlap with a ligand orbital. After summing the effects of all the ligands, the total energy shift was then typically correlated with that observed in the electronic spectrum of the complex. In principle, however, perturbation theory can also be used to estimate the mixing coefficients that result from the interaction. If the AOM could be developed in this manner, it should allow ligand field theory to be extended in a number of ways. One important application is in the interpretation of magnetic superexchange interactions between metal ions (Section 9.11.3). These occur via the unpaired spin density transferred to bridging ligands, and the magnetic coupling is often observed to be sensitive to the bridging angle. The AOM should provide a powerful method of estimating the changes in ligand mixing coefficient, and hence superexchange coupling magnitudes, in a series of structurally related complexes. Probably the most direct readily accessible method of measuring the mixing of metal and ligand orbitals in complexes is from the superhyperfine coupling constants obtained from EPR spectroscopy (Section 10.3.2.2). Again, the AOM should provide an ideal framework within which to compare the ligand unpaired spin densities in a series of complexes. Finally, the AOM might be used to rationalize the intensities of d–d transitions, because these are derived largely from transition moment integrals involving ligand components of the ground and excited state wavefunctions (Section 8.1.1).

Some research has been done in each of the above areas, and the results look promising. It is still too early, however, to know to what extent the AOM may be expanded beyond the way in which it is largely used at present, namely as a simple energy parametrization scheme. What does seem clear is that these areas of extension are well worth exploring.

3.6   CALCULATION OF ELECTRONIC SPECTRA AND MAGNETIC PROPERTIES USING COMPUTER PROGRAMS BASED ON THE AOM

A major advantage of an approach such as the AOM, which parametrizes the ligand field in terms of individual metal–ligand interactions, is that it can readily be used to form the basis of a general computer program to calculate properties derived from the energy levels and wavefunctions of a metal complex. This would be much harder using a global crystal field model, for which the ligand field parameters are a function of the symmetry of the complex, which is often, formally at least, low.

The first ligand field program to be widely used, *CAMMAG*, was developed by Gerloch and co-workers (23). Basically, the positions and orientations of the ligands in the complex are defined using information from a crystal structure analysis. Each ligand is assigned a set of bonding parameters. These, together with the d configuration of the metal and appropriate Racah and spin–orbit coupling parameters, are used to calculate the energy levels and associated wavefunctions of the metal in the complex. The energy levels may be compared with the observed electronic spectrum of the complex. The wavefunctions are used to estimate the molecular g values and their orientation both in a defined molecular coordinate system and with respect to the crystal axes. If desired, the molecular and crystal magnetic susceptibilities may also be calculated at specified temperatures. The second version of the program, *CAMMAG II*, also provides estimates of the relative intensities of the d–d transitions if the complex is noncentrosymmetric.

It should be noted that in *CAMMAG II* the underlying cause of the metal–ligand interaction is ascribed to a rather different origin from that which forms the basis of the conventional AOM. Rather than relating the interactions to metal–ligand orbital overlap, the electron density of the ligands is partitioned into cells, which perturb the d electrons of the metal. Mathematically, however, the cellular ligand field (CLF) is represented by parameters that are analogous to those of the conventional AOM (7). The conceptual arguments underlying the formulation of the CLF and the interpretation of the metal–ligand bonding parameters derived using this model are presented in detail elsewhere (22–24).

Computer programs such as *CAMMAG* provide a powerful means of exploring the relationships between experimental observables such as optical transition energies and magnetic properties, and the metal–ligand σ- and π-bonding parameters in a complex. They are well suited not only to calculations involving complexes of known crystal structure but also to investigating the effect that a particular change in geometry has on the energy levels of a complex. For instance, the effect of a particular vibration may be explored by carrying out a series of calculations in which the positions of the ligands in a 'dummy' complex are moved progressively along the normal coordinate of interest. This latter application is particularly useful in estimating vibronic coupling coefficients (25–27).

In the *CAMMAG* programs, the free ion term wavefunctions are used as the basis set in the calculation. The eigenfunctions are, therefore, linear combinations of these wavefunctions, which means that—except in special cases (e.g. d^1 and d^9 metal ions) — the d orbital occupancy associated with each energy level is not readily available. Another computer program, *AOMX*, developed by Adamsky and co-workers (28), uses the real d orbitals as the basis of the calculations. This is a particular advantage if one desires to correlate the d electron density with some property, for instance the metal–ligand bond length. At present, *AOMX* calculates just the energy levels and associated wavefunctions of the metal ion and is basically designed to facilitate the interpretation of the d-d spectra of transition-metal complexes. It includes a procedure to allow the 'best-fit' bonding parameters to be derived from a particular assignment of the band energies.

REFERENCES

1. Schäffer C. E., and Jørgensen, C. K., *Mol. Phys.*, 1965, **9**, 401.

2. Schönherr, T. *Topics Curr. Chem.*, 1997, **191**, 88.

3. Larson, E., and LaMar, G. N., *J. Chem. Educ.*, 1974, **51**, 633.

4. Richardson, D. E., *J. Chem. Educ.*, 1993, **70**, 372.

5. Smith, D.W., *Inorg. Chim. Acta*, 1977, **22**, 107.

6. Ceulemans, A., Beyens D., and Vanquickenborne, D. G., *Inorg. Chim. Acta*, 1982, **61**, 199.

7. Schäffer, C. E., *Inorg. Chim. Acta*, 1995, **240**, 5814.

8. Kasai, P. H., Whipple, E. A., and Weltner, W., *J. Chem. Phys.* 1966, **44**, 2581.

9. Beattie, I. R., Brown, J. M., Crozet, P., Ross, A. J., and Yiannopoulou, A., *Inorg. Chem.* 1997, **36**, 3207.

10. Clack, D. W., Mingdan, C., and Warren, K. D., *J. Mol. Struct.*, 1987, **153**, 323.

11. Bermejo, M., and Pueyo, L., *J. Chem. Phys.*, 1983, **78**, 854.

12. Smith, D. W., *Struct. Bond.*, 1972, **12**, 50.

13. Deeth, R. J., and Gerloch, M., *Inorg. Chem.*, 1985, **24**, 1754.

14. Bencini, A., Benelli, C., and Gatteschi, D., *Coord. Chem. Rev.*, 1984, **60**, 131.

15. Lever, A. P. B., *Inorganic Electronic Spectroscopy*, 2nd ed., Elsevier, Amsterdam, The Netherlands,1984, chap. 9.

16. Kennedy, B. J., Murray, K. S., Hitchman, M. A., and Rowbottom, G., *J. Chem. Soc. Dalton Trans.*, 1987, 825.

17. Schönherr, T., Itoh, M., and Urushiyama, A., *Bull. Chem. Soc. Jpn.*, 1995, **68**, 2271.

18. McDonald, R. G., Riley, M. J., and Hitchman, M A., *Inorg. Chem.*, 1988, **27**, 894.

19. Vanquickenborne, L. G., and Ceulemans, A., *Inorg. Chem.*, 1981, **20**, 796.

20. Schäffer, C. E., and Yamatera, H., *Inorg. Chem.* 1991, **30**, 2840.

21. Ceulemans, A., and Vanquickenborne, L.G., *Pure Appl. Chem.*, 1990, **62**, 1081.

22. Bridgeman, A. J., and Gerloch, M., *Prog. Inorg. Chem.*, 1996, **45**, 179–281.

23. Gerloch, M., *Magnetism and Ligand-Field Analysis*, Chapman-Hall, London, 1972.

24. Reidel, D., in Avery, J., eds., *Understanding Molecular Properties*, J. Avery, J. P. Dahl and A .E. Hanien, Editors. D. Reidel (Publisher), New York, 1987, p. 111.

25. Bacci, M., *Chem. Phys. Lett.*, 1978, **58**, 537.

26. Bacci, M., *Chem. Phys.* 1979, **40**, 244.

27. Warren, K. D., *Struct. Bond.*, 1984, **57**, 119.

28. theochem.uni-duesseldorf.de/~heribert/aomx/aomxhtml/aomxhtml.html.

CHAPTER 4

THE ORIGIN AND CALCULATION OF Δ

It was pointed out in Chapter 2 that, although the parameter Δ or 10 Dq provides a useful way of representing the splitting of the d orbitals in cubic symmetry, in fact its primary origin cannot be the electrostatic interaction of central metal d (or f) orbitals with the surrounding set of ionic point charges or dipoles. Chapter 3 made some attempt to relate it to covalency in metal–ligand bonding. We now examine this issue a little more carefully, although still not in depth, because that would require a diversion into theoretical chemistry at a level we cannot contemplate. We shall try to make some general observations about the d orbital splitting without becoming involved in the mechanics of the calculations. Gerloch and Slade (1) should be consulted for a lucid and more extensive account.

4.1 CALCULATIONS BASED ON ELECTROSTATIC INTERACTIONS

The first calculations of Dq were remarkably successful. Shortly after the formulation of crystal field theory, a calculation on the $Cr(H_2O)_6^{3+}$ ion in chrome alum using a hydrogenic 3d wavefunction on the Cr^{3+} ion and treating the water molecules as point charges yielded $Dq_{oct}(calc)$ $\sim 1000\,cm^{-1}$ (2). The experimental value is $1700\,cm^{-1}$. A little later, a treatment based on point dipoles applied to the $Cu(H_2O)_6^{2+}$ ion in $K_2Cu(SO_4)_2 \cdot 6H_2O$ gave a similar degree of agreement with experiment (3). That study of a system in which the ligand is uncharged but possesses a permanent dipole moment largely avoids the difficulty of the contribution from ligands outside the first coordination sphere, because the effect of the permanent dipoles falls off much more rapidly with distance than does that of charges. Moreover, the effects of induced dipole terms, which are also important, decrease even more rapidly. On this basis, it was accepted for some time that the crystal field approach adequately covered the basic features of metal–ligand d orbital interactions, even though there were obviously some awkward points remaining.

Progress in computational aspects of theoretical chemistry soon showed that the initial successes were illusory. As soon as Hartree-type self-consistent wavefunctions became available for the first transition series ions, it was obvious that the earlier values obtained for $\overline{r^4}$, say, were grossly in error. The self-consistent field 3d radial wavefunctions show maxima at little over half the radial extent that pertains to hydrogenic functions. The value of $\overline{r^4}$ that they yield is an order of magnitude less. Dq, then, was found to be calculated as an order of magnitude lower than that found by experiment on this "improved" basis (4).

There was worse to come. A higher level of calculation replaced the effective water molecule charges or dipoles, permanent plus induced, with a fuller consideration of the Coulomb attractions of the metal 3d electrons to the oxygen nuclei and repulsions by the oxygen electrons (5). This procedure reversed the sign of the calculated Dq, making it about $-500 \, \text{cm}^{-1}$: The dominant term was found to be the attraction to the oxygen cores! The development of further improved metal 3d wavefunctions saved the situation from absurdity but did not restore it to credibility (6). Self-consistent wavefunctions of the Hartree–Fock type are even more contracted than those of the Hartree type, and use of them removed the dominance of the attractions of the 3d electrons for the oxygen cores. However Dq, although calculated as positive, was still more than an order of magnitude lower than the experimental value. It should be pointed out that the outermost portion of an atomic orbital is the least accurately defined, and it is this section that is critical in determining the interaction with ligand charges or dipoles.

A further step was to allow for modification of the central metal d orbital wavefunctions by the potential of the ligand charges or dipoles. The results of such treatments are not clear, being subject to the danger of some circularity of argument. Certainly a confident prediction of Dq is not available from this procedure.

The electrostatic approach to the calculation of Dq thus failed (7), and by inference, classical electrostatic interactions are not the primary basis of the splitting of the d orbitals in transition-metal ions. Crystal field theory, and so ligand field theory derived from it, cannot be justified, except on the empirical grounds that it provides a useful and effective framework for summarizing many symmetry-related d orbital splitting effects.

4.2   ONE-ELECTRON MOLECULAR ORBITAL CALCULATIONS

The history of calculations of Dq by the simpler forms of MO theory has almost as checkered a career as that of the crystal field approach. A brief outline of simple molecular orbital processes applied to transition-metal complex ions was given in Section 3.1.1, and we follow on from that here. We neglect the possibility of ligand–orbital–ligand–orbital overlap. The general form of a bonding molecular orbital is:

$$\phi_i = (1 + 2\lambda_i S_i + \lambda_i^2)^{-1/2}(\phi_{iM} + \lambda_i \phi_{iL}) \tag{4.1}$$

For an octahedral complex, $i = e_g$ for σ-bonding and $i = t_{2g}$ for π bonding interactions. λ_i is the mixing coefficient for bonding of the type specified by i; and for the present purposes, ϕ_{iM} can be taken to be a metal d orbital set of symmetry i, and ϕ_{iL} an appropriate combination of ligand p orbitals, also of symmetry i. Corresponding to this bonding molecular orbital there is an antibonding companion:

$$\phi_i^* = (1 - 2\lambda_i S_i + \lambda_i^2)^{-1/2}(\lambda_i \phi_{iM} - \phi_{iL}) \tag{4.2}$$

The mixing coefficient λ is a measure of the covalency of bonding. For $\lambda = 0$, the molecular orbitals are pure atomic orbitals located on the metal atom for ϕ and on the ligand atoms for ϕ^*, and the bonding is entirely ionic in nature. For $\lambda = 1$, there is equal contribution to the molecular orbitals from each source, and the bonding is entirely covalent in nature. One may approximately identify the covalent character in a bonding MO with the quantity:

$$\lambda^2/(1 + \lambda^2) \tag{4.3}$$

The molecular orbitals are arranged to be, as nearly as possible, eigenfunctions of a Hamiltonian operator which includes the kinetic energy operators for each of the atoms of the molecule, Coulomb potentials between all the nuclei and their valence electrons, and electron exchange terms. Again, we do not need to inquire into its specific nature here. In this treatment, the molecular orbitals formed by combinations of atomic orbitals have well-defined energies. The term *one electron* is often used in reference to treatments at this level.

The results of calculations based on this level of approximation are most immediately the energies and compositions of a set of bonding and antibonding molecular orbitals. The energies may be divided into contributions of three types as follows:

1. The electrostatic interactions as discussed above, together with the primary features of the formation of molecular orbitals, principally the exchange integral between the metal and the ligand orbitals.
2. Overlap—often referred to in the physics literature on this subject as non orthogonality—of metal and ligand orbitals. Note that appreciable covalency in the bonding does not necessarily require a large degree of overlap as defined by the overlap integral S_i.

3. Covalency—mixing of the metal and ligand orbitals, as defined by non zero values of λ_i—is required to differentiate the model from the purely ionic case.

As far as the one-electron level of molecular orbital theory is concerned, the ion most frequently studied—the test case—is the NiF_6^{4-} cluster, which appears in $KNiF_3$ and NiF_2. It has been chosen because a great deal is known about its magnetic, spectral and other physical properties, and it is of simple chemical formulation and has relative few electrons to take into account. The most obvious step is to take the calculated energies of the t_{2g}^* and e_g^* molecular orbitals of the metal ion and equate the difference to $10\,Dq$ as defined earlier (Chapters 2 and 3). On this basis, a value of $6300\,cm^{-1}$ was obtained (7), made up of contributions approximately as follows:

Crystal field / exchange integral, etc.	$-2700\,cm^{-1}$
Overlap	3700
Covalency	5300
Total	6300

This is to be compared with the experimental observation for the quantity of $7250\,cm^{-1}$, as taken from the $^3A_{2g} \rightarrow {}^3T_{2g}$ spectral transition. The agreement appears to be quite satisfactory.

The definition of Δ as the $e_g^* - t_{2g}^*$ separation is, however, an over-simplification. A more careful consideration of this separation reveals that it is a complex quantity. The point is that an occupied antibonding molecular orbital exerts an influence on the bonding and nonbonding valence electrons through its partner of the same spin in the corresponding bonding molecular orbital, and through the coupling between those electrons. The influence exerted is found to be distinctly different for the t_{2g}^* and the e_g^* orbitals. The transition $t_{2g}^* \rightarrow e_g^*$ then involves some rearrangement of the bonding and nonbonding valence orbital system, with consequent energy changes. When this effect is taken into account, a distinctly different picture emerges: Δ for the NiF_6^{4-} cluster is calculated to be only $2800\,cm^{-1}$, made up by these contributions:

Crystal field / exchange integral etc.	$-3600\,cm^{-1}$
Overlap	5700
Covalency	700
Total	2800

giving a rather poor agreement with experiment.

The application of the method to other MF_6 clusters, particularly those from Mn^{2+}, Fe^{3+}, and Cr^{3+} ions, yields about the same degree of success: Δ is calculated to be half or less of the experimental value. Minor variations of the procedure outlined have been tried without substantially improving the position.

Claims of close agreement in reproducing experiment are made from time to time in individual cases. There is, however, always a lack of generality, and some special feature in the approximations or empirical data act to produce agreement for the case in hand.

4.3 ALL-ELECTRON MOLECULAR ORBITAL CALCULATIONS

It is now technically possible to calculate the eigenvalues of the complete Hamiltonian of a system as complex as a first transition series MF_6^{n-} cluster using wavefunctions of good quality. By *the complete Hamiltonian*, we mean one that includes the kinetic energy operators of each atom, potential terms for the Coulomb attraction of each nucleus for every electron, repulsion between each electron pair, all exchange terms, and possibly some relativistic effects. Such a calculation is referred to as *ab initio*, or *all electron*. Also included here may be less rigorous calculations of the "frozen-core" type, in which the innermost electrons in closed shells are treated as units invariant to bonding influences.

In most studies, such a calculation is performed just for the ground configuration. Many—perhaps millions—of other configurations arise from an open shell, and they are mixed to an appreciable extent with the ground configuration under the Hamiltonian. This constitutes configurational interaction (CI).

Ab *initio–level* calculations may be considered under five headings:

1. *Restricted.* Here the radial parts of the wavefunctions for α- and β-spin electrons in a given orbital are defined to be the same. Unless CI is treated properly (see below) restricted calculations for open-shell systems, such as transition-metal complexes, are much less satisfactory than the unrestricted type, and this type of calculation is not recommended.

2. *Unrestricted.* Here the radial parts of the wavefunctions of an orbital are allowed to differ for α- and β-spin electrons. Unrestricted treatments include some, but not necessarily very much, of the possible CI.

3. *Hartree–Fock.* In this approximation, the solution to the Schrödinger equation for the multi electron atom takes place in the approximation that it must correspond to a single configuration.

4. *Density Functional.* Here advantage is taken of the fact that there must be a unique relationship between the ground state energy and the electron density (i.e., essentially the wavefunction), but it is not necessary to know exactly what that relationship (functional) is to solve the Schrödinger's equation to a good approximation. In particular, the assumption of a single configuration is not necessary. The energy–density relationship, however, is *not* required for excited states, and these are not as well reproduced. The density functional treatment includes some, perhaps much, of the CI absent in restricted theory.

5. *Configurational Interaction*. In general, a proper solution to Schrödinger's equation requires the participation of many different configurations. This is because the motions of electrons are correlated—if an electron is in one part of a molecule, the tendency of other electrons to venture into that region is affected. To deal with CI, it is necessary to perform a calculation that includes in some way at least some of the non-ground-state configurations, and the treatment can be computationally extremely demanding.

With the exception of restricted forms, all-electron calculations reproduce the relative energy levels of simple transition-metal complex ions with reasonable success. They do *not* however, retain the energies of the molecular orbitals as well-defined quantities.

The relevance of all-electron calculations to the crystal-field parameter then raises the question of definition. For example, if Δ for the NiF_6^{4-} cluster is defined on the basis of the $^3A_{2g} \rightarrow {}^3T_{2g}$ spectral transition energy, then the experimental value can readily be compared with a calculation. It is equally appropriate, however, to define Δ by comparing other term energy separations with experiment. Of course, different degrees of agreement are obtained, and it is a matter of choice as to which is to be accepted. The point is that when examined at sufficient depth, Δ is not a well-defined quantity. It is only in the simpler models of the bonding in complexes that it has a clear interpretation. Nevertheless, accepting some uncertainty in definition, it is found that unrestricted Hartree–Fock (UHF) calculations reproduce the equivalent of Δ in the simpler complex ions quite well. For the NiF_6^{4-} ion, a value of $8020 \, cm^{-1}$ has been quoted (8), agreeing with the experiment within about 10%. For the $CoCl_4^{2-}$ ion, the calculated value of $3500 \, cm^{-1}$ may be compared with an experimental one of $3000 \, cm^{-1}$. More recent density functional calculations do better ($7400 \, cm^{-1}$ for the NiF_6^{4-} ion) because they include the CI, and agreement within 10% is not uncommon for other systems (9).

Molecules may be treated by the methods of density functional theory (DFT) at greater depth because of its efficiency, and within the limits of definition of Δ, quite respectable reproduction of experimental values is obtained. For the CrF_6^{3-} ion, this technique gives a value of about $15,000 \, cm^{-1}$ for Δ (cf. the observed value of $15800 \, cm^{-1}$), and for the $Fe(CN)_6^{3-}$ ion $38,000 \, cm^{-1}$ (cf. $35000 \, cm^{-1}$ from experiment). Quite large systems, such as $M(bpy)_3^{n+}$ complexes, can also be handled, and useful estimates of d orbital splittings are still obtained.

To summarize, the parameter Δ has a clear significance in simple models of the chemical bonding in a complex, such as crystal field theory or the empirical and approximate molecular orbital treatments, but not in all-electron calculations for which the molecular orbital energies are not well defined. On the other hand, calculations based on the simpler models fail to reproduce experimental observations, whereas the laborious, sophisticated treatments succeed quite well in this respect, insofar as the relationship between the definition of Δ in the experimental and calculational frameworks allows.

4.4 SYMMETRIES LOWER THAN CUBIC

Calculations on a nonempirical basis are straightforward for complexes of lower than cubic symmetry, although they are more time-consuming. Quite good estimates of effective d orbital splittings are obtained, and DFT is the method of choice. Molecules as large as the tetragonal metal phthalocyanines have been handled successfully (10).

The effects of a small departure from cubic symmetry present problems, because then the separations between the d orbitals that correspond to the lifting of the degeneracy of, say, a T-type term are so small relative to the total energy. Calculations of this type are not difficult to carry out as a perturbation in the empirical crystal field framework by the introduction of parameters such as Cp (Section 2.7). The difficulty is to provide a nonempirical evaluation of, say, Cp.

Some attempts have been made to deduce say the trigonal ligand field component in a trigonally distorted octahedral complex using *ab initio* methods. Success has been limited, however, because the d orbital separations are so small compared with the overall bonding energies of the complexes. They reproduce the spectral transitions fairly well, but here the question of the relationship of the definition of the crystal field parameter sets in terms of calculated term energies presents even greater problems than in cubic symmetry. Dealing with the magnetic properties of, say, cubic field T ground state complexes is rarely successful. No attempt to reproduce say the Dq-Ds-Dt parameters of crystal field theory by an all-electron calculation has been really successful when the geometry is close to but not precisely octahedral.

Of course, on its entirely empirical basis, the AOM has shown promise in accounting for the energies of d and f orbitals in low-symmetry systems (Chapter 3).

4.5 f ELECTRON SYSTEMS

Calculation of f orbital splittings is on an even less satisfactory basis than for d orbital systems. The purely electrostatic approach immediately runs into the difficulty of dealing with the shielding effects of the valence electrons. Atomic wavefunctions are developed to minimize the energy of the system, so that they are more accurate in the inner regions where they maximize near the nucleus and contribute more to the energy than in the tail, far from the nucleus. It is this tail of the s and p valence electrons, however, that is responsible for the polarizability that provides the shielding of, say, the 4f electrons of a lanthanide ion.

Ab initio molecular orbital methods can produce some semiquantitative account of F orbital splittings. It seems that the important contributions occur via covalent interactions involving d orbitals, with these coupling to the f orbitals, which are only slightly lower in energy. The splittings are small, however, and the symmetries of ligand distribution are low, so that comparison between theory and experiment is scarcely possible. Nevertheless, it seems that, in principle at least, all-electron calculations provide terms of the right type to account for f orbital splitting effects (11).

4.6   REAL ELECTRON DENSITY DISTRIBUTION

The conventional thrust of ligand field theory is the production of sets of energy levels that account for those deduced from experimental studies of the physical properties of complexes. The models that produce such a set of energy levels, viz, CF or AOM, are openly empirical in nature and in various ways are manifestly in conflict with the processes of quantum mechanics. Those models are taken to act on a set of central metal effective d orbitals. These latter, while often having a sound quantum mechanical origin, are likely to have other orbitals, of metal as well as of associated ligand character, mixed into them. Consequently, the presumption has been that the three-dimensional electron densities that are predicted from the effective d orbitals under the action of a CF or an AOM Hamiltonian cannot be expected to have much resemblance to those actually physically present. That this is a somewhat pessimistic view was already demonstrated in the discussion of the t_{2g} orbital of a CrF_6^{3-} ion in Section 1.5.

More recently, further evidence has become available that the valence electron densities predicted by ligand field models may be taken with some degree of seriousness. A ligand field model for the Fe^{2+} ion in ammonium ferrous Tutton salt, FeTS, $[(NH_4)_2Fe(SO_4)_2 \cdot 6H_2O]$ has been developed on the basis of experimental studies involving an unusually wide set of physical techniques. The model predicts the t_{2g} components to have the populations given in Table 4.1, and consequently, the electron density to be cigar shaped and oriented along a direction defined by the polar angles θ and ϕ given in the table. θ is the angle measured from one of the Fe–O bonds and ϕ is the angle in the plane normal to that bond, and measured from another Fe–O bond.

High-quality X-ray diffraction experiments are now able to map, with worthwhile accuracy, the electron density around atoms in complexes. For FeTS, they can be interpreted to give valence d orbital populations and polar angles that describe the orientation of the associated electron density (12); the values obtained are quoted in Table 4.1. It can be seen that there is a pleasing agreement between them and their ligand field model counterparts. This study indicates

TABLE 4.1. Parameters Defining the d Orbital Populations in Electrons and the Orientation of the Electron Density in the Ammonium Ferrous Tutton Salt

Orbital	Experiment	Ligand Field Model
d_{xy}	1.06	1.24
d_{xz}	1.46	1.57
d_{yz}	1.06	1.19
θ	37°	41°
ϕ	135°	139°

that, in accounting for observed energy levels, ligand field theory may be able to predict the shape and magnitude of the distribution of the d valence electrons in a quite useful manner.

REFERENCES

1. Gerloch, M., and Slade, R.C., *Ligand Field Parameters*, Cambridge University Press, UK, 1973; chap. 6.
2. Van Vleck, J. H., *J. Chem. Phys.*, 1939, **7**, 72.
3. Polder, D., *Physica*, 1942, **9**, 709.
4. Ilse, F. E., and Hartman, H., *Z. Phys. Chem.*, 1951, **197**, 239.
5. Kleiner, W. H., *J. Chem. Phys.*, 1952, **20**, 1784.
6. Freeman, A. J., and Watson, R. E., *Phys. Rev.*, 1960, **120**, 1711.
7. Sugano, S., and Schulman, R. G., *Phys. Rev.*, 1963, **130**, 517.
8. Yamaguchi, T., Shibuya, S., Suga, S., and Shin, S., *J. Phys. C, Solid State Phys.*, 1982, **15**, 2641.
9. Pavao, A. C., and da Silva, J. P. B, *J. Phys. Chem. Solids*, 1989, **50**, 669.
10. Reynolds, P. A., and Figgis, B. N., *Inorg. Chem.*, 1991, **30**, 2995.
11. Watson, R. E., and Freeman, A. J., *Phys. Rev.*, 1967, **156**, 251.
12. Figgis, B. N., Sobolov, A. N., Young, D. M., Schultz, A. J., and Reynolds, P. A., *J. Amer. Chem. Soc.*, 1998, **120**, 8715.

CHAPTER 5

ENERGY LEVELS OF TRANSITION METAL IONS

5.1 INTRODUCTION

In Chapters 2 and 3, the effect of certain arrangements of ligand atoms on the energies of the d orbitals of a central metal ion was investigated. It was shown that, irrespective of the nature of the interaction, the effect of the ligand field is to remove the degeneracy of the d orbitals. For metal ions with a single d electron, this alone is sufficient to account for the existence of what is perhaps the most striking feature of transition metal complexes—their attractive colors. For instance, the magenta of the octahedral $Ti(H_2O)_6^{3+}$ ion is caused by absorption of light of energy $\sim 20,000 \, cm^{-1}$ associated with excitation of the electron from the lower energy t_{2g} to the higher energy e_g^* set of orbitals (1).

The energies of the electronic transitions of most metal ions with more than one electron cannot be explained fully in this way. Thus the d^2 $V(H_2O)_6^{3+}$ ion absorbs at 17,800 and 25,700 cm^{-1} and it is impossible to explain these transitions merely in terms of excitation of the electrons from the t_{2g} to the e_g^* set of orbitals. Although the lower energy transition may plausibly be the result of promotion of just one electron between these orbitals, the only other excitation that this simple model would allow involves the simultaneous promotion of both electrons. This would occur at twice the energy of the first transition, which is incompatible with what is observed. Similar, or even more pronounced, complications occur for most many-electron transition metal ions.

In addition, certain aspects of the magnetic properties of complexes cannot be explained solely in terms of the relative energies of the d orbitals. For instance, it is hard to see why the d^5 complex $Fe(H_2O)_6^{3+}$ should be strongly paramagnetic, with five unpaired electrons, while the similar d^6 complex $Co(H_2O)_6^{3+}$ is diamagnetic, with all six electrons completely filling the t_{2g} set of orbitals.

Clearly, some additional factor must be included to explain the properties of transition ions having several d electrons. The most important aspect so far neglected is interelectron repulsion. This repulsion depends on the way in which the electrons are distributed among the d orbitals and is the reason why it is energetically favorable for two electrons to be placed in separate orbitals with parallel spins, rather than occupying a single orbital with paired spins (Hund's rule). Moreover, the spatial orientation of the d orbitals means that even when different orbitals are involved, the repulsion between two electrons depends on which pair of orbitals they occupy. Thus less repulsion occurs for the configuration $(d_{xy})^1(d_{z^2})^1$ than for the arrangement $(d_{xy})^1(d_{x^2-y^2})^1$ This means that a transition-metal ion with more than one d electron has a number of energy levels *even when the degeneracy of the d orbitals is not removed by a set of ligands.*

For first-row transition-metal ions, the effects of interelectron repulsion are, in fact, comparable to, or even larger than those owing to the interaction of the ligands with the d orbitals. Before tackling the effects of both interactions together, it is, therefore, convenient to consider in some detail the energy levels expected for a free transition-metal ion. Although chemists might think that such considerations lack a practical application, free transition ions have long been studied by physicists, particularly astronomers, and the theoretical model used to explain the spectra of these metal ions was in fact first developed by them.

5.2 FREE TRANSITION IONS

It is necessary at this stage to establish clearly the significance of two expressions that enter into the discussion frequently from now on (1). Great confusion can result if the difference between them is not appreciated.

> *Configuration.* A configuration is the assignment of a given number of electrons to a certain set of orbitals. Thus when there are two electrons in the d orbital set, we have the configuration d^2; if there are four electrons in the d set, two in the e_g^* set, and two in the t_{2g} set, the configuration is d^4; $t_{2g}^2 \cdot e_g^{*2}$.
>
> *Term*: A term is an energy level, or set of levels, of a system. The system may be specified by a configuration. Each configuration, in general, gives rise to a number of sets of energy levels and hence to a number of terms.

As is well known, it can be shown that there is little interaction between the electrons in an unfilled shell of an atom and those in filled shells, apart from the shielding of the nuclear charge by the latter. The electrons within an unfilled shell of a free ion are subject to two main perturbations. First, the electrons in a shell repel each other, and the energy of a configuration depends on how the

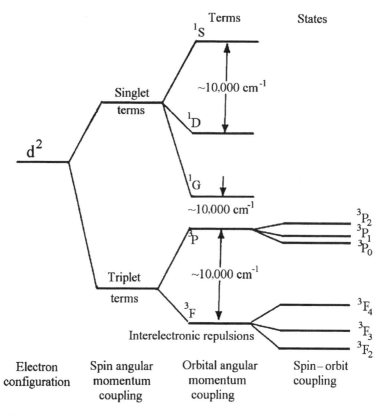

Figure 5.1. The splitting of a configuration into terms by interelectronic repulsions and of those into states by spin–orbit coupling. d^2 is used as the example.

electrons occupy the orbitals of the shell relative to one another. This is because the degree to which the electrons are brought into proximity is determined in that way. Second, there is the interaction of electron spin and orbital angular momenta through the spin–orbit coupling mechanism (Section 1.6). The interelectronic repulsion perturbation serves to break the highly degenerate energy levels of the unperturbed configuration down into a number of terms, as shown in Figure 5.1. The energy separation between adjacent terms, of free transition metal ions is generally of the order of $10,000\ cm^{-1}$. Because this is an order of magnitude larger than the spin–orbit coupling energies for first transition series ions, attention is focused first on the electron repulsion perturbation. It should be mentioned that the interelectronic repulsion effects include electron correlation and exchange interactions, which arise from the operation of the Pauli exclusion principle.

The terms that arise from electron repulsions are specified by wavefunctions that are eigenfunctions of a hydrogen-like Hamiltonian for the set of electrons, with an additional component owing to the interelectronic repulsions:

$$\mathbf{H}_0 = \sum_{i=1}^{n} (\mathbf{H}_i^0 + \sum_{j=1}^{n} e^2/r_{ij}) \quad (j \neq i) \tag{5.1}$$

5.3 FREE ION TERMS

A term is a set of energy levels that arise from a configuration. In general, it is degenerate, and the wavefunctions that correspond to the degeneracy are specified by a set of quantum numbers that are related to those already employed to specify the wavefunctions of single electrons. A term is characterized by: (a) its total orbital angular momentum L, which is zero or a positive integer, corresponding to l for a single electron, and (b) its total spin angular momentum S, which is zero or a positive integer or half-integer, corresponding to the quantum number s for a single electron. A term is labeled by means of its values of L and S using the nomenclature $^{(2S+1)}X$, where for the first four values X is a capital letter analogous to the lower case letter used for an orbital having a similar angular momentum:

$$L = 0 \quad 1 \quad 2 \quad 3 \quad 4 \quad 5 \quad 6 \cdots$$
$$X = S \quad P \quad D \quad F \quad G \quad H \quad I \cdots$$

The quantity $(2S + 1)$ is known as the *multiplicity* of the term. It is the number of values of M_S (equivalent to m_s) which occur, and is equal to the number of unpaired electrons present, plus one. Terms are named according to the value of S as follows:

S =	Singlet	Doublet	Triplet	Quartet	Quintet	Sextet
	0	$\frac{1}{2}$	1	$1\frac{1}{2}$	2	$2\frac{1}{2}$

Just as the z component of orbital angular momentum for a single electron l is specified by the quantum number m, so the z component of the total orbital angular momentum of the term L is given by the quantum number M_L. M_L is a number running in integral steps from L to $-L$. Similarly, the z component of total spin angular momentum for the term S_z is specified by the quantum number M_S, running from S to $-S$. Because there are $(2L + 1)$ values of M_L and $(2S + 1)$ values of M_S in each term, the total degeneracy of the term is $(2L + 1)(2S + 1)$. Each value of M_L occurs $(2L + 1)$ times, and each value of M_S occurs $(2S + 1)$ times in the term. For instance, a ^{3}P term is ninefold degenerate.

The wavefunctions specified by M_L are linear combinations of the Slater determinantal wavefunctions formed from the single electron wavefunctions. We write the wavefunction in which one electron has quantum number m_1, the next m_2, another m_3, and so on, as the antisymmetrised Slater determinant

$$(m_1, m_2, m_3, \ldots) = n!^{-1/2} \begin{vmatrix} (m_1)^1 & (m_1)^2 & (m_1)^3 & \cdots \\ (m_2)^1 & (m_2)^2 & (m_2)^3 & \cdots \\ & & & \cdots \\ (m_3)^1 & (m_3)^2 & (m_3)^3 & \cdots \\ & & & \cdots \\ & & & \cdots \end{vmatrix} \qquad (5.2)$$

where the superscript indicates the numbering of the electrons. For example, the wavefunction of a two-electron system, of one electron with $m = 2$ and the other with $m = -1$ is:

$$(2, -1) = 2^{-1/2} \begin{vmatrix} (2)^1 & (2)^2 \\ (-1)^1 & (-1)^2 \end{vmatrix}$$

$$= 2^{-1/2}[(2)^1(-1)^2 - (-1)^1(2)^2] \qquad (5.3)$$

The wavefunction $(m_1, m_2 \cdots)$ is associated with a value of M_L, which is the sum of the m values of the component electrons

$$M_L = \sum_{i=1}^{n} m_i \qquad (5.4)$$

Thus the wavefunction $(2, -1)$ belongs to an M_L value of 1.

If the wavefunction specified by M_L is written as $\{L, M_L\}$, the fact that it is a linear combination of the single electron orbital wavefunctions may be put in the form:

$$\{L, M_L\} = C_0(m_1, m_2, \ldots) + C_1(m'_1, m'_2, \ldots) + \cdots \qquad (5.5)$$

where, of course, Eq. 5.4 holds for each of the single electron product functions concerned. $\{L, M_L\}$ is normalized, so that

$$C_0 C_0^* + C_1 C_1^* + \cdots = 1 \qquad (5.6)$$

By our definition of L, the observable value of L_z is M_L (h/2π units):

$$\overline{L_z} = \int \{L, M_L\}^* \mathbf{L}_z \{L, M_L\} d\tau$$
$$= \langle L, M_L | \mathbf{L}_z | L, M_L \rangle \qquad (5.7)$$
$$= L_z \cdot h/2\pi$$

In the last relationship the "bra" and "ket" notation $\langle \Psi_1 | \alpha | \Psi_2 \rangle$ has been introduced to represent integrals of the form:

$$\int \psi_1^* \alpha \psi_2 \, d\tau \qquad (5.8)$$

to save the trouble of writing the integral and complex conjugate signs. These integrals, made up of an operator α taken between two wavefunctions ψ_1 and ψ_2, occur frequently as entries in secular determinants in quantum mechanics and are referred to as the *matrix elements* of the operator α.

The spin wavefunctions are treated in exactly the same manner as the orbital wavefunctions. In the wavefunction:

$$(m_{s_1}, m_{s_2}, \ldots) \, (m_{s_i} = \pm \tfrac{1}{2}) \qquad (5.9)$$

the total spin quantum number is, therefore, given by:

$$M_S = \sum_{i=1}^{n} m_{s_i} \qquad (5.10)$$

To differentiate readily between orbital and spin wavefunctions for terms, the spin wavefunctions are written as $[S, M_S]$. They are linear combinations of single electron spin product wavefunctions:

$$[S, M_S] = C_0(m_{s_1}, m_{s_2}, \ldots) + C_1(m'_{s_1}, m'_{s_2}, \ldots) + \cdots \qquad (5.11)$$

and

$$\langle S, M_S | \mathbf{S}_z | S, M_S \rangle = M_S \cdot h/2\pi$$
$$= \overline{S_z} \qquad (5.12)$$

The sets of wavefunctions for a term specified by M_L are orthonormal. The same statement applies for the sets specified by M_S:

$$\langle L, M_L | L, M'_L \rangle \quad \delta(M_L, M'_L) \qquad (5.13)$$
$$\langle S, M_S | S, M'_S \rangle \quad \delta(M_S, M'_S) \qquad (5.14)$$

The process of finding the linear combinations of single electron product wavefunctions that correspond to $\{L, M_L\}$ is, in the general case, a matter of some complexity. Standard works on atomic spectroscopy can be consulted for elegant methods (1–3). The same holds for the $[S, M_S]$ wavefunctions also. Certain particular combinations, however, can be seen by inspection. Consider the configuration d^2. There is only one way in which it is possible to obtain a value of $M_L = 3$ in d^2, and that is by having one electron with $m = 2$ and the other with $m = 1$, so that

$$\{3, 3\} = (2, 1) \tag{5.14}$$

Similarly, for the same configuration,

$$[1, -1] = (-\tfrac{1}{2}, -\tfrac{1}{2}) \tag{5.15}$$

This matter will be developed in more detail in the next section.

The process of finding what terms arise from a given configuration can be undertaken in the way illustrated below for d^1 and d^2. The first step is to write down all the possible combinations of single electron wavefunctions that are compatible with the Pauli exclusion principle, together with the M_L and M_S values to which they correspond. The second step is to start with the highest value of M_L that occurs and to say that there must be a term of multiplicity the number of times it appears, and with L having the same value $M_L(\text{max})$. Because the term possesses all values of M_L from L down to $-L$—i.e., from $M_L(\text{max})$ to $-M_L(\text{max})$—$(2S + 1)$ times, the number of times $M_L(\text{max})$ occurs is subtracted from all the values of M_L found. (Similarly, each time M_S occurs for the term, it is subtracted from the appropriate entry in the M_S set). From the remaining sets of M_L values, the process is repeated until there is nothing left. Of course, due regard must be paid to the fact that only certain multiplicities can arise from a given configuration: doublets from d^1, triplets and singlets from d^2, quartets and doublets from d^3, etc.

5.4.1 d^1

There are only 10 wavefunctions that can be written for one electron in the d orbital set. Because there is only one electron with $L = l = 2$ and $S = s = \tfrac{1}{2}$

$$\{2, M_L\} = (m) \tag{5.16}$$

$$[\tfrac{1}{2}, M_S] = (m_s) \tag{5.17}$$

Writing an upward-pointing arrow to represent an electron with $m_S = \tfrac{1}{2}$ and a downwards-pointing arrow for $m_S = -\tfrac{1}{2}$, the method of finding the term wavefunctions can be illustrated thus:

$m =$	2	1	0	-1	-2
	↑				

$M_L = 2$ $M_S = \frac{1}{2}$

Each value of M_L from 2 down to -2 occurs twice (corresponding to the two orientations of the arrow), so the configuration leads to nothing but a ^2D term, because $M_L\,(\text{max}) = 2$, and occurs twice.

$$d^1 \rightarrow {}^2D \tag{5.18}$$

5.4.2 d²

There are 45 ways of arranging two electrons, with spin, in the five d orbitals. A few of them are given below, together with the M_L and M_S values to which they lead.

$m =$	2	1	0	-1	-2	M_L	M_S
	↑	↑				3	1
	↑		↑			2	1
			↑		↓	-2	0
			↓		↑	-2	0
	↑↓					4	0

Note that although the third and fourth are distinct wavefunction representations, the last is *not* different from:

↓↑				

Because of the Slater determinantal relationship of Eq. 5.2, the M_L values of these arrangements occur the following numbers of times:

$$
\begin{aligned}
M_L = \pm 4 \quad & 1 \text{ time (each)}\\
= \pm 3 \quad & 4 \text{ times}\\
= \pm 2 \quad & 5 \text{ times}\\
= \pm 1 \quad & 8 \text{ times}\\
= 0 \quad & 9 \text{ times}\\
\hline
\text{Total} = \; & 45 \text{ components}
\end{aligned}
$$

The M_S values are found to be distributed as

$$M_S = \pm 1 \quad 10 \text{ times (each)}$$
$$= \quad 0 \quad 25 \text{ times}$$
$$\text{Total} = \quad 45 \text{ components}$$

Because $M_L = 4$ occurs only once, there must be a ^1G term from the configuration. We subtract one from each M_L value. The term leads to $M_S = 0$ only, and $(2L + 1) = 9$: we subtract 9 from the $M_S = 0$ set. There remains:

$$M_L = \pm 3 \quad 3 \text{ times}$$
$$= \pm 2 \quad 4 \text{ times}$$
$$= \pm 1 \quad 7 \text{ times}$$
$$= \pm 0 \quad 8 \text{ times}$$
$$M_S = \pm 1 \quad 10 \text{ times}$$
$$= \pm 0 \quad 16 \text{ times}$$

The maximum value of M_L remaining is 3, occurring 3 times, and a ^3F term is indicated. (The alternative of three ^1F terms makes it impossible to complete the assignment of the terms of the configuration). Subtracting 3 from each M_L and 7 from each M_S value:

$$M_L = \pm 2 \quad 1 \text{ time}$$
$$= \pm 1 \quad 4 \text{ times}$$
$$= \pm 0 \quad 5 \text{ times}$$
$$M_S = \pm 1 \quad 3 \text{ times}$$
$$= \quad 0 \quad 9 \text{ times}$$

The maximum M_L value now left is 2, once, and there must be a ^1D term. The requisite quantities, after subtracting 1 from each M_L value and 5 from $M_S = 0$, are:

$$M_L = \pm 1 \quad 3 \text{ times}$$
$$= \quad 0 \quad 4 \text{ times}$$
$$M_S = \pm 1 \quad 3 \text{ times}$$
$$= \quad 0 \quad 4 \text{ times}$$

M_L (max) is now 1, occurring 3 times, so that there is a ^3P term, which leaves

$$M_L = 0 \quad 1 \text{ time}$$
$$M_S = 0 \quad 1 \text{ time}$$

TABLE 5.1. Terms That Arise from d^n Configurations

Configurations	Terms
d^1, d^9	^2D
d^2, d^8	$^3F, {}^3P, {}^1G, {}^1D, {}^1S$
d^3, d^7	$^4F, {}^4P, {}^2H, {}^2G, {}^2F, 2X{}^2D, {}^2P, ({}^2D$ occurs twice$)$
d^4, d^6	$^5D, {}^3H, {}^3G, 2X{}^3F, {}^3D, 2X{}^3P, {}^1I, 2X{}^1G, {}^1F,$
d^5	$^6S, {}^4G, {}^4F, {}^4D, {}^4P, {}^2I, {}^2H, 2X{}^2G, 2X{}^2F, 3X{}^2D, {}^2P, {}^2S$

which corresponds to a ^1S term. Thus:

$$d^2 \rightarrow {}^1G + {}^1D + {}^1S + {}^3F + {}^3P \tag{5.19}$$

5.4.3 d^3-d^9

The argument for the higher configurations carried out in the same way as for d^2 is rather laborious. The results are given in Table 5.1. It is not necessary to work out the results for the configurations d^6 to d^9, because these can be obtained directly from the previous ones. We use the fact that, formally, holes in a full shell are equivalent to positrons (positive electrons), which repel each other in exactly the same way as do the electrons in a less-than-half-full shell. Consequently, the splitting of the configuration d^{10-n} $(n < 5)$ into terms takes the same form as does that for d^n.

Having found what terms arise from a configuration, it is now necessary to know something about the order in which they lie in energy. The most important point—which term lies lowest of all—is decided by the application of Hund's rules. In effect these rules sum up the tendency of electrons to share as little common space as possible, so that the repulsions between them are minimized. Hund's rules state that

1. Of the terms of a configuration, those with maximum multiplicity tend to lie lowest; this is a statement that the energy required to force the electrons to pair off spins is the major feature of the system.
2. Of the terms with maximum multiplicity, that with the largest value of L lies lowest.

With the aid of Hund's rules it is easy to find the ground term of any configuration. One just fills up the boxes (employed above) to have as many unpaired electrons as possible and to have the largest value of M_L consistent with this and with the exclusion principle. The ground terms of the configurations d^1 to d^9 are derived in this manner in Table 5.2.

The order in which other terms lie above the ground term and their separations from each other and from the ground term cannot be predicted in a simple manner (1–3). Some indication of the procedure by which this is done was given by Ballhausen (4). The following paragraphs summarize the results.

TABLE 5.2. The Ground Terms for d^n Configurations

Ground Configuration.	Term	Maximum M_L and M_S					M_L	M_S
		$m=2$	1	0	−1	−2		
d^1	^{2}D	↑					2	$\frac{1}{2}$
d^2	^{3}F	↑	↑				3	1
d^3	^{4}F	↑	↑	↑			3	$1\frac{1}{2}$
d^4	^{5}D	↑	↑	↑	↑		2	2
d^5	^{6}S	↑	↑	↑	↑	↑	0	$2\frac{1}{2}$
d^6	^{5}D	↑↓	↑	↑	↑	↑	2	2
d^7	^{4}F	↑↓	↑↓	↑	↑	↑	3	$1\frac{1}{2}$
d^8	^{3}F	↑↓	↑↓	↑↓	↑	↑	3	1
d^9	^{2}D	↑↓	↑↓	↑↓	↑↓	↑	2	$\frac{1}{2}$

It is possible to develop the theory of the interelectronic repulsions within a configuration to give the energies of the terms above the ground term. For our purposes, the energies are a function of two parameters: the *electron repulsion parameters*. The two parameters may be chosen in either of two completely equivalent ways: either in that of Condon and Shortley (2) or in that of Fano and Racah (3). The two choices lead to the *Condon–Shortley* parameters F_2 and F_4 or to the *Racah* parameters B and C, respectively. These pairs of parameters are sufficient if attention is restricted to energy differences involving d electrons. If, for example, f electrons are considered other parameters to extend the set must be introduced. The two sets of repulsion parameters are linearly related:

$$B = F_2 - 5F_4 \qquad (5.20)$$

$$C = 35F_4 \qquad (5.21)$$

A list of some of the relevant energies for transition-metal ions is given in Appendix 6.

The Racah set of parameters has a small advantage in that the separations between terms of the same multiplicity within a configuration involve only its parameter B. Separations between terms of different multiplicities, however, involve both B and C. With the Condon–Shortley parameters, separations between terms, even of the same multiplicity, in general are functions of both F_2 and F_4. Racah parameters are employed here.

A complete account of the energies of the terms that arise from a configuration requires a knowledge of the radial wavefunctions of the electrons concerned and, consequently, depends on their principal quantum number. For many purposes, it is useful to know the qualitative features of the distribution of the terms, and these may be obtained if the approximate value of the ratio C/B or (F_4/F_2) is known. For the first-transition series ions, C/B is around 4.0, and B is about $1000 \, \text{cm}^{-1}$. Using this information, Figure 5.2 was constructed to

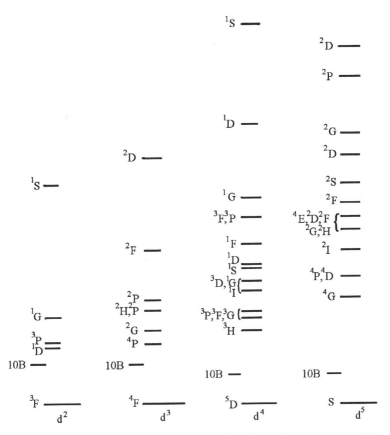

Figure 5.2. The relative energies of terms from d^n configurations, assuming $C/B = 4.7$. The diagrams for d^{10-n} are the same as for d^n.

illustrate the composition of the d^n configurations. In Figure 5.3, the manner in which the terms of the d^2 configuration vary in relative energy with the ratio C/B is outlined. The available data on the values of B and of the ratio C/B for transition element ions are summarized in Table 5.3 (5).

5.4   TERM WAVEFUNCTIONS

To proceed from a knowledge of how the single electron d orbitals are influenced by a ligand field to the effect of the ligand field on the terms of a free ion, it is necessary to know how the term wavefunctions are composed of single electron wavefunctions. We inquire only briefly into how some of the simpler combinations may be obtained. The results for the more difficult cases are listed elsewhere (1–3).

The single electron wavefunctions that give rise to various values of M_L and of M_S are listed in Table 5.4 for the configuration d^2. From the table, it is possible

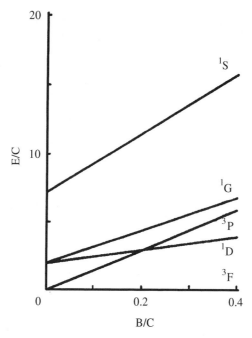

Figure 5.3. The relative energies of the terms from d^2 as the ratio C/B varies.

to deduce immediately that the wavefunction for the component of the ^1G term with $M_L = 4$ is given by:

$$\{4,4\}[0,0] = (2\tfrac{1}{2}, 2 - \tfrac{1}{2}) \tag{5.22}$$

Here, for example, $(2\tfrac{1}{2}, 2 - \tfrac{1}{2})$ means the wavefunction specified by one electron with $m = 2$ and $m_S = \tfrac{1}{2}$ and another electron with $m = 2$ and $m_S = -\tfrac{1}{2}$. Similarly, the wavefunctions of the components of the ^3F term with $M_L = 3$ and $M_L = 2$, both with $M_S = 1$, are:

$$\{3,3\}[1,1] = \left(2\tfrac{1}{2}, 1\tfrac{1}{2}\right) \tag{5.23}$$

$$\{3,2\}[1,1] = \left(2\tfrac{1}{2}, 0\tfrac{1}{2}\right) \tag{5.24}$$

These results follow because there is only the one combination of the single electron wavefunctions that can lead to these term quantum numbers. It is also obvious that the wavefunctions of both the ^1G and ^3F terms are composed of the pair $(2\tfrac{1}{2}, 1 - \tfrac{1}{2})$ and $(2 - \tfrac{1}{2}, 1\tfrac{1}{2})$. Similarly, the $M_L = 1, M_S = 1$ wavefunctions of the ^3F and ^3P terms are made up of $(2\tfrac{1}{2}, -1\tfrac{1}{2})$ and $(1\tfrac{1}{2}, 0\tfrac{1}{2})$.

TABLE 5.3. The Racah Parameters of Electron Repulsion for Transition-Element Free Ions [a,b]

Element	Charge									
	0		1+		2+		3+		4+	
Ti	**380,**	2.4	**583,**	3.4	**714,**	3.7				
Zr	**250,**	7.9	**450,**	3.9	**540,**	3.0				
Hf	**280**	—	**440,**	3.4	—	—				
V	**436,**	2.4	**585,**	4.2	**760,**	3.8	**886,**	4.0		
Nb	**300,**	8.0	**260,**	7.7	**530,**	3.8	**604,**	4.1		
Ta	**350,**	3.7	**480,**	3.8	—	—	**562,**	4.3		
Cr	**790,**	3.2	**655,**	4.1	**796,**	4.2	**933,**	4.0	**1038,**	4.1
Mo	**460,**	3.9	**440,**	4.5	—	—	—	—	**680**	—
W	**370,**	3.1	—	—	—	—	—	—	—	—
Mn	**720,**	4.3	**680,**	4.6	**859,**	4.1	**950,**	4.3	**1088,**	4.1
Tc	—	—	—	—	—	—	—	—	—	—
Re	**850,**	1.4	**470,**	4.0	—	—	—	—	—	—
Fe	**805,**	4.4	**764,**	4.5	**897,**	4.3	**1029,**	4.1	**1122,**	4.2
Ru	**600,**	5.4	**670,**	3.5	**620,**	6.5	—	—	—	—
Os	—	—	—	—	—	—	—	—	—	—
Co	**780,**	5.3	**798,**	5.5	**989,**	4.3	**1080,**	4.2	**1185,**	4.2
Rh	—	—	—	—	—	—	—	—	—	—
Ir	—	—	—	—	—	—	—	—	—	—
Ni	**1025,**	4.1	**1040,**	4.2	**1042,**	4.4	**1149,**	4.2	**1238,**	4.2
Pd	—	—	—	—	**830,**	3.2	—	—	—	—
Pt	—	—	—	—	—	—	—	—	—	—
Cu	—	—	**1220,**	4.0	**1240,**	3.8	—	—	—	—
Ag	—	—	—	—	—	—	—	—	—	—
Au	—	—	—	—	—	—	—	—	—	—

[a] First transition series are from Brorson and Schäffer (5).
[b] The energies in boldface are B in cm^{-1}; the figure following B is the ratio C/B. For the second- and third-transition series, the values listed are not reliable.

In fact, the wavefunctions for the orbital and spin parts of the complete wavefunction can be separated out. The complete orbital combinations for an F term from d^2, which is one of the principal results desired, are:

$$\{3, \pm 3\} = \pm(\pm 2, \pm 1) \tag{5.25}$$

$$\{3, \pm 2\} = (\pm 2, 0) \tag{5.26}$$

$$\{3, \pm 1\} = (2/5)^{1/2}(\pm 1, 0) + (3/5)^{1/2}(\pm 2, \mp 1) \tag{5.27}$$

$$\{3, 0\} = (4/5)^{1/2}(1, -1) + 5^{-1/2}(2, -2) \tag{5.28}$$

TABLE 5.4. The Single Electron Wavefunctions That Can Give Rise to Particular Values of M_L and M_S, for the configuration d^2

M_L	$M_S \to 1$	0	-1
4		$(2\frac{1}{2}, 2-\frac{1}{2})$	
3	$(2\frac{1}{2}, 1\frac{1}{2})$	$(2\frac{1}{2}, 1-\frac{1}{2}), (2-\frac{1}{2}, 1\frac{1}{2})$	$(2-\frac{1}{2}, 1-\frac{1}{2})$
2	$(2\frac{1}{2}, 0\frac{1}{2})$	$(2\frac{1}{2}, 0-\frac{1}{2}), (2-\frac{1}{2}, 0\frac{1}{2}), (1\frac{1}{2}, 1-\frac{1}{2})$	$(2-\frac{1}{2}, 0-\frac{1}{2})$
1	$(2\frac{1}{2}, -1\frac{1}{2}),$ $(1\frac{1}{2}, 0\frac{1}{2})$	$(2\frac{1}{2}, -1-\frac{1}{2}), (2-\frac{1}{2}, -1\frac{1}{2}),$ $(1\frac{1}{2}, 0-\frac{1}{2}), (1-\frac{1}{2}, 0\frac{1}{2})$	$(2-\frac{1}{2}, -1-\frac{1}{2}),$ $(1-\frac{1}{2}, 0-\frac{1}{2})$
0	$(2\frac{1}{2}, -2\frac{1}{2}),$ $(1\frac{1}{2}, -1\frac{1}{2})$	$(2\frac{1}{2}, -2-\frac{1}{2}), (2-\frac{1}{2}, -2\frac{1}{2}), (1\frac{1}{2}, -1-\frac{1}{2}),$ $(1-\frac{1}{2}, -1\frac{1}{2}), (0\frac{1}{2}, 0-\frac{1}{2})$	$(2-\frac{1}{2}, -2-\frac{1}{2}),$ $(1-\frac{1}{2}, -1-\frac{1}{2})$
-1	$(-2\frac{1}{2}, 1\frac{1}{2}),$ $(-1\frac{1}{2}, 0\frac{1}{2})$	$(-2\frac{1}{2}, 1-\frac{1}{2}), (-2-\frac{1}{2}, 1\frac{1}{2}), (-1\frac{1}{2}, 0-\frac{1}{2}),$ $(-1-\frac{1}{2}, 0\frac{1}{2})$	$(-2-\frac{1}{2}, 1-\frac{1}{2}),$ $(-1-\frac{1}{2}, 0-\frac{1}{2})$
-2	$(-2\frac{1}{2}, 0\frac{1}{2})$	$(-2\frac{1}{2}, 0-\frac{1}{2}), (-2-\frac{1}{2}, 0\frac{1}{2})$	$(-2-\frac{1}{2}, 0-\frac{1}{2})$
-3	$(-2\frac{1}{2}, -1\frac{1}{2})$	$(-2\frac{1}{2}, -1-\frac{1}{2}), (-2-\frac{1}{2}-1\frac{1}{2})$	$(-2-\frac{1}{2}, -1-\frac{1}{2})$
-4		$(-2\frac{1}{2}, -2-\frac{1}{2})$	

Similarly, the spin wavefunctions for a triplet term from d^2 are:

$$[1, \pm 1] = (\pm \tfrac{1}{2}, \pm \tfrac{1}{2}) \tag{5.29}$$

$$[1, 0] = 2^{-1/2}[(\tfrac{1}{2}, -\tfrac{1}{2}) + (-\tfrac{1}{2}, \tfrac{1}{2})] \tag{5.30}$$

Because the pairs $(\pm 1, 0)$ and $(\pm 2, \mp 1)$ must be completely employed between the ^{3}F and ^{3}P terms of d^2, it follows that the sum of the squares of the coefficients with which they appear in the $\{L, 1\}$ wavefunctions of ^{3}F and ^{3}P must be unity. Hence, for the P term,

$$[1, \pm 1] = \pm (3/5)^{1/2}(\pm 1, 0) + (2/5)^{1/2}(\pm 2, \mp 1) \tag{5.31}$$

Also, $\{1, 0\}$ of the P term is:

$$\{1, 0\} = 5^{-1/2}(1, -1) - (4/5)^{1/2}(2, -2) \tag{5.32}$$

In the same way, it follows that the spin wavefunction for a singlet term of d^2 is:

$$[0, 0] = 2^{-1/2}[(\tfrac{1}{2}, -\tfrac{1}{2}) - (-\tfrac{1}{2}, \tfrac{1}{2})] \tag{5.33}$$

5.5   SPIN–ORBIT COUPLING

Each of the terms of a configuration is $(2L + 1)(2S + 1)$-fold degenerate. The second perturbation that acts within an unfilled shell of electrons of a free ion is the coupling between the spin and orbital angular momenta of the electrons (Section 1.6.3). In most of the transition elements, the spin–orbit coupling perturbation is less than that owing to electronic repulsions. It is sufficient to see how spin–orbit coupling affects each of the terms of the configuration and to ignore the way in which the action of spin–orbit coupling in one term may influence its action in another term. This is the basis of the *Russell–Saunders* coupling scheme: Spin–orbit coupling is considered as a small perturbation on the terms set up by electron repulsions.

The Russell–Saunders coupling scheme is valid for most lighter elements, for which the effects of spin–orbit coupling are small. In particular, it applies quite well to the first-transition series ions; however, for heavy elements, spin–orbit coupling becomes more important and as may be seen from Table 5.3, electron repulsion parameters decrease somewhat. Consequently, for second- and, more particularly, third-transition series elements, the approximation implicit in the Russell–Saunders coupling scheme ceases to be good, and the scheme breaks down to at least some extent.

Another possible relationship between electron repulsion and spin–orbit coupling perturbations is that in which the spin-orbit coupling is much more important than the former. Then, instead of a configuration being split into terms as the first approximation, it is split into levels determined by spin–orbit coupling. The second step in this alternative relationship is to treat the electron repulsions as a perturbation on the spin–orbit coupling levels. Such an arrangement is known as the $j - j$ coupling scheme; it is the direct reverse of the Russell–Saunders scheme. No ion of interest to ligand field theory comes anywhere near to conforming to the $j - j$ coupling scheme, even for the actinide elements. Among the heavier transition-element ions, many do not conform to either coupling scheme and require something in between. To them an *intermediate* coupling scheme is said to apply. The only coupling scheme to be dealt with in detail here is the Russell–Saunders scheme, but occasional reference will be made to the intermediate coupling scheme.

In the Russell–Saunders coupling scheme, each term is split up into a number of *states*, which are specified by the total angular momentum quantum number J. Each state is $(2J + 1)$-fold degenerate. The wavefunctions for the state are specified according to their z component J_z of total angular momentum by the quantum number M_J. M_J, of course, runs in integral steps from J to $-J$. For example, if $J = 2\frac{1}{2}$, $M_J = 2\frac{1}{2}, 1\frac{1}{2}, \frac{1}{2}, -\frac{1}{2}, -1\frac{1}{2}, -2\frac{1}{2}$. The $(2J + 1)$-fold degeneracy of a state is lifted upon the application of a magnetic field so that the components become separated.

Just as the total angular momentum of the state is specified by J, so it is possible to say that the orbital and spin angular momenta, l and s, of each individual electron are coupled together to give a total angular momentum for the electron j. j_z, the z component of the single electron total angular momentum, is specified by the quantum number, m_j, which proceeds in integral steps from j to $-j$. If, as outlined in the definition of the $j - j$ coupling scheme, spin–orbit coupling is the most important perturbation acting on a configuration, it is the coupling between the total angular momenta of the individual electrons that defines primary splitting of the energy of the configuration. For that reason, it is called the $j - j$ coupling scheme.

Two parameters are in common use to describe the action of spin–orbit coupling. The first of these is the *single-electron* spin–orbit coupling parameter ζ, which measures the strength of the interaction between the spin and orbital angular momenta of a single electron of the configuration; it is a property of the *configuration*. The operator that corresponds to perturbation by spin–orbit coupling is, as far as a single electron is concerned, ζ times the scalar product between the spin and orbital angular moment vectors of the electron:

$$\zeta l \cdot s \tag{5.34}$$

The parameter, ζ is related to fundamental atomic parameters:

$$\zeta = (Z_{eff} e^2 / 2m^2 c^2) \overline{r^{-3}} \tag{5.35}$$

It is a positive quantity.

For use in connection with terms it is more convenient to deal with a parameter that is a property of the term itself; and for this purpose, the parameter λ is introduced. The operator that corresponds to perturbation by spin–orbit coupling within a term is λ times the scalar product between the spin and orbital angular momenta of the term:

$$\lambda L \cdot S \tag{5.36}$$

λ and ζ are related:

$$\lambda = \pm \zeta / 2S \tag{5.37}$$

The plus sign holds for a shell less than half-full because ζ is by definition positive. It includes (from Eq. 5.35) the charges on the nucleus and electron; and because a more-than-half-full shell of electrons may be regarded as a set of positively charged holes in a filled shell, the negative sign in Eq. 5.37 holds for a more-than-half-full shell.

Each term is split into states by spin–orbit coupling, and successive states are specified by values of J differing by unity. Given this, it is easy to show that the number of states that arise from a term is $(2S + 1)$ or $(2L + 1)$, whichever is the smaller. It also follows that J ranges from $|L - S|$ to $|L + S|$. The energy separation between a state specified by J and the one specified by $(J + 1)$ is:

$$\Delta E_{J,J+1} = \lambda(J + 1) \tag{5.38}$$

For a shell less than half full, λ is positive, and the lowest value of J, $|L - S|$, lies lowest in energy.

To specify a particular state that arises from a term, the value of J is added as a subscript to the term symbol. The full symbol for a state is thus $^{(2S+1)}L_J$ The $J = 1$ state from the ^3P term is, then, $^3P_1$.

The splitting arrangement just outlined—a term split into a set of states of increasing value of J—is said to make up a *normal* multiplet. When dealing with configurations that correspond to shells more than half full, λ is negative. This has the effect of reversing the order in the splitting diagram of the same term of the equivalent less-than-half-full configuration. The resulting multiplet, which has the highest value of J lowest in energy, is said to be *inverted*.

The foregoing relationships are summarized in Figure 5.4. The spin–orbit coupling parameters for a number of transition-element free ions are given in Table 5.5. The parameter ζ is listed, because λ can then be derived for the various terms of any configuration via Eq. 5.37. For instance, in d^2, $\lambda = \zeta/2$; in d^8

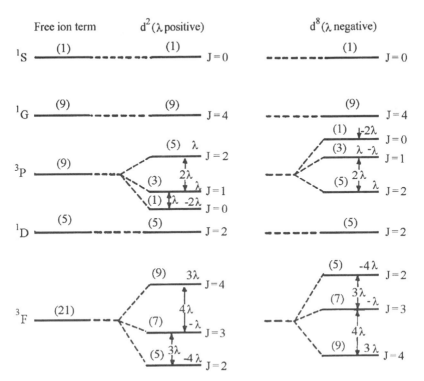

Figure 5.4. The splitting of the terms from d^2 and d^8 by spin–orbit coupling.

TABLE 5.5. Free Ion, Single-Electron Spin–Orbit Coupling Parameters (ζ_{nd}) for Transition Elements[a]

Metal	0	Charge					
	0	1+	2+	3+	4+	5+	6+
Ti	70	90	123	155			
Zr	—	(300)	(400)	(500)			
Hf	—	—	—	—			
V	95	135	170	210	250		
Nb	—	(420)	(610)	(800)	—		
Ta	—	—	—	(1400)	—		
Cr	135	185	230	275	335	380	
Mo	—	—	(670)	800	(850)	(900)	
W	—	—	(1500)	(1800)	(2300)	(2700)	
Mn	190	255	300	355	415	475	540
Tc	—	—	(950)	(1200)	(1300)	(1500)	(1700)
Re	—	—	(2100)	(2500)	(3300)	(3700)	(4200)
Fe	275	335	400	460	520	590	665
Ru	—	—	—	(1250)	(1400)	(1500)	(1700)
Os	—	—	—	(3000)	(4000)	(4500)	(5000)
Co	390	455	515	580	650	715	790
Rh	—	—	—	—	(1700)	(1850)	(2100)
Ir	—	—	—	—	(5000)	(5500)	(6000)
Ni	—	565	630	705	790	865	950
Pd	—	(1300)	(1600)	—	—	—	—
Pt	—	(3400)	—	—	—	—	—
Cu	—	—	830	890	960	1030	1130
Ag	—	—	(1800)	—	—	—	—
Au	—	—	(5000)	—	—	—	—

[a] Values are given in em^{-1}; Values in parentheses are estimates.

$\lambda = -\zeta/2$. For the ^3F ground term of d^2, $J = |3 - 1| = 2$ lies lowest, with $J = 3$ higher by 3λ and $J = 4$ higher again by 4λ. In Figure 5.4, the center of gravity of the states from each term remains the same as that for the unperturbed term, because the spin–orbit coupling is a small perturbation; i.e., if W_J is the energy of the state specified by J, for any term:

$$\sum (2J + 1)W_J = 0 \qquad (5.39)$$

The state wavefunctions [J,M_J] are expressed as linear combinations of the term wavefunctions $\{L, M_L\}[S, M_S]$. Methods by which they may be obtained are set out in standard works on the theory of atomic spectroscopy (1–3).

REFERENCES

1. Eyring, H., Walter, J., and Kimball, G. E., *Quantum Chemistry*, Wiley, New York, 1944, chap. 9.
2. Condon, E. U., and Shortley, G. H., *The Theory of Atomic Spectra*, Cambridge University Press, UK 1957.
3. Fano, U., and Racah, G., *Irreducible Tensorial Sets*, Academic Press, New York, 1958.
4. Ballhausen, C. J., *Introduction to Ligand Field Theory*, McGraw-Hill, New York, 1962, chap. 2.
5. Brorson, M., and Schäffer, C. E., *Inorg. Chem.* 1988, **27**, 2522.

CHAPTER 6

EFFECT OF LIGAND FIELDS ON THE ENERGY LEVELS OF TRANSITION IONS

6.1 THE EFFECT OF A CUBIC LIGAND FIELD ON S AND P TERMS

The compositions of free ion term wavefunctions were described in Chapter 5, expressed as functions of single electron wavefunctions. It is now possible to proceed with the examination of the effect of a ligand field on a polyelectron configuration. The general problem is to set up the secular determinant whose perturbing Hamiltonian is the sum of the ligand field, interelectronic repulsion, and spin–orbit coupling operators. It acts on the set of all the wavefunctions that can be written for the configuration under consideration.

This general problem is, however, much too complicated to have a simple solution. Although it has been studied in full (1), the approach we consider initially is to introduce the approximation that the ligand field is a small perturbation compared to the electron repulsion effects within the configuration but a large one compared to spin–orbit coupling. This is the *weak field* approximation; the ligand field sorts out the various wavefunctions of a free ion term but does not alter them. In other words, the effect of the ligand field on a term is not influenced by what it does to any other term. The introduction of the weak field approximation allows the general features of the effects of the ligand field to be presented in an easily visualized form. It is not a valid approximation for any but the simplest applications of ligand field theory, but it can readily be improved on when required. Ligand fields that cause splittings about as large as and much larger than the interelectronic repulsions are referred to as *medium* and *strong*, respectively.

In the weak field approximation, one sets up the secular determinant for the action of the ligand field potential operator, $\mathbf{V}_{oct}$, say, on the wavefunctions of the term being considered. Spin–orbit coupling is neglected for the time being. The matrix elements:

$$\langle L, M_L | \mathbf{V}_{oct} | L, M'_L \rangle \tag{6.1}$$

for example, are required.

6.1.1 S Terms

An S term is orbitally nondegenerate. Because the ligand field potential is concerned only with the orbital part of a wavefunction, it does not affect an S term.

6.1.2 P Terms

The orbital wavefunctions for a P term from the d^2 configurations are Eqs. 5.31 and 5.32;

$$\{1, \pm 1\} = \pm (3/5)^{\frac{1}{2}} (\pm 1, 0) + (2/5)^{\frac{1}{2}} (\pm 2, \mp 1)$$
$$\{1, 0\} = -(1/5)^{\frac{1}{2}} (1, -1) + (4/5)^{\frac{1}{2}} (2, -2)$$

Take first the matrix element

$$\langle 1, 1 | \mathbf{V}_{oct} | 1, 1 \rangle = \int [(3/5)^{\frac{1}{2}} (1, 0) + (2/5)^{\frac{1}{2}} (2, -1)]^* \mathbf{V}_{oct} [(3/5)^{\frac{1}{2}} (1, 0)$$
$$+ (2/5)^{\frac{1}{2}} (2, -1)] d\tau$$
$$= \int [(3/5)(1, 0)]^* \mathbf{V}_{oct} (1, 0) - (6^{\frac{1}{2}}/5)(1, 0)^* \mathbf{V}_{oct} (2, -1)$$
$$- (6^{\frac{1}{2}}/5)(2, -1)^* \mathbf{V}_{oct} (1, 0) + (2/5)(2, -1)^* \mathbf{V}_{oct} (2, -1)] d\tau \tag{6.2}$$

Now $\mathbf{V}_{oct}$ acts on the coordinates of each electron independently, so that we may suppose that it can be divided into components that act on the coordinates of a particular electron:

$$\mathbf{V}_{oct} = \mathbf{V}_{oct}^1 + \mathbf{V}_{oct}^2 + \cdots \tag{6.3}$$

The superscripts indicate the numbering of the electrons. Because we are at the moment dealing with only a two-electron system, the expansion has been carried far enough. It follows from this property of $\mathbf{V}_{oct}$ that:

$$\int (m, m')^* \mathbf{V}_{oct}(m'', m''') d\tau = 0 \tag{6.4}$$

unless:

$$m = m'' \quad \text{and} \quad \int (m')^* \mathbf{V}^2_{oct}(m''') d\tau = 0 \tag{6.5}$$

and/or:

$$m' = m''' \quad \text{and} \quad \int (m)^* \mathbf{V}^1_{oct}(m'') d\tau = 0 \tag{6.6}$$

In other words, the m values of one of the electrons must be the same on each side of $\mathbf{V}_{oct}$ in the matrix element, and the matrix element of $\mathbf{V}_{oct}$ with the wavefunctions of the other electron must be nonzero in at least one of the two possible cases. These relationships hold simply because $\mathbf{V}^1_{oct}$ (etc). cannot change the m value of more than one electron.

Adding the effects of $\mathbf{V}_{oct}$ on both electrons, the result may be put in the form:

$$\int (m, m')^* \mathbf{V}_{oct}(m'', m''') d\tau = \int (m')^* (m''') d\tau_2 \cdot \int (m)^* \mathbf{V}^1_{oct}(m'') d\tau_1$$
$$+ \int (m)^* (m'') d\tau_1 \cdot \int (m'_1)^* \mathbf{V}^2_{oct}(m'''_1) d\tau_2 \tag{6.7}$$

But the integrals that appear in this expression have already been evaluated in the electrostatic crystal field formalism (Section 2.3). Applying those results to Eq. 6.2:

$$\int (-1, 0)^* \mathbf{V}_{oct}(-1, 0) d\tau = \int (1, 0)^* \mathbf{V}_{oct}(1, 0) d\tau$$
$$= \int (1)^* \mathbf{V}_{oct}(1) d\tau_1 + \int (0)^* \mathbf{V}_{oct}(0) d\tau_2$$
$$= -4Dq + 6Dq = 2Dq \tag{6.8}$$

$$\int (\pm 1, 0)^* \mathbf{V}_{oct}(\pm 2, \mp 1) d\tau = \int (\pm 2, \mp 1)^* \mathbf{V}_{oct}(\pm 1, 0) d\tau = 0 \tag{6.9}$$

$$\int (\pm 2, \mp 1)^* \mathbf{V}_{oct}(\pm 2, \mp 1) d\tau = Dq - 4Dq = -3Dq \tag{6.10}$$

So that:

$$\langle 1,1| \mathbf{V}_{oct}|1,1\rangle = [(3 \times 2)/5]Dq + [(2 \times -3)/5]Dq = 0 \qquad (6.11)$$

which is the result we required. Also:

$$\int (1,-1)\,^{*}\mathbf{V}_{oct}(1,-1)d\tau = -8Dq \qquad (6.12)$$

$$\int (2,-2)\,^{*}\mathbf{V}_{oct}(2,-2)\,d\tau = 2Dq \qquad (6.13)$$

and so:

$$\langle 1,\pm 1| \mathbf{V}_{oct}|1,\pm 1\rangle = -8Dq/5 + 8Dq/5 = 0 \qquad (6.14)$$

Following exactly the same procedure with the cross terms between different M_L wavefunctions, we find in total

$$\langle 1,0|\mathbf{V}_{oct}|1,0\rangle = \langle 1,\pm 1|\mathbf{V}_{oct}|1,\pm 1\rangle = \langle 1,0|\mathbf{V}_{oct}|1,\pm 1\rangle$$
$$= \langle 1,\pm 1|\mathbf{V}_{oct}|1,\mp 1\rangle = 0 \qquad (6.15)$$

All the matrix elements are zero, so that *the crystal field potential, does not split the P term.*

The above argument was developed specifically for the ^3P term from the d^2 configuration. The d^3 configuration also, for example, gives rise to a P term, ^4P. It is possible to find the three-electron wavefunctions that give rise to the ± 1 and 0 M_L values of a P term and to proceed as above to show that all the matrix elements of $\mathbf{V}_{oct}$ within the ^4P term also are zero. It is simpler, however, to remark that, as far as the orbital properties of the terms with *maximum multiplicity* are concerned, the two "holes" in a half-filled d shell that result from the d^3 configuration are equivalent, with a change of charge sign, to the two electrons of a d^2 configuration (Section 6.2). Thus the ^4P term, like the ^3P term, is unaffected by a weak ligand field.

The ligand field potential employed above was that for an octahedral ligand arrangement; i.e., $\mathbf{V}_{oct}$; however, because the matrix elements of the potentials from a tetrahedron or a cube of ligands, $\mathbf{V}_{tet}$ or $\mathbf{V}_{cube}$, differ from those of $\mathbf{V}_{oct}$ only by a numerical constant (Section 2.5), they likewise are without influence on an S or a P term in the weak field approximation.

6.2 THE EFFECT OF A CUBIC LIGAND FIELD ON D TERMS

We have seen (Section 5.4) that the orbital wavefunctions of the ^2D term of d^1, $\{2,\pm 2\}$, $\{2,\pm 1\}$ and $\{2,0\}$, are identical to the single electron orbital wavefunctions (± 2), (± 1) and (0), respectively. Consequently, the ^2D term of d^1 must be split in exactly the same way as are the members of the d orbital set. In the electrostatic crystal field model, the wavefunctions $\{2,0\}$ and $2^{-1/2}$ $(\{2,2\} + \{2,-2\})$ occur at energy 6Dq, and the wavefunctions $\{2,\pm 1\}$ and $2^{-1/2}(\{2,2\} - \{2,-2\})$ lie at energy -4Dq. It is easy to obtain the orbital splitting of the ^2D term of the d^9 configuration from that for d^1. We merely note that d^9 corresponds to a single positive hole in a filled d^{10} shell. The positive hole behaves like a positron (the positive analogue of an electron). The expression for the electrostatic crystal field potential (Section 2.2) involves linearly the charge on the electron e. Consequently, changing the charge on the electron reverses the sign of the matrix elements of $\mathbf{V}_{oct}$ with the d electron term wavefunctions. Thus the ^2D term from the d^9 configuration is split by a crystal field in exactly the opposite manner to that for the ^2D term from d^1: At energy -6Dq there are the wavefunctions $\{2,0\}$ and $2^{-1/2}(\{2,2\} + \{2,-2\})$, and at energy 4Dq, the wavefunctions are $\{2,\pm 1\}$ and $2^{-1/2}(\{2,2\} - \{2,-2\})$.

Alternatively, the reversal in sign of the splitting of the ^2D term of d^9 relative to that of d^1 may be seen by noting that the expression for Dq contains e (Section 2.3). Properly, then, one should write:

$$Dq(d^9) = -Dq(d^1) \qquad (6.16)$$

and leave the splittings as 6Dq and -4Dq, but the convention given above is that normally adopted.

Although the above reasoning is based on the electrostatic crystal field approach, an identical result is obtained if the metal–ligand interaction is covalent in nature. The form of the splitting depends on the symmetry properties of the term wavefunctions, not on the nature of the interaction. The sign of the splitting is reversed for d^9, because for the latter the vacancy in the filled d shell occurs in the antibonding e_g^* orbital set, whereas the single electron of the d^1 configuration occupies the nonbonding or only weakly antibonding t_{2g} orbital set.

The other D terms that arise from d electron configurations are the ^5D terms of d^4 and d^6. The effect of the ligand field on ^5D (d^4) may be obtained by the laborious procedure of writing out the term orbital wavefunctions as products of the four single electron wavefunctions. It is much simpler, however, to note that, just as the d^9 configuration may be considered to be a hole in the filled d^{10} shell and the d^3 configuration may be considered to be two holes in the d^5 half-shell, so the d^4 configuration, may be considered to be a hole in the d^5 half-shell (Section 5.3). The argument only holds provided as many electrons as possible are unpaired (maximum multiplicity). But the ^5D term of d^4 is one in which all the electrons are unpaired. Consequently, this term is split in exactly the same way as is the ^2D term of d^9: upside down relative to the splitting of the ^2D term of d^1. Continuing the argument in this manner, the ^5D term of d^6 is one with the maximum multiplicity, and may be considered to be four holes in the filled d^{10} shell, and so is split in the opposite manner as the ^5D term of d^4; this means that it is split in the same way as the ^2D term of d^1.

These splitting relationships for D ground terms are summarized as follows:

1. The splitting pattern of the D term does not depend on its multiplicity.
2. When $n < 5$, the splitting of d^{n+5} is the same as for d^n, the splittings for d^{5-n} and d^{10-n} are identical, and are the reverse of that for d^n.

It was pointed out in Section 1.5 that the p orbital set is not split by an octahedral ligand field; likewise, of course, an s orbital cannot be split by a ligand field. Including these observations with those on d orbital splitting, we see that there is a one-to-one correspondence of the effect of a ligand field on an orbital set and on the term specified by the same letter.

6.3 THE EFFECT OF A CUBIC LIGAND FIELD ON F TERMS

Before treating the F term wavefunctions in the same way that we have dealt with the S, P, and D term wavefunctions, it is instructive to examine the manner in which f orbitals are affected by a ligand field. Consider how the real forms of the f orbitals, (Fig. 1.3) relate to the ligand positions of an octahedral complex (Fig. 1.5). Irrespective of whether the interaction is electrostatic or antibonding covalent, the orbital that is directed through the octahedral faces f_{xyz} is the one that is farthest from the ligands and hence has least destabilization. The set directed along the axes $f_{x(5x^2-3r^2)}$, $f_{y(5y^2-3r^2)}$, and $f_{z(5z^2-3r^2)}$ are closest to the ligands and are most destabilized. The remaining set—$f_{z(x^2-y^2)}$, $f_{x(y^2-z^2)}$, and

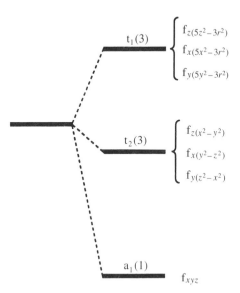

Figure. 6.1. The splitting of the f orbital set by an octahedral ligand field. The degeneracy of each set is indicated in parentheses.

$f_{y(z^2-x^2)}$, involves orbitals that are symmetrically directed toward the ligands to an extent that is intermediate between those of the two other sets. Thus the splitting of the orbitals is as shown in Figure 6.1. If the correlation between the splitting of an orbital set and the term with the same letter as the term symbol holds, we expect $\mathbf{V}_{oct}$ to split an F term as shown in Figure 6.1.

Much of the necessary detail for finding the effect of $\mathbf{V}_{oct}$ on the orbital wavefunctions of the ^{3}F term of d^2 was given in connection with the P terms (Section 6.1). It is necessary merely to use the different coefficients with which the single electron wavefunctions appear in $\{3, \pm1\}$ and $\{3,0\}$ of the F term as compared to $\{1, \pm1\}$ and $\{1,0\}$ of the P term (Section 5.4) to obtain the corresponding matrix elements. For example, in the electrostatic formalism:

$$\langle 3, 1|\mathbf{V}_{oct}|3, 1\rangle = -[(3 \times 3)/5]\mathrm{Dq} + [(2 \times 2)/5]\mathrm{Dq} = -\mathrm{Dq} \tag{6.17}$$

$$\langle 3, 0|\mathbf{V}_{oct}|3, 0\rangle = -[(8 \times 4)/5]\mathrm{Dq} + [(2 \times 1)/5]\mathrm{Dq} = -6\,\mathrm{Dq} \tag{6.18}$$

For dealing with the F term the following additional results concerning the matrix elements of single electron product functions are required. For example, they are obtained exactly as outlined for $\int(1,0)^*\mathbf{V}_{oct}(1,0)\mathrm{d}\tau$ of the P term. Referring to Sections 5.4 and 6.1:

$$\int(\pm2, \pm1)\,^*\mathbf{V}_{oct}(\pm2, \pm1)\mathrm{d}\tau = \mathrm{Dq} - 4\,\mathrm{Dq} = -3\,\mathrm{Dq} \tag{6.19}$$

$$\int(\pm2, 0)\,^*\mathbf{V}_{oct}(\pm2, 0)\mathrm{d}\tau = \mathrm{Dq} + 6\,\mathrm{Dq} = 7\,\mathrm{Dq} \tag{6.20}$$

$$\int(\pm2, 0)\,^*\mathbf{V}_{oct}(\mp2, 0)\mathrm{d}\tau = 5\,\mathrm{Dq} \tag{6.21}$$

$$\int(\pm2, \pm1)\,^*\mathbf{V}_{oct}(\mp2, \pm1)\mathrm{d}\tau = 5\,\mathrm{Dq} \tag{6.22}$$

From these and the expressions for the F term orbital wavefunctions, we obtain:

$$\langle 3, \pm3|\mathbf{V}_{oct}|3, \pm3\rangle = -3\,\mathrm{Dq} \tag{6.23}$$

$$\langle 3, \pm2|\mathbf{V}_{oct}|3, \pm2\rangle = 7\,\mathrm{Dq} \tag{6.24}$$

$$\langle 3, \pm1|\mathbf{V}_{oct}|3, \pm1\rangle = -\mathrm{Dq} \tag{6.25}$$

$$\langle 3, 0|\mathbf{V}_{oct}|3, 0\rangle = -6\,\mathrm{Dq} \tag{6.26}$$

$$\langle 3, \pm3|\mathbf{V}_{oct}|3, \mp1\rangle = 15\tfrac{1}{2}\,\mathrm{Dq} \tag{6.27}$$

$$\langle 3, \pm2|\mathbf{V}_{oct}|3, \mp2\rangle = 5\,\mathrm{Dq} \tag{6.28}$$

All the other matrix elements of the series are zero; for example, $\langle 3, \pm3|\mathbf{V}_{oct}|3, \pm2\rangle$.

We are now in a position to set up the secular determinant for the action of V_{oct} on the F orbital wavefunctions, which is, then:

	$\{3, 3\}$	$\{3, 2\}$	$\{3, 1\}$	$\{3, 0\}$	$\{3, -1\}$	$\{3, -2\}$	$\{3, -3\}$
$\{3, 3\}$	$-3\,Dq-E$	0	0	0	$15^{1/2}\,Dq$	0	0
$\{3, 2\}$	0	$7\,Dq-E$	0	0	0	$5\,Dq$	0
$\{3, 1\}$	0	0	$-Dq-E$	0	0	0	$15^{1/2}\,Dq$
$\{3, 0\}$	0	0	0	$-6\,Dq-E$	0	0	0
$\{3,-1\}$	$15^{1/2}\,Dq$	0	0	0	$-Dq-E$	0	0
$\{3,-2\}$	0	$5\,Dq$	0	0	0	$7\,Dq-E$	0
$\{3,-3\}$	0	0	$15^{1/2}\,Dq$	0	0	0	$-3\,Dq-E$

$= 0$

$$(6.29)$$

This determinant reduces immediately to the sub-determinants:

$$
\begin{array}{cc}
\{3, \pm 3\} & \{3, \mp 1\}
\end{array}
$$
$$
\begin{vmatrix}
-3\,Dq - E & 15^{1/2}Dq \\
15^{1/2}\,Dq & -Dq - E
\end{vmatrix} = 0
\qquad (6.30)
$$

$$
\begin{array}{cc}
\{3, 2\} & \{3, -2\}
\end{array}
$$
$$
\begin{vmatrix}
7\,Dq - E & 5\,Dq \\
5\,Dq & 7\,Dq - E
\end{vmatrix} = 0
\qquad (6.31)
$$

and

$$\{3, 0\} \qquad -6\,Dq - E = 0 \qquad (6.32)$$

the solutions are:

$$E = -6\,Dq \text{ occurring 3 times}$$
$$E = 2\,Dq \text{ occurring 3 times}$$
$$E = 12\,Dq \text{ occurring 1 time}$$

The corresponding wavefunctions obtained from the sets of secular equations are as follows: at $-6Dq$:

$$24^{-1/2}[15^{1/2}\{3, 3\} - 3\{3, -1\}]$$
$$\{3, 0\}$$
$$24^{-1/2}[15^{1/2}\{3, -3\} - 3\{3, 1\}] \qquad (6.33)$$

at 2Dq:

$$24^{-1/2}[3\{3,3\} + 15^{1/2}\{3,-1\}]$$
$$2^{-1/2}[\{3,2\} - \{3,-2\}]$$
$$24^{-1/2}[3\{3,-3\} + 15^{1/2}\{3,1\}] \tag{6.34}$$

at 12Dq:

$$2^{-1/2}[\{3,2\} + \{3,-2\}] \tag{6.35}$$

Thus the ^{3}F term of d^2 is split by $\mathbf{V}_{oct}$, as shown in Figure 6.2.

The center of gravity of the F term is at $(3\times-6+3\times2+1\times12)Dq=0$ relative to the unperturbed term. That is, the center of gravity is not shifted by this electrostatic crystal field, a result required by the definition of $\mathbf{V}_{oct}$. The splitting illustrated in Figure 6.2 is of the same form as that of the splitting of the f orbital set (Fig. 6.1), but reversed in sign, showing that the equivalence between the splittings of orbital sets and of terms with the same letter symbol is the same, apart possibly from the sign. In fact, in general, the sign is the same for l even, opposite for l odd.

A consideration of the equivalence between holes in filled and half-filled shells and electrons in a less-than-half-filled shell, carried out as for the D terms

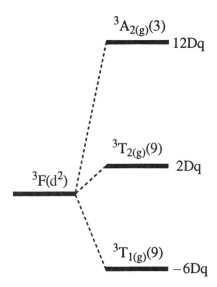

Figure. 6.2. The effect of an octahedral ligand field on the ^{3}F term of d^2. The numbers in parentheses represent the total degeneracy, including the spin. The energy of each term is given on the right side, in the electrostatic crystal field formalism.

in Section 6.2, shows that the same relationship holds. The splitting diagram is independent of the multiplicity for the F term, provided it is of the maximum multiplicity for the configuration. For d^{n+5}, the diagram is identical to that of d^n. The diagram is inverted relative to that of d^n for d^{5-n} and d^{10-n}. Thus the diagram shown in Figure 6.2 is maintained for the ^4F term of d^7 but is inverted for the ^4F term of d^3 and the ^3F term of d^8.

6.4   THE EFFECT OF A CUBIC LIGAND FIELD ON G, H AND I TERMS

Although G, H, and I terms do not occur as ground terms of d electron configurations, the results for the splitting of these terms are required in connection with spectra, in which excited terms are involved. G, H and I terms may be dealt with in the same manner as the terms previously treated. The splitting diagrams for the terms are shown on the left side of the correlation diagrams discussed in Section 6.7. The relationship $d^{5+n} \equiv d^n$, $d^{5-n} \equiv d^{10-n}$ inverted relative to d^n cannot always be used to obtain the results for excited terms, because there may not be an equivalence between holes and electrons within the appropriate configurations when there is a change in multiplicity.

The splitting of terms by V_{oct} having been determined, that by V_{tet} and V_{cube}, may be obtained from the fact that the matrix elements of these potentials with the d electron wavefunctions are respectively $-4/9$ and $-8/9$ times those of V_{oct} (Sections 2.5 and 3.2.3). Consequently, in the ligand field from a tetrahedron or a cube, the splitting diagrams are inverted relative to those for V_{oct} and the splitting is changed by the appropriate factor.

6.5   STRONG-FIELD CONFIGURATIONS

Up to now, the ligand field has been considered to be merely a perturbation acting on the free ion terms resulting from interelectronic repulsions. Such a treatment is not, however, generally valid. The separations between free ion terms are on the order of 10^4 cm^{-1}, whereas the electronic spectra of d^1 ions show that the splitting of the d orbitals is similar to this energy. We now consider the problem of a free ion subject to interelectronic repulsion and ligand field effects of comparable magnitude.

The first step is to examine a situation exactly contrary to the earlier one. We now assume that the interelectronic repulsions are small compared to the ligand field effects. In other words, the splitting of the free ion terms is large compared to the separation between them. A ligand field that is large enough to cause splittings of such magnitude is said to be *strong*. The strength of a ligand field is, then, measured in relationship to the magnitude of the interelectronic repulsions present in the ion. From the point of view of wavefunctions for the system, the effect of a strong ligand field is that *strong-field configurations,* rather than free ion {L,M_L} wavefunctions, become the natural basis for a calculation. A strong-field configuration is obtained by assigning each d electron to either a t_{2g} or an e_g^* orbital in octahedral stereochemistry. The orbitals are t_2^* or e for tetrahedral stereochemistry. Thus if there are n electrons assigned to t_{2g} orbitals and m to e_g^* orbitals, the corresponding strong-field configuration is $t_{2g}^n \cdot e_g^{*m}$.

For the configurations d^1 and d^9, there are no interelectronic repulsions, so the question of the relative strength of the ligand field does not arise. With two d electrons d^2, both the electrons in an octahedral strong-field complex go into the t_{2g} orbital set to form the lowest lying configuration. The strong-field configuration is t_{2g}^2. In the strong octahedral field, the excitation of an electron from the lowest configuration to the next higher lying configuration leads to the excited strong-field configuration $t_{2g}^1 \cdot e_g^{*1}$: excitation of both electrons gives the still higher lying e_g^{*2} strong-field configuration.

Feeding d electrons into strong-field systems results in the first six occupying the t_{2g} orbital set. The next four go into the e_g^* orbital set. In tetrahedral complexes, the first four electrons go into the e set, then six into the t_2^* set. The strong-field ground configurations in both octahedral and tetrahedral stereo-chemistry are shown in Table 6.1, in which three spaces represent the t_{2g} orbital set and two spaces the e_g^* set. The excited strong-field configurations are obtained by removing one or more electrons from the t_{2g} orbital set and placing them in the e_g^* set. For each electron so excited, the energy of the excited strong-field configuration is increased by Δ_{oct}. Exactly equivalent statements hold for the e and t_2^* orbitals in the tetrahedral case.

Of course, it is possible to describe the strong-field configurations as linear combinations of the free ion $\{L, M_L\}$ wavefunctions (1,2), but the combinations are different from those obtained for weak ligand field terms (Sections 6.1–4). For our purposes, there is no particular point in finding the strong-field $\{L, M_L\}$ combinations. Calculations involving the strong field configurations can usually be performed by using the single electron (m) wavefunctions, of which the t_{2g} and e_g^* orbitals are simple combinations (Section 1.5).

6.6   TRANSITION FROM WEAK TO STRONG LIGAND FIELDS

We now investigate the region where ligand fields and interelectronic repulsions are of comparable magnitude. A ligand field that causes free ion term splittings that are comparable to the separations between those terms is said to be of medium strength.

Just as free-ion configurations are split into terms by interelectronic repulsions so, if interelectronic repulsions are treated as perturbations on strong-field configurations, these latter are also split into terms. Such terms, for octahedral or tetrahedral stereochemistry, are called cubic-field terms to distinguish them from the free ion terms. The method of finding what terms arise from a strong-field configuration resembles that employed for deducing the free ion terms of a d^n configuration (Section 5.3); however, the process can be facilitated by the use of group theory (Section 1.4). The rule is, If the terms that

TABLE 6.1. Strong-Field Ground Configurations and Terms for Various Numbers of Electrons in Octahedral and Tetrahedral Stereochemistries

Stereochemistry	Octahedral Ground Configuration		Ground Term	Tetrahedral Ground Configuration			Ground Term
	t_{2g}	e_g			e	t_2	
d^1 — t_{2g}^1	↑		$^2T_{2g}$	e^1	↑		^2E
d^2 — t_{2g}^2	↑ ↑		$^3T_{1g}$	e^2	↑ ↑		$^3A_2$
d^3 — t_{2g}^3	↑ ↑ ↑		$^4A_{2g}$	e^3	↑↓ ↑		^2E
d^4 — t_{2g}^4	↑↓ ↑ ↑		$^3T_{1g}$	e^4	↑↓ ↑↓		$^1A_1$
d^5 — t_{2g}^5	↑↓ ↑↓ ↑		$^2T_{2g}$	$e^4 t_2^1$	↑↓ ↑↓	↑	$^2T_2$
d^6 — t_{2g}^6	↑↓ ↑↓ ↑↓		$^1A_{1g}$	$e^4 t_2^2$	↑↓ ↑↓	↑ ↑	$^3T_1$
d^7 — $t_{2g}^6 e_g^1$	↑↓ ↑↓ ↑↓	↑	^2E_g	$e^4 t_2^3$	↑↓ ↑↓	↑ ↑ ↑	$^4A_2$
d^8 — $t_{2g}^6 e_g^2$	↑↓ ↑↓ ↑↓	↑ ↑	$^3A_{2g}$	$e^4 t_2^4$	↑↓ ↑↓	↑↓ ↑ ↑	$^3T_1$
d^9 — $t_{2g}^6 e_g^3$	↑↓ ↑↓ ↑↓	↑↓ ↑	^2E_g	$e^4 t_2^5$	↑↓ ↑↓	↑↓ ↑↓ ↑	$^2T_2$

arise from a given configuration are known, then the terms that arise from the configuration with one additional electron must have the symmetry labels of the irreducible representations that are contained in the direct product between the symmetry labels of the initial terms and the added electron. Now, with only one d electron, there are no interelectronic repulsions, so the symmetry labels of the terms of d^1 must be the same as those of the electron. As shown in Section 5.3, the term and electron wavefunctions are then identical. Thus the t_{2g} and e_g^* electrons of d^1 in O_h give rise, respectively, to $^2T_{2g}$ and ^2E_g terms.

To get the terms of t_{2g}^2 we need to take the direct product between T_{2g} (the symmetry label of the term of t_{2g}^1) and T_{2g} (the symmetry label of the added electron). We have:

$$T_{2g} \times T_{2g} = T_{1g} + T_{2g} + E_g + A_{1g} \tag{6.36}$$

so that t_{2g}^2 gives rise to terms, one of each of the symmetries T_{1g}, T_{2g}, E_g, and A_{1g}. Similarly, $t_{2g}^1 \cdot e_g^{*1}$, an excited configuration of d^2, leads to T_{2g} and T_{1g} terms $(T_{2g} \times E_g = T_{1g} + T_{2g})$. The second excited configuration e_g^{*2} leads to the terms E_g, A_{1g}, and A_{2g} $(E_g \times E_g = E_g + A_{1g} + A_{2g})$.

Along with the symmetry labels of the terms that arise from a configuration, we also must know their multiplicities. The full method of finding the multiplicities of the terms of a cubic-field configuration is complicated and is given, for example, by Griffith (2), but some of the arguments can be illustrated here. In Table 6.1 the three t_{2g} orbitals are represented by a set of three spaces. Such a set of three spaces might also represent the set of p orbitals. If the methods of Section 5.3 are followed, it is found that two electrons may be put into three spaces in 15 ways. Thus there are 15 wavefunctions for t_{2g}^2. A T-type term is of threefold orbital degeneracy, an E-type term of two fold orbital degeneracy, and an A-type term orbitally nondegenerate. Only triplet or singlet multiplicities can arise from two electrons. If the multiplicities of the four terms of t_{2g}^2, are, say, a, b, c, and d, the terms are:

$$^aT_{1g} + {}^bT_{2g} + {}^cE_g + {}^dA_{1g} \tag{6.37}$$

where a, b, c, and d = either 1 or 3. We must have:

$$3a + 3b + 2c + d = 15 \tag{6.38}$$

Eq. 6.38 expresses the fact that the 15 wavefunctions of t_{2g}^2 are made up of the sum of the products of the orbital and spin degeneracies of the individual terms. The relationship can be satisfied only if c = d = 1 and one of a and b is 3, the other 1. That is, one of the T-type terms must be a triplet, the other three terms are singlets. Further arguments (3), not presented here, lead to the conclusion that the T_{1g} term is the triplet. Thus:

$$t_{2g}^2 \rightarrow {}^3T_{1g} + {}^1T_{2g} + {}^1E_g + {}^1A_{1g} \tag{6.39}$$

For the configuration $t_{2g}^1 \cdot e_g^{*1}$, it is possible to write 24 wavefunctions. For each of the six ways of putting an electron in the t_{2g} orbital set, there are four ways of placing one in the e_g^* set. Hence, if a and b are the multiplicities of the T_{1g} and T_{2g} terms, we must have:

$$^aT_{1g} + {}^bT_{2g} = 24 \tag{6.40}$$

which is satisfied, for instance, by $a = b = 4$. This result is unacceptable as it stands, because we have stated that a and b must be either 3 or 1. The interpretation to be placed on $a = b = 4$ is that there are T_{1g} and T_{2g} terms with $a = b = 3$ and with $a = b = 1$. In fact:

$$t_{2g}^1 \cdot e_g^{*1} \rightarrow {}^3T_{1g} + {}^3T_{2g} + {}^1T_{1g} + {}^1T_{2g} \tag{6.41}$$

Placing two electrons in the two spaces that represent the e_g^* orbital set can be accomplished in six ways. Proceeding as before:

$$^aE_g + {}^bA_{1g} + {}^cA_{2g} = 6 \tag{6.42}$$

For this to hold, we must have $a = 1$, one of b and $c = 3$, the other $= 1$. It can be shown that, in fact, it is the A_{2g} term that is the triplet. Hence:

$$e_g^{*2} \rightarrow {}^3A_{2g} + {}^1E_g + {}^1A_{1g} \tag{6.43}$$

We have dealt with all the strong-field configurations that arise from d^2. The total number of wavefunctions taken up by these configurations $15 + 24 + 6 = 45$, which is, as it must be, the number that was found Section 5.3 for d^2.

Which term of a strong-field configuration lies lowest is determined by an appeal to Hund's rules, after they have been slightly modified. In a configuration, terms with the highest multiplicity lie lowest. Of terms with equal multiplicity, terms with highest orbital degeneracy $(T > E > A)$ tend to lie lower. This appeal to Hund's rules fixes the ground term of the t_{2g}^2 configuration as $^3T_{2g}$; however, it does not, for instance, distinguish between the $^3T_{1g}$ and $^3T_{2g}$ terms of $t_{2g}^1 \cdot e_g^{*1}$.

The terms of the d^3 strong field configurations can be obtained from those of t_{2g}^2, $t_{2g}^1 \cdot e_g^{*1}$, and e_g^{*2} by taking the direct product of the symmetry labels with T_{2g} and with E_g. The terms of the configurations t_{2g}^3, $t_{2g}^2 \cdot e_g^*$, and $t_{2g}^1 \cdot e_g^{*2}$ and of the configurations $t_{2g}^2 \cdot e_g^{*1}$, $t_{2g}^1 \cdot e_g^{*2}$, and e_g^{*3}, respectively, follow. The results for a t_{2g} or e_g^* subshell more than half full can be obtained by considering it is an equivalent number of holes (see Section 5.3). Thus $e_g^{*3} \equiv e_g^{*1}$, for example. The ground terms for the various strong-field ground configurations of d^n are given in Table 6.1.

The accent in the last few paragraphs has been on the case of octahedral stereochemistry. Exactly the same arguments apply to finding the terms from strong-field tetrahedral configurations. Indeed, on dropping the g subscripts, it is noticed that all the strong-field configurations for d^2 in a tetrahedral ligand field, e^2, $e^1 \cdot t_2^{*1}$ and t_2^{*2} have been dealt with under the octahedral heading. The ground terms for strong-field configurations in tetrahedral stereochemistry are included in Table 6.1.

We now have studied the free ion: on the one hand, subject to interelectronic repulsions with the ligand field as a perturbation (weak-field case), and on the other hand, subject to a ligand field with interelectronic repulsions as a perturbation (strong-field case). To follow the transition from weak to strong fields we need to appeal to the so-called noncrossing rule. The operator for the energy of a system, the Hamiltonian **H**, can be shown to be necessarily totally symmetric to the operations of the relevant group. Consequently, it transforms as $A_{1(g)}$ in the group to which the system belongs. (This is, in a way, a restatement that the energy of a system must be independent of direction.) It immediately follows from the arguments of group theory that integrals of the form $< \psi_1 | \mathbf{H} | \psi_2 >$ may be non zero if ψ_1 and ψ_2 have the same symmetry labels. From this, it may be demonstrated that if energy is plotted on the vertical coordinate and a perturbing potential as the horizontal coordinate, the energy levels that correspond to wavefunctions that have the same symmetry labels never cross. The application of the non crossing rule is made obvious in the figures presented in the next section.

6.7 CORRELATION DIAGRAMS

We now show how the weak-field splittings discussed in Sections 6.1 to 6.4 correlate with the strong-field terms developed in Section 6.5. We present diagrams that show the transition from a weak to a strong ligand field, starting on the left side with the free ion terms (zero ligand field) and finishing on the right side with strong-field configurations (infinite ligand field). As soon as the ligand field becomes non vanishing, the symmetry is defined (O_h or T_d for our arguments). Each symmetry label, including multiplicity, must occur with equal frequency on the left and right hand sides of the diagram. Each label is joined up from side to side, observing the noncrossing rule. The procedure resembles the well- known correlation diagrams for the formation of a united atom from the separated atoms of a heteronuclear or homonuclear diatomic molecule. The resultant diagrams give a qualitative idea of how the energy levels of the system behave as the strength of the ligand field increases. For d^2 and d^8, all 45 wavefunctions on each side of the diagram are accounted for and the correlation is complete. Furthermore, these diagrams have been drawn roughly to scale on the extreme left and right sides.

For the configurations d^3 to d^7, there are so many wavefunctions that it complicates the diagrams too much to include them all. Consequently, only the portions of lower energy on each side of the diagram are given. Correlation lines going upward without a specific ending point indicate that the levels concerned reach high energy on the other side of the diagram. Two diagrams must be created for each d^n configuration because of the reversal of signs of the splitting of orbitals and free ion terms by O_h and T_d. At the same time, however, it is possible to incorporate the reversal of sign of the splitting for holes in the filled d manifold in the pair of diagrams. In that way, the conjugate configurations (d^{10-n}) are included. The relevant diagrams are given in Figures. 6.3 to 6.9. Although those figures are, for the most part, only qualitative, they illustrate some interesting points when taken in conjunction with Table 6.1.

6.7.1 d^1, d^9: Any Stereochemistry

Because there are no interelectronic repulsions in d^1 or d^9, the d wavefunctions are sorted out into t_{2g} and e_g^* types in O_h, for instance, by even a small ligand field. Increasing the magnitude of the ligand field produces no new result.

6.7.2 d^2 (O_h), d^8 (T_d)

For d^2 (O_h), d^8 (T_d), the ground term remains $^3T_{1(g)}$ for any strength of the ligand field (Fig. 6.3). No cubic ligand field, no matter how strong, can force pairing of the two unpaired electrons of, for example, octahedral V^{3+}. Indeed, $(NH_4)V(SO_4)_2 \cdot 12H_2O$ and similar compounds possess two unpaired electrons. Two unpaired electrons are also exhibited by $(Et_4N)_2NiCl_4$ and other tetrahedral nickel compounds (Chapter 9).

6.7.3 d^2 (T_d), d^8 (O_h)

Because the lowest term in Figure. 6.4 remains $^3A_{2(g)}$ for all strengths of the ligand field, no cubic field can pair off the two unpaired electrons of tetrahedral V^{3+} or octahedral Ni^{2+} complexes. Tetrahedral d^2 complexes are rare. Complexes of Ni^{2+}, such as $(NH_4)_2Ni(SO_4)_2 \cdot 6H_2O$, possess two unpaired electrons.

6.7.4 d^3 (O_h), d^7 (T_d)

In Figure. 6.5, the lowest term is the $^4A_{2(g)}$ at all strengths of the ligand field. All Cr^{3+} (octahedral) and Co^{2+} (tetrahedral) complexes are expected to retain three unpaired electrons in any cubic ligand field. Compounds such as $(NH_4)Cr(SO_4)_2 \cdot 12H_2O$ and Cs_2CoCl_4, indeed possess three unpaired electrons.

6.7.5 d^3 (T_d), d^7 (O_h)

For d^3 (T_d) and d^7 (O_h), there is a change in the ground term as the ligand field strength increases: $^4T_{1(g)} \rightarrow \, ^2E_{(g)}$. Thus a sufficiently strong cubic ligand field reduces the three unpaired electrons of tetrahedral Cr^{3+} or octahedral Co^{2+} to one. Tetrahedral Cr^{3+}, as $PCl_4^+ \cdot CrCl_4^-$ is an example of the three-unpaired-electron form. Because tetrahedral ligand fields are weaker than octahedral fields, spin-paired tetrahedral Cr^{3+} complexes are not likely to exist. Octahedral

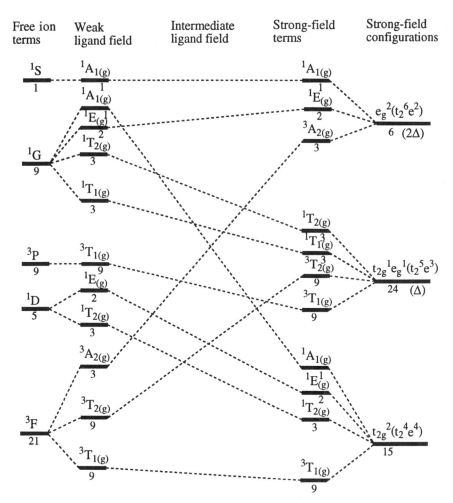

Figure. 6.3. Correlation for free ion terms with strong-field configuration for d^2 in O_h (V_{oct}^{3+}) and d^8 in T_d (Ni_{tet}^{2+}). The diagram is not to scale, except on the far right side. T_d symmetry does not require the g subscripts. The numbers under a term indicate the total degeneracy.

Co^{2+} is known in the three-unpaired-electron $[(NH_4)_2Co(SO_4)_2 \cdot 6H_2O]$ and one unpaired-electron $[K_2BaCo(NO_2)_6]$ forms.

6.7.6 d^4 (O_h), d^6 (T_d)

Increasing ligand field strength causes a change of ground term in d^4 (O_h), d^6 (T_d): $^5E_{(g)} \rightarrow {}^3T_{1(g)}$ (Fig. 6.7). Compounds of octahedral Mn^{3+} and Cr^{2+} exist as both high-spin (spin-free with four unpaired electrons) and low-spin (spin-paired with two unpaired electrons) species. Examples are $Mn(acetylacetonate)_3$

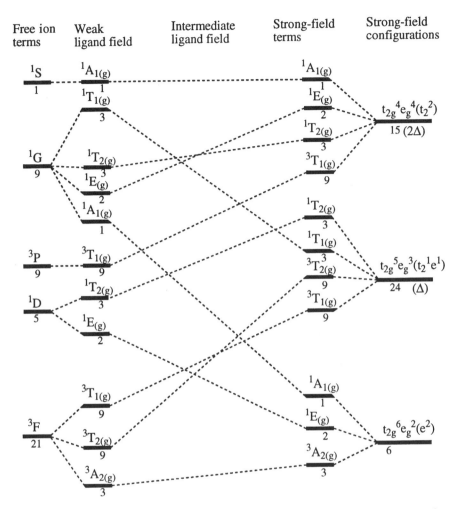

Free ion terms	Weak ligand field	Intermediate ligand field	Strong-field terms	Strong-field configurations

Figure. 6.4. Correlation diagram for free ion terms with strong-field configurations for d^2 in T_d (V_{oct}^{3+}) and d^8 in O_h (Ni_{oct}^{2+}). The diagram is not to scale, except on the far right side. The g subscript does not apply in T_d symmetry. The numbers under terms indicate their total degeneracy.

and $CrCl_2 \cdot 6H_2O$ (high spin), and $K_3Mn(CN)_6$ and $K_4Cr(CN)_6 \cdot 3H_2O$, (low spin). In the case of tetrahedral Fe^{2+}, the ligand field does not reach sufficient magnitude to force spin pairing, and compounds of only four unpaired electrons are known, as in $(Et_4N)_2FeCl_4$.

6.7.7 d^4 (T_d), d^6 (O_h)

For d^4 (T_d) and d^6 (O_h), the transition $^5T_{2(g)} \rightarrow {}^1A_{1(g)}$ takes place as the ligand field strength increases. Both Fe^{2+} and Co^{3+} in octahedral coordination give rise to compounds of both the high-spin (four unpaired electrons) and low-spin

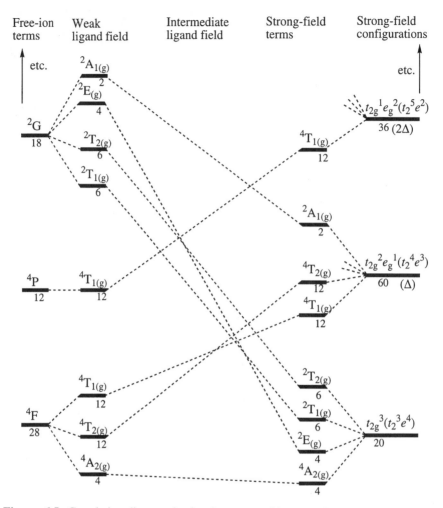

Figure. 6.5. Correlation diagram for free ion terms with strong-field configurations for d^3 in O_h (Cr_{oct}^{3+}) and d^7 in T_d (Co_{tet}^{2+}). The diagram is not to scale, except on the far right side. For T_d terms drop the g subscripts. The numbers under a term indicate the total degeneracy.

(no unpaired electrons) types. $(NH_4)_2Fe(SO_4)_2·6H_2O$ and Li_3CoF_6 are examples of the former and $K_4Fe(CN)_6·3H_2O$ and $Co(NH_3)_6·Cl_3$ of the latter spin type. The few tetrahedral d^4 complexes known are high spin, as expected from the small ligand field splittings of the stereochemistry.

6.7.8 d^5 (O_h and T_d)

For d^5 (O_h and T_d), increasing ligand field strength results in the ground term changing from $^6A_{1(g)}$ to $^2T_{2(g)}$. Fe^{3+} and Mn^{2+} in octahedral compounds are

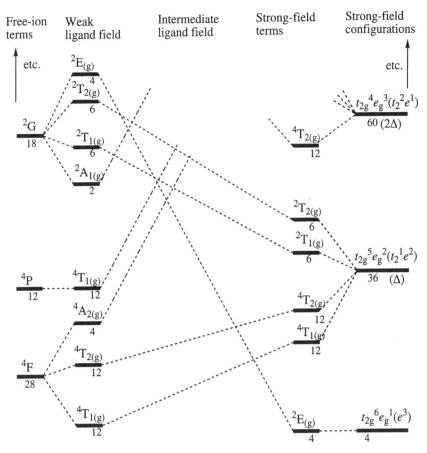

Free-ion terms	Weak ligand field	Intermediate ligand field	Strong-field terms	Strong-field configurations

Figure. 6.6. Correlation diagram for free ion terms with strong-field configurations for d^3 in T_d (Cr^{3+}_{tet})) and d^7 in O_h, (Co^{2+}_{oct}). The diagram is not to scale, except on the far right side. The g subscript is deleted for T_d symmetry. The numbers under a term indicate the total degeneracy.

each known as both high-spin (five unpaired electrons) and low-spin (one unpaired electrons) species. Examples are $(NH_4)Fe(SO_4)_2 \cdot 12H_2O$ and $(NH_4)_2Mn(SO_4)_2 \cdot 6H_2O$ (high spin) and $K_3Fe(CN)_6$ and $K_4Mn(CN)_6 \cdot 3H_2O$ (low spin). Tetrahedral Mn^{2+} and Fe^{3+} complexes, such as $(Et_4N)_2MnCl_4$ and $(Et_4N)FeCl_4$, are known in the high-spin form only. The ligand field strength does not become large enough to force the transition to the $^2T_2$ term.

6.8 TANABE–SUGANO DIAGRAMS

In addition to the above qualitative treatment of the transition from weak to strong ligand fields, it is also necessary to have quantitative results available for

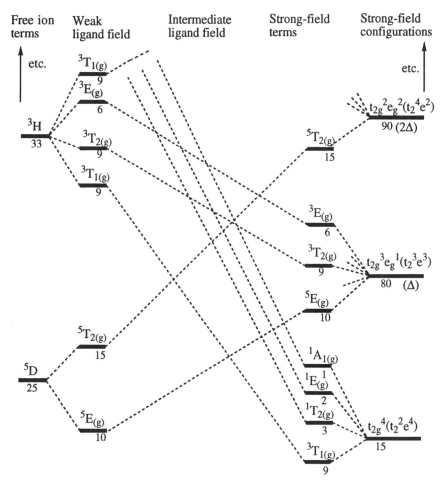

Figure. 6.7. Correlation diagram for free ion terms with strong-field configurations for d^4 in O_h (Mn^{3+}, Cr^{2+}_{oct}) and d^6 in T_d (Fe^{2+}_{tet}). The diagram is not to scale except on the far right side. The g subscript is deleted for T_d symmetry. The numbers under a term indicate the total degeneracy.

the interpretation of spectra. The calculation of all the energy levels of a d^n system in the presence of both interelectronic repulsions and ligand fields of medium strength is quite tedious, and the results are normally presented only for certain relationships between the interelectronic repulsion parameters (3,4). It is not possible here to indicate the general methods involved in the calculations. The results are presented in the form of the so-called Tanabe–Sugano diagrams (Figs. 6.10 to 6.16) (4,5). In the Tanabe–Sugano diagrams, the energies of the levels of a d^n system are plotted as the vertical coordinate, and the ligand field strength Δ is the horizontal coordinate. The energy unit is generally taken as the Racah parameter B. This has the advantage that the same diagram may be used

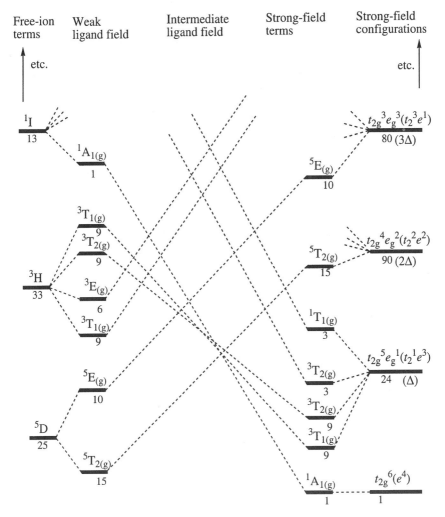

Figure. 6.8. Correlation diagram for free ion terms with strong-field configurations for d^4 in T_d (Mn^{3+}, Cr^{2+}_{tet}) and d^6 in O_h (Fe^{2+}, Co^{2+}_{oct}). The diagram is not to scale, except on the far right side. The g subscript is deleted for T_d symmetry. The numbers under a term indicate the total degeneracy.

for complexes with different values of B. Two parameters—B and C—are required to describe the interelectronic repulsions for a d electron system. The C/B ratio must, therefore, be specified and the diagrams have been compiled for the values of the ratio that are considered most likely for the most common first-transition series ions of the configuration concerned (see Table 5.3). The value of C/B is given for each diagram. Note that this ratio affects only the relative energies of excited states that have a different spin multiplicity from the ground term. Many of the results required later concern only terms of the same

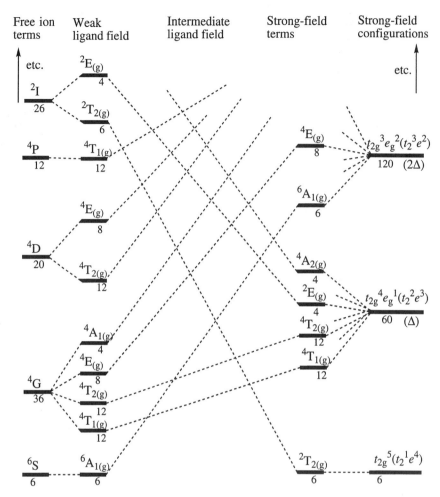

Figure. 6.9. Correlation diagram for free ion terms with strong-field configurations for d^5 in O_h and T_d (Mn^{2+}, $Fe^{3+}_{oct,\,tet}$). The diagram is not to scale, except on the far right side. The g subscript is deleted for T_d symmetry. The numbers under a term indicate the total degeneracy.

multiplicity as the ground term, and these are unaffected by the value of C. Note also that the same diagram applies for any metal ion giving rise to the configuration in question, for example V^{2+} and Cr^{3+}.

In these diagrams the zero of energy is *always* taken as that of the lowest term. Hence, when there is a change in the ground term, the diagram is discontinuous. The discontinuity always takes the form of an increase in the slope of term energies above a critical value of Δ/B.

The Tanabe–Sugano diagrams have been drawn up primarily for use with ligand fields of O_h symmetry. Provided it is noted that the C/B ratio may vary

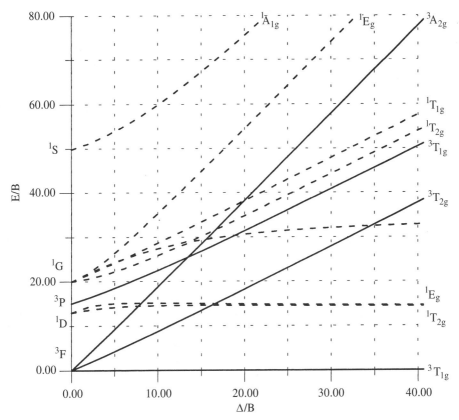

Figure 6.10. The energy diagram for d^2. $B = 886\,\text{cm}^{-1}$. $C/B = 4.42$.

somewhat with the number of d electrons, the same diagrams may be used for tetrahedral ligand fields. We note the equivalence $d^{10-n} \equiv d^n$ (Figs. 6.10 to 6.16). For example, the diagram for d^2 may be used for tetrahedral d^8 (Ni^{2+}).

For the interpretation of many spectral and magnetic results, simplified diagrams of the Tanabe-Sugano type may suffice. In them only the ground and low-lying higher levels of the same multiplicity are considered. In such diagrams it is also possible to demonstrate the carry over from O_h to T_d symmetry. The behaviour of D ground terms is illustrated in Figure 6.17 and of F ground terms in Fig. 6.18.

Ground D terms occur for the configurations d^1, d^9, d^4, and d^6. For these configurations, there is no other term among the lower levels of the same symmetry label and multiplicity as the ground term, provided that the ligand field is not strong. Consequently, the splitting diagram for D terms in weaker fields consists of two intersecting straight lines, crossing at zero ligand field.

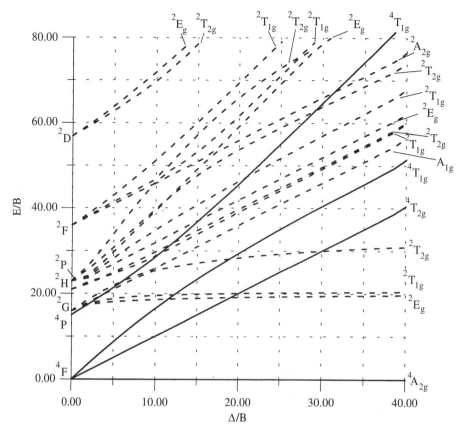

Figure 6.11. The energy diagram for d^3. $B = 933 \, \text{cm}^{-1}$. $C/B = 4.5$.

Where the ligand field term is of the F type, d^2, d^3, d^7, and d^8, a $T_{1(g)}$ term arises both from the ground F term and from the P term that lies immediately above it. These two $T_{1(g)}$ terms have the same multiplicity, $^3T_{1(g)}$ or $^4T_{1(g)}$, depending on the configuration. We concentrate for the moment on the $^3T_{1(g)}$ terms of d^2 in octahedral stereochemistry. We label them according to the free-ion terms to which they belong: $^3T_{1(g)}$ (F) and $^3T_{1(g)}$ (P). These two terms meet the symmetry label requirement laid down earlier for the existence of the matrix element $\langle ^3T_{1(g)} (F)|\mathbf{H}|^3T_{1(g)}(P)\rangle$, because $T_{1g} \times T_{1g}$ contains A_{1g}. Here the term symbols represent the whole set of nine wavefunctions. Consequently, these two terms interact with each other, and we examine the interaction as a function of the strength of the ligand field. That the ligand field changes the composition of the term during the transition from the free ion to the strong field configuration can be seen by comparing the free ion term wavefunctions (Section 5.4) and the fact that the $^3T_{1(g)}$ (F) term becomes the t_{2g}^2 configuration, and the $^3T_{1(g)}$ (P) term $t_{2g}^1 \cdot e_g^{*1}$ in the strong field limit. This system provides an elementary example of configuration interaction at work.

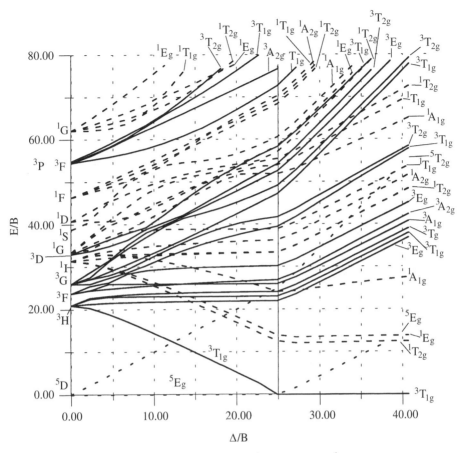

Figure 6.12. The energy diagram for d^4. $B = 796 \, \text{cm}^{-1}$. $C/B = 4.6$.

To find the energies of the $^3T_{1(g)}$ terms of d^2 in a medium-strength ligand field, we set up the secular determinant for the Hamiltonian operator between the two free ion terms. Now this operator differs from the free ion operator (Section 5.2) by only the addition of $\mathbf{V}_{\text{oct}}$:

$$\mathbf{H} = \mathbf{H_0} + \mathbf{V}_{\text{oct}} \tag{6.44}$$

In the absence of the ligand field, we know that the $^3T_{1(g)}$ (F) term has the same energy as the other components of the F term, which we define as zero. The $^3T_{1(g)}$ (P) term lies higher, at 15B (Fig. 5.2).

$$\langle \, ^3T_{1(g)} \, (F)_0 \, |\mathbf{H_0}|^3T_{1(g)} \, (F)_0 \rangle = \mathbf{0} \tag{6.45}$$

$$\langle \, ^3T_{1(g)} \, (P)_0 \, |\mathbf{H_0}|^3T_{1(g)} \, (P)_0 \rangle = 15B \tag{6.46}$$

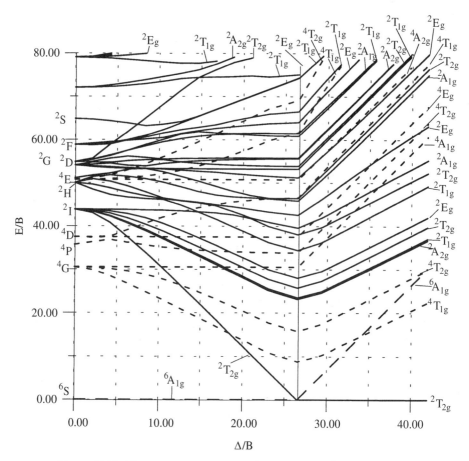

Figure 6.13. The energy diagram for d^5. $B = 859\,\text{cm}^{-1}$. $C/B = 4.48$.

In Section 6.1 it was shown that V_{oct} does not have any effect on a P term. It splits an F term, reducing the energy of the $^3T_{1(g)}$ (F) term by 6Dq. Hence:

$$\langle\,^3T_{1(g)}\,(F)_0\,|V_{oct}|\,^3T_{1(g)}\,(F)_0\,\rangle = -6Dq \qquad (6.47)$$

$$\langle\,^3T_{1(g)}\,(P)_0\,|V_{oct}|\,^3T_{1(g)}\,(P)_0\,\rangle = 0 \qquad (6.48)$$

The secular determinant required for **H** acting on the wavefunctions is:

$$
\begin{array}{cc}
^3T_{1(g)}(F)_0 & ^3T_{1(g)}(P)_0
\end{array}
$$
$$
\begin{vmatrix}
-6Dq - E & X \\
X & 15B - E
\end{vmatrix} = 0 \qquad (6.49)
$$

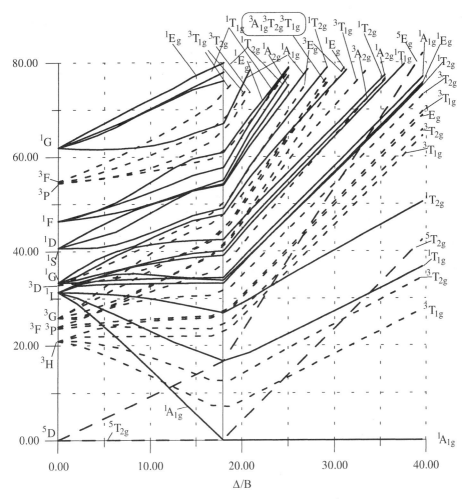

Figure 6.14. The energy diagram for d^6. $B = 1080\,\text{cm}^{-1}$. $C/B = 4.42$.

Here, the off-diagonal element $X = \langle\,^3T_{1(g)}\,(F)_0|V_{oct}|^3T_{1(g)}\,(P)_0\,\rangle$ can be evaluated by direct methods (Section 6.3); however, the following stratagem gives it easily (3). Suppose the interelectronic repulsions are negligible. Then the energies given by the determinant must be those of t_{2g}^2 ($-8Dq$), to which the $^3T_{1(g)}$ (F) term correlates, and of $t_{2g}^1 \cdot e_g^{*1}$ ($2Dq$), to which the $^3T_{1(g)}$ (P) term correlates. Omitting 15B from the determinant (Eq. 6.49) and expanding it we get the equation:

$$E^2 + 6DqE - X^2 = 0 \tag{6.50}$$

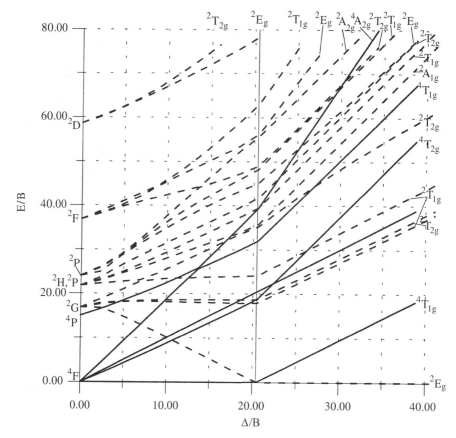

Figure 6.15. The energy diagram for d^7. $B = 986\,\text{cm}^{-1}$. $C/B = 4.63$.

which can give the required energies of $-8Dq$ and $2Dq$ only if:

$$X = 4Dq \tag{6.51}$$

The secular determinant for the interaction of the two $^3T_{1(g)}$ terms of d^2 is obtained by replacing X in Eq. 6.49 with $4Dq$. On expanding the resultant determinant, we obtain:

$$E^2 + (6Dq - 15B)E - 16(Dq)^2 - 90DqB = 0 \tag{6.52}$$

The roots of this equation are the energies of the two $^3T_{1(g)}$ terms. They, together with the other term energies from the ^3F term, are plotted as a function of the strength and sign of the ligand field in Figure. 6.18. It is seen that the result is to introduce curvature into the behavior of the two $^3T_{1g}$ terms. The other terms, which do not interact, form straight lines.

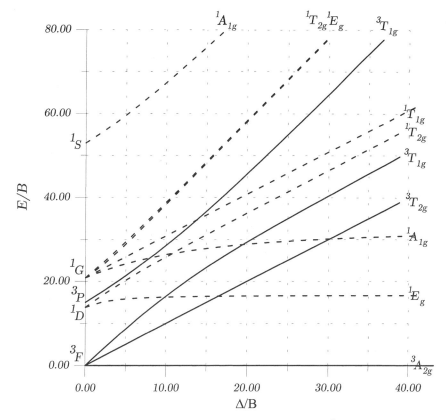

Figure 6.16. The energy diagram for d^8. $B = 1042\,cm^{-1}$. $C/B = 4.71$.

The argument used for d^2 can readily be extended to the $^3T_{1g}$ terms of d^8 and to the $^4T_{1g}$ terms of d^3 and d^7, and they are included in Figure. 6.18. A special sign convention must be observed for use with Eq. 6.52 and Figure 6.13 and 6.14. Δ is positive for an orbital splitting of the form of d^1 and d^2 in octahedral stereochemistry and negative for their inverted orbital splitting patterns. The reasons for the convention can be seen from the arguments given in Sections 6.2 to 6.4, which lead to the inversion patterns.

6.9 SPIN-PAIRING ENERGIES

A question of considerable interest is: At what value of Δ does spin pairing occur in d^n systems? A simple answer to the question cannot be given, because the relative energies of the states depend on the balance between Δ and the interelectron repulsion parameters, and the latter are reduced from their free ion values on complex formation by an amount that varies from one compound to

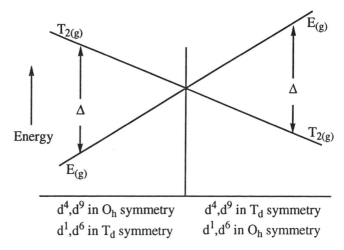

| d^4, d^9 in O_h symmetry | d^4, d^9 in T_d symmetry |
| d^1, d^6 in T_d symmetry | d^1, d^6 in O_h symmetry |

Figure 6.17. The relationship between the cubic field term energies and the lower lying terms of d^n systems with D ground terms.

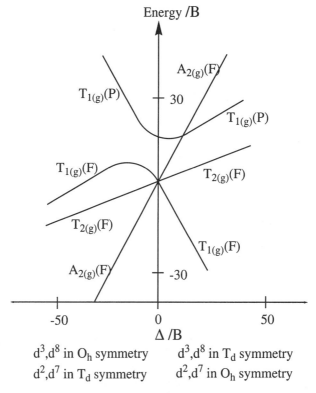

| d^3, d^8 in O_h symmetry | d^3, d^8 in T_d symmetry |
| d^2, d^7 in T_d symmetry | d^2, d^7 in O_h symmetry |

Figure 6.18. The relationship between the cubic field term energies of the lower lying terms of d^n systems with F ground terms.

another. The reduction can sometimes be estimated from the electronic spectrum, as will be discussed in Section 8.4.2.

As a first approximation, the correlation diagrams shown in Figures 6.3 to 6.9 provide a useful guide to the possible spin ground states of complexes. The d configurations for which two spin ground states are possible are those for which the term of the weak-field limit correlates with an excited configuration in the strong-field limit. Considering just octahedral stereochemistry, as Δ for tetrahedral complexes is almost always too low to force spin pairing, it is apparent that metal ions with d^4, d^5, d^6, and d^7 configurations may have either high-spin or low-spin ground states. Moreover, inspection of the energies of the strong-field configurations shows that for d^4 and d^7, the low-spin state is stabilized by Δ, whereas for d^5 and d^6 it is stabilized by 2Δ. The terms of the weak-field limit, however, show that the difference in spin multiplicity between the two spin states is $\Delta S = 1$ for the former configurations, but $\Delta S = 2$ for the latter. If it is assumed that Δ_s represents a constant energy for the pairing of two electron spins, then the same condition applies to each configuration: Spin pairing will occur, as a first approximation, when $\Delta > \Delta_s$. In general (as will be discussed in Chapter 9 in connection with the magnetic properties of complexes), observations confirm this. Low-spin complexes tend to occur with ligands that produce large Δ values, such as cyanide, and for trivalent metal ions that have larger d orbital splittings than divalent metal ions. Moreover, second and third row transition ions are predominantly low-spin, and these have both higher Δ values and lower values of Δ_s, because of lower interelectron repulsion energies, than do first row metal ions.

From the Tanabe–Sugano diagrams, it is possible to obtain a more quantitative measure of the ease of spin pairing. In Figures. 6.12 to 6.15 it is seen that the discontinuity corresponding to the change of ground term takes place at:

Figure	Number of d Electrons	Δ/B
6.12	4	26
6.13	5	27
6.14	6	18
6.15	7	20

It may be seen that spin pairing occurs most easily for d^6 and d^7 (Fe^{2+}, Co^{3+}, and Co^{2+}, respectively) and rather less easily for d^4 (Cr^{2+}, Mn^{3+}) and d^5 (Mn^{2+}, Fe^{3+}). The results can be related to the particular stability associated with configurations involving filled and half-filled shells. Thus low-spin d^6 has a completely filled t_{2g} orbital set, whereas high-spin d^5 has half-filled t_{2g} and e_g^* sets. The difference between the spin pairing energy of the d^5 and d^6 configurations explains the observation that the d^6 $Co(H_2O)_6^{3+}$ ion is low-spin, whereas the d^5 $Fe(H_2O)_6^{3+}$ ion is high-spin. The ligand field of the water molecule is sufficient to cause spin pairing for the former complex but not for the latter.

REFERENCES

1. Gerloch, M., *Magnetism and Ligand Field Analysis*, Chapman-Hall, London, 1972.
2. Griffith, J. S., *The Theory of Transition-Metal Ions*, Cambridge University Press, UK 1961, chap. 9.
3. Ballhausen, C. J., *Introduction to Ligand Field Theory*, McGraw-Hill, New York 1962, chap. 4.
4. Tanabe, Y., and Sugano, S., *J. Phys. Soc. Jpn.*, 1954, **9**, 753, 766.
5. McClure, D. S., *Solid State Physics*, Vol. 9, Academic Press, New York 1950.

CHAPTER 7

INFLUENCE OF THE d CONFIGURATION ON THE GEOMETRY AND STABILITY OF COMPLEXES

The geometry and bond energy of any molecule are determined by its electronic structure (1). In particular, bond strengths and lengths are related to the numbers of electrons occupying bonding and antibonding molecular orbitals. For coordination complexes, the bulk of the bond energy is derived from the interactions between the metal ns and np orbitals and the ligand orbitals, where n pertains to the valence shell. The bonding MOs derived from these interactions are mainly ligand in character and are filled with electrons, whereas the corresponding antibonding MOs are mainly metal in character and are empty. The molecular orbitals involving the d electrons are much less influential in determining the energy, with those of σ symmetry being the most important. Again, the bonding MOs are mainly ligand based and are filled; however, in this case the antibonding MOs, which are again mainly metal in character, are often partially occupied. This latter occupancy has a destabilizing influence and causes a lengthening of the metal–ligand bonds. Although energetically these effects are always a small fraction of the total bond energy, the d electron configuration of the metal does have an important and sometimes critical influence on several aspects of transition-metal chemistry, and these are considered in this chapter.

7.1   DEPENDENCE OF THE GEOMETRY OF A COMPLEX ON ITS d CONFIGURATION

7.1.1   Non-degenerate Electronic States

Consider two electron configurations of a transition-metal complex that differ in the distribution of m electrons among the d orbitals; these orbitals are separated in energy by an amount E. For example, in its ground state, a planar Pt^{2+} complex has the configuration $(d_{z^2})^2(d_{xz})^2(d_{yz})^2(d_{xy})^2(d_{x^2-y^2})^0$, whereas an excited state has the configuration $(d_{z^2})^2(d_{xz})^2(d_{yz})^2(d_{xy})^1(d_{x^2-y^2})^1$. Here, $m = 1$ and E is simply the energy difference between the d_{xy} and $d_{x^2-y^2}$ orbitals. If the metal–ligand bond length for the complex in the first configuration is r_0, the change to the second configuration generally produces a uniform increase in all metal–ligand bonds; i.e., an expansion Q in the totally symmetric stretching mode of the complex. This produces an energy increase $f Q^2/2$, where f is the force constant of that vibration. If the energy separation E is inversely proportional to some power n of the bond length, the total energy difference between the complex in the two configurations is:

$$U = f Q^2/2 + mE[(r_0 + CQ)/r_0]^{-n} + S \qquad (7.1)$$

$$= f Q^2/2 + mE[1 + CQ/r_0]^{-n} + S \qquad (7.2)$$

where C is a normalization coefficient and S represents any contribution owing to interelectron repulsion terms. Provided CQ/r_0 is small, the expansion of the second term in Eq. 7.2 yields the approximation:

$$U \approx f Q^2/2 + mE[1 - (nCQ)/r_0] + S \qquad (7.3)$$

The displacement in Q corresponding to the energy minimum of the second electron configuration is obtained by setting $dU/dQ = 0$, and because, to a first approximation, S does not depend on Q, this gives:

$$Q_{min} \approx mnEC/fr_0 \qquad (7.4)$$

The change in each bond length is given by $\delta r_0 = CQ_{min}$, and for the totally symmetric vibrational mode $C^2 = 1/N$, where N is the number of ligands. Thus:

$$\delta r \approx mnE/Nfr_0 \qquad (7.5)$$

Note that both E and f refer to the complex in the initial electron configuration.

   Energy, usually in the form of light, must be provided to convert a complex from one electron configuration to another. An exception is the case in which the complex occurs in two available spin states where thermal energy is sometimes sufficient (see Section 9.10). Measurement of light absorption, as the electronic spectrum, provides an important method of determining the bond length difference between a complex in two different electronic configurations. The way in which this is achieved, which involves analyzing the vibrational structure associated with the electronic excitation, is discussed in Section 8.1.3. Bond length differences derived in this way for electronic transitions between different d orbitals in complexes with a range of coordination numbers and shapes are shown in Table 7.1. Both theoretical considerations and the pressure dependence of the optical spectra of transition metal compounds suggest that the d orbital energy separation depends inversely on about the fifth or sixth power of the bond distance (Section 3.2.5). Values of bond length changes calculated for the relevant excited states using Eq. 7.5 and assuming an inverse fifth-power dependence are also shown in Table 7.1.

It can be seen that the bond length change associated with the transfer of one or more electrons between two d orbitals with differing antibonding character is described remarkably well by Eq 7.5. The influence of antibonding energy is well illustrated by the bond lengths in the different states of the unusual linear ion NiO_2^{3-}. The bonding in species of this type was discussed in Section 3.2.2; and because the Ni^+ ion has the d^9 electron configuration, the electronic spectrum can be interpreted directly in terms of the energy differences between the d orbitals. The second excited state is at approximately twice the energy of the first; in agreement with Eq 7.5, the spectral analysis suggests that the Ni–O bonds lengthen by twice as much in this higher state as in the lower one. For the same reason, somewhat greater bond length changes are expected in the excited states of complexes formed by ligands that are high in the spectrochemical series compared to complexes formed by ligands that are low in this series.

TABLE 7.1 Comparison of the Calculated Changes in Metal–Ligand Bond Length Accompanying Excitation of d Electrons to More Antibonding Orbitals with Those Derived from Electronic Spectra[a]

Complex	Shape	N	m	f (mdyne $Å^{-1}$)	E (cm^{-1})	r_0 (Å)	δr_0 Calculated	δr_0 Spectra
NiO_2^{3-}	Linear	2	1	4.27	6,500	1.759	0.045	0.048
		2	1	4.27	12,500	1.759	0.086	0.085
$NiCl_4^{2-}$	Tet	4	1.75	1.78	3,045	2.245	0.03	0.03
$CuCl_4^{2-}$	Planar	4	1	1.59	12,150	2.27	0.085	0.096
		4	1	1.59	16,500	2.27	0.114	0.116
$PdCl_4^{2-}$	Planar	4	1	2.01	21,491	2.29	0.12	0.11
$PbBr_4^{2-}$	Planar	4	1	1.70	20,051	2.46	0.12	0.10
$Co(NH_3)_6^{3+}$	Oct.	6	1	2.49	24,000	1.97	0.08	0.07
		6	~2	2.49	24,000	1.97	0.16	0.12

[a]See text for details.

The results for the tetrahedral $NiCl_4^{2-}$ complex are less well defined, because the electronic states involved in the transition are both orbitally degenerate, and hence subject to Jahn–Teller distortions. In this case, however, any changes in geometry associated with Jahn–Teller coupling are likely to be small (Section 7.1.2.2, and it seems clear that the electronic transition is accompanied by a minimal change in bond length. This is consistent with the small d orbital splitting associated with the tetrahedral stereochemistry. Here, the number of electrons involved in the transition is nonintegral because, owing to configurational interaction effects, the terms involved in the electronic transition do not correspond to simple strong-field d configurations (Section 5.4). Much larger bond length changes occur for the planar Cu^{2+} complex, which has the same number of ligands and a similar vibrational force constant but considerably higher antibonding energies in the excited states.

An inverse correlation is expected between the number of ligands N and the bond length change. The results for the planar Pd^{2+} complexes and the lower energy single-electron transition of the octahedral $Co(NH_3)_6^{3+}$ complex, for which the excited state energies are fairly similar, show that this trend is indeed followed. The bond length change derived from the spectrum of the last-mentioned complex involves an orbitally degenerate excited state, which means that a correction must be made for a distortion in the Jahn–Teller active vibration. The excited states of the NiO_2^{3-} ion are also orbitally degenerate, but a problem does not arise here because the Jahn–Teller theorem does not apply to linear molecules. It is apparent that, despite the small number of ligand atoms, the distortions in the two excited states are smaller than, or comparable to, those in the octahedral and planar complexes. This is partly owing to the much smaller excitation energies, but the high force constant of the metal–ligand bond is also an important contributing factor.

The influence that the number of electrons m involved in the electronic rearrangement has on the bond length may be seen by comparing the changes derived from the first- and second-electronic transitions of the $Co(NH_3)_6^{3+}$ ion, because the latter transition approximates to a two-electron jump. As expected from Eq. 7.5, the bond length increase approximately follows the number of electrons involved in the transition.

7.1.2 Degenerate States: The Jahn–Teller Effect

The Jahn–Teller theorem, which is based on group theoretical arguments, states that any nonlinear molecule or ion having an orbital degeneracy is electronically unstable and distorts in way that removes that degeneracy (2,3). Because it is based solely on symmetry arguments, the theorem indicates only that a particular geometry is unstable. It says nothing about the *size* of the distortion that acts to remove the orbital degeneracy. This latter depends basically on the balance of two competing factors: the lowering in electronic energy and the rise in nuclear potential energy when the molecule distorts along the normal coordinate of the Jahn–Teller active vibration. To a first approximation, the total energy U

as a function of the displacement Q in the normal coordinate is, therefore, given by:

$$U = -VQ + f/2Q^2 \tag{7.6}$$

where V and f are the electronic coupling constant and force constant of the vibration, respectively. Here it is assumed that the electronic energy varies linearly with the distortion, which takes the form of a simple harmonic function. This coupling of the vibrational and electronic wavefunctions (vibronic coupling) produces an energy minimum when $dU/dQ = 0$, which occurs when:

$$Q_{min} = V/f \tag{7.7}$$

with a Jahn–Teller stabilization energy E_{JT} given by:

$$E_{JT} = -V^2/2f \tag{7.8}$$

A significant displacement and stabilization is thus expected only when the electronic coupling constant V is large. For transition-metal complexes, this mainly occurs when the orbital degeneracy involves d orbitals that interact strongly with the ligands via σ bonding; i.e., when an odd number of electrons occupies the e_g^* orbital set of an octahedral complex. It is also seen, to a lesser extent, when any number other than 3 or 6 electrons occupy the t_2^* orbital set of a tetrahedral complex. The former case applies to metal ions with high-spin d^4 (Cr^{2+}, Mn^{3+}), low-spin d^7(Co^{2+}, Ni^{3+}), and d^9 (Cu^{2+}, Ag^{2+}, Ni^+) electron configurations, and the latter largely to the d^8 and d^9 metal ions Ni^{2+} and Cu^{2+}.

7.1.2.1 Octahedral Complexes.

For an octahedral complex with orbital degeneracy in the antibonding e_g^* orbitals, only one normal vibration is Jahn–Teller active, that of ε_g symmetry. The forms of the two components of this vibration, Q_θ and Q_ε, are shown in Figure 7.1a. In the first order, distortions in the two components are energetically equivalent. The way in which the energy depends on the ε_g coordinate is, therefore, conventionally represented by a three-dimensional diagram, the so-called Mexican hat potential surface (Fig. 7.1b). The energy minimum is a circular trough of radius Q_{min}, and the ligand positions are defined by an angular function ϕ, such that:

$$Q_\theta = Q\cos\phi \quad \text{and} \quad Q_\varepsilon = Q\sin\phi \tag{7.9}$$

For the geometry defined by any particular value of ϕ, the electronic wavefunction containing the electron giving rise to the orbital degeneracy is given by:

$$\Psi = d_{x^2-y^2}\cos\phi/2 + d_{z^2}\sin\phi/2 \tag{7.10}$$

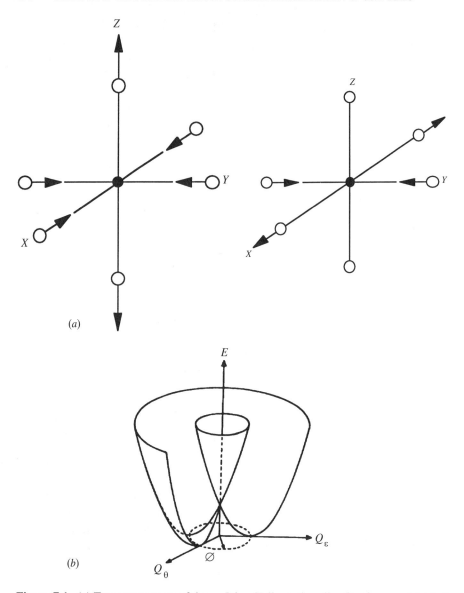

Figure 7.1. (*a*) Two components of the ε_g Jahn–Teller active vibration for an octahedral complex. (*b*) The Mexican hat potential surface of an octahedral complex undergoing a first-order Jahn–Teller distortion as a result of an orbital degeneracy in the e_g^* orbital set.

The normal coordinate is related to the ligand displacements from their mean positions by:

$$Q_{min} = (2\delta x^2 + 2\delta y^2 + 2\delta z^2)^{1/2} \tag{7.11}$$

An approach to derive the displacement in the totally symmetric α_1 mode associated with the transfer of one or more electrons between two nondegenerate orbitals was described in Section 7.1.1. An exactly analogous procedure may be used to estimate the approximate displacement in the Jahn–Teller active mode (4). This gives the linear Jahn–Teller coupling constant as:

$$V \approx nmE'C/r_0 \tag{7.12}$$

Where, m is the difference between the number of electrons in the split components of the e_g^* orbitals and C is a normalization coefficient, which in this case is $12^{-1/2}$. The energy E' is the antibonding energy of the e_g^* orbitals in the undistorted complex. The AOM provides the relationship $E' = 3e_\sigma$ (Section 3.2.1), so that this model represents a particularly convenient way of estimating Jahn–Teller coupling coefficients (5). If π bonding is negligible, $E' = \Delta$, the energy separation between the e_g^* and t_{2g} orbitals. The bond length of the undistorted complex is r_0, and it is assumed that the energy of the e_g^* orbitals depends inversely on the nth power of the bond length. Substitution into Eq. 7.12 yields the relationship:

$$Q_{min} \approx nmE'/12^{1/2}fr_0 \tag{7.13}$$

The values of Q_{min} estimated for various complexes using Eq. 7.13 are compared with those observed experimentally in Table 7.2. Here, n is assumed equal to 5, as for distortions in the α_1 mode (Section 7.1.1) and in every case $m = 1$, because the examples concern an orbital degeneracy involving just a single electron. The antibonding energies generally have an uncertainty of 5 to 10% when the complex involves π-antibonding ligands.

In the case of Cu^{2+} complexes, the Jahn–Teller distortions may be measured directly by X-ray structural analysis. The distortions calculated using Eq. 7.13, while of the right order of magnitude, are $\sim 25\%$ smaller than those observed experimentally. For the $Cr(NH_3)_6^{3+}$ and PtF_6^{2-} complexes, in which the distortions occur in excited states, they are known less accurately because they

TABLE 7.2 Calculated and observed Jahn–Teller Distortions in Six-Coordinate Complexes[a]

				Q (Å)	
Complex	f (mdyne Å^{-1})	E'(cm^{-1})	r_0 (Å)	Calculated	Observed
$Cu(H_2O)^{2+}$	0.57	11,500	2.10	0.28	0.39
$Cu(NO_2)_6^{4-}$	0.64	12,000	2.17	0.24	0.31
$Cr(NH_3)_6^{3+}$	1.72	22,300	2.16	0.17	0.16
PtF_6^{2-}	1.91	$\sim 41,000$	1.91	0.16	0.18

[a]See text for details.

are derived from an analysis of the vibrational structure observed in the electronic spectra (Section 8.1.3.3). For these examples, agreement between the distortions calculated using Eq. 7.13 and those derived from the spectra is rather good. The distortions for these higher-valent metal ions are considerably smaller than those for the divalent ones. This is because the distortion depends on the ratio of the linear coupling constant V and the vibrational force constant f. The former is dominated by the σ antibonding energy E'. Although E' rises progressively with an increasing oxidation state of the metal, this effect is more than counterbalanced by the increase in the force constant of the ε_g vibration (Table 7.2).

So far, the magnitude, but not the nature of the distortion, has been discussed. At this level of approximation, the ligands may be thought of as moving along the two components of the ε_g vibration in a cyclical fashion, accompanied by a concomitant change in the electronic wavefunction. In fact, quantum mechanically, the vibronic wavefunctions are specified by a set of quantum numbers J analogous to those used to describe the rotational energy levels of a molecule (2,3). In practice, higher order effects invariably act to discriminate between the two ε_g components. The Q_ε component carries the complex into a lower symmetry point group (D_{2h}) than the Q_θ component (D_{4h}), and the so-called epikernel principle may be used to show that the higher symmetry distortion is energetically more favorable (6). This may produce either a tetragonally elongated or a tetragonally compressed geometry. In considering which is likely to be favored, it may be noted that, unlike the Q_ε normal coordinate, the Q_θ distortion involves *unequal* ligand displacements—the axial ligands move twice as far as the in-plane ligands. Because of this, as discussed below, three different factors act to discriminate between the compressed and elongated forms of the tetragonal distortion (3,4).

In a first-order treatment, the Jahn–Teller coupling constant V is derived assuming that the d orbital energies vary linearly with bond length r. In fact, they vary approximately as r^{-5}. Because the ligands along the Z axis move twice as far as those along the X and Y axes, the d_{z^2} orbital is affected more than $d_{x^2-y^2}$ (Fig. 7.2a). Because only a single electron occupies the d_{z^2} orbital for a tetragonal compression, the net effect is to stabilize this form of the distortion relative to an elongation. Second, on distortion from O_h to D_{4h} symmetry, configuration interaction occurs between the d_{z^2} and metal 4s orbitals, both of which transform as a_{1g} in the latter point group. This causes a progressive stabilization of the d_{z^2} orbital, irrespective of whether the distortion is a compression or an elongation. Because of the higher occupancy of the orbital for the latter case, however, it is favored by d-s mixing, as may be seen in Figure 7.2b.

It is comparatively straightforward to obtain order-of-magnitude estimates of these two competing factors. The first may be calculated by extending the power expansion of V as a function of r, and the second from the d_{z^2} transition energy observed in the electronic spectra of planar complexes, in which the effect of d-s mixing is substantial (Section 3.2.2). The two effects are found to be comparable in magnitude.

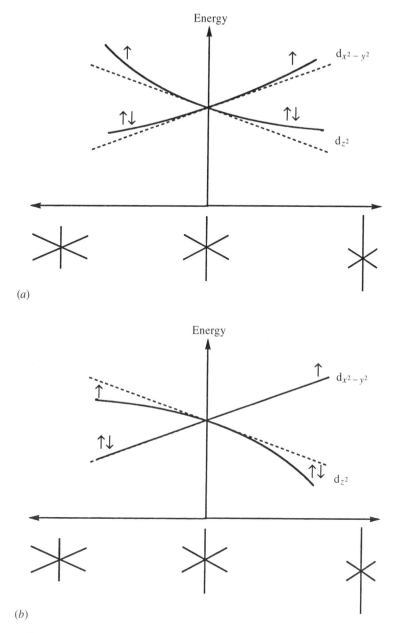

Figure 7.2. Effect of elongated and compressed tetragonal distortions on the $d_{x^2-y^2}$ and d_{z^2} orbital energies. The d occupancy corresponds to the d^9 configuration of Cu^{2+}. *Dashed line*, first-order linear dependence; *solid line*, effects of higher-order electronic terms (a) and configuration interaction between the d_{z^2} and metal 4s orbitals (b).

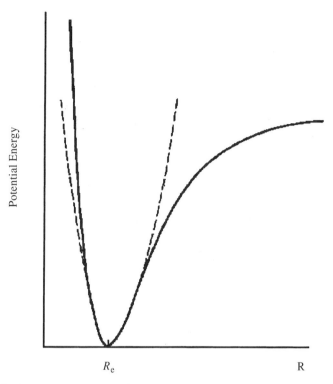

Figure 7.3. Comparison of the Morse function (*solid line*) with simple parabolic behavior (*dashed line*) for the energy change caused by the stretching and bending of a metal–ligand bond.

The third factor influencing the sense of the distortion is the anharmonicity of the ε_g vibration. The motion of each individual ligand is described better by a Morse potential than by the simple harmonic function used in the first-order Jahn–Teller approach (Fig. 7.3). This means that the total energy rises more steeply for bond compression than for bond extension. For this reason, anharmonicity of the concerted ligand motions of the ε_g vibration energetically favors the tetragonal elongation over the compression, while leaving the Q_ε component unaffected. Estimates of the anharmonicity coefficient suggest that this effect should be comparable in magnitude to the other two factors. It should be noted that the Morse potential applies only for terminal ligands. When the ligands bridge to another metal ion, the anharmonicity corrections to the vibrational potential are rather different, and the above arguments no longer apply.

Thus while two of the higher order effects favor the tetragonally elongated geometry, only one favors the tetragonally compressed one. Overall, the tetragonal elongation might, therefore, be expected to be more stable. In agreement with this, almost all complexes with a ground state having an odd

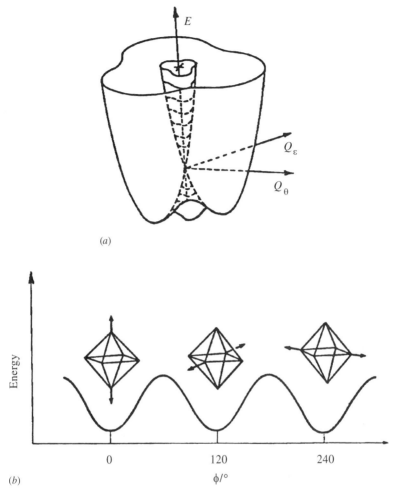

Figure 7.4. (a), Warping of the Mexican hat potential surface owing to the influence of higher order effects. (b) Energy change associated with motion of the ligands around the trough of the warped Mexican hat potential surface. The three minima correspond to tetragonally elongated octahedral geometries.

number of electrons in the e_g^* orbitals exhibit this geometry. The potential surface of such a complex, after including the higher order terms, is shown in Figure 7.4a. The trough in the Mexican hat is warped to give three minima, corresponding to tetragonal elongations of the metal–ligand bonds along Z, X, and Y, and higher-energy saddlepoints occur, corresponding to axial compressions of the metal–ligand bonds along these three directions. Figure 7.4b shows complex's energy dependence on the relative contributions of the two components of Q to the vibronic wavefunction, as specified by ϕ. Calculations suggest that in a complex such as the $Cu(H_2O)_6^{2+}$ ion the energy difference between the minima and saddlepoints is $\sim 600\,cm^{-1}$, and experimental measurements of the EPR spectra of compounds containing this species confirm the estimate (3,4). In agreement with the arguments of the preceeding paragraph, considerably smaller warping energies are sometimes observed for Cu^{2+} ions in continuous lattices involving bridging ligands. In general, therefore, although the complexes are strongly stabilized by the Jahn–Teller distortion and stabilization energies E_{JT} typically range from ~ 1500 to $\sim 2500\,cm^{-1}$, the forms of the distortion differ in energy by only a few hundred cm^{-1}.

The above treatment applies when the six ligands are identical, and in that situation, the three energy minima are symmetry equivalent. In practice, even in a cubic crystal, random strains lower the symmetry at any particular lattice site, so that overall an equal number of complexes are localized in each of the three energy minima. The *average* ligand positions determined by X-ray diffraction then define a regular octahedral geometry. Such a situation occurs in the compound $[Cu(H_2O)_6]SiF_6$, in which the six Cu–O bonds are crystallographically equivalent. The EPR spectrum of the compound shows an isotropic signal at room temperature, but on cooling to below $\sim 40\,K$ this changes to the anisotropic signal characteristic of a complex with the tetragonally elongated octahedral geometry. This behavior is quite consistent with a potential surface such as that shown in Figure 7.4. Thus at room temperature each complex exchanges between the three possible conformations more rapidly than the EPR time scale ($\sim 10^{-9}$ s), so an average signal is observed. On cooling, each complex becomes "frozen" into one of the energy minima, and the spectrum becomes that of a tetragonally elongated complex, with the orientations of the **g** tensors randomly distributed over the three Cu–O bond directions.

Compounds involving six crystallographically equivalent ligands are rather rare. Generally, even when six ligands of the same kind bond to the metal, the crystal site symmetry is lower than cubic. Lattice forces then act to make the three minima in the potential surface different. Provided that the energies arising from the lattice forces are larger than thermal energies, the complex becomes localized in the lowest energy minimum. This usually corresponds to a tetragonally elongated geometry, often with an additional orthorhombic component induced by the lattice forces. Occasionally, however, the compressed tetragonal geometry is more stable. The X-ray crystal structure and spectroscopic properties generally reflect the local coordination geometry, although when the perturbations owing to the lattice are comparable to thermal energies, X-ray analysis and the EPR spectrum are likely to suggest a geometry averaged over the thermal population of the three wells. Moreover, under these circumstances, this geometry changes as a function of temperature. On the other hand, the band positions observed in the electronic spectrum are not averaged in this way, because electronic transitions occur far more rapidly than the time scale of molecular vibrations.

In mixed-ligand complexes, the potential surface is perturbed by the difference in σ-bonding strength of the ligands (3,4). For a complex of stoichiometry ML_4X_2, if X is a weaker σ donor than is L, the energy minimum corresponding to the M–X long bonds is stabilized relative to the other two minima. This is the case for a complex such as $Cu(NH_3)_4Cl_2$, which has four short equivalent Cu–NH_3 bonds and two long Cu–Cl bonds. If, on the other hand, X is a much stronger σ donor than is L, the ligand field asymmetry is dominant, and a potential surface corresponding to the unusual tetragonally compressed geometry occurs. This is the case for the $Cu(NH_3)_2Cl_4^{2-}$ ion in $Cu(NH_3)_2Cl_2$, which has short axial bonds to the ammonia ligands and four approximately equal bonds to the weaker, bridging, chloride ions. When X is a marginally stronger σ donor than is L, the ligand field anisotropy may not be sufficient to completely overcome the tendency of the metal ion to adopt a tetragonally elongated geometry, and two minima corresponding to an orthorhombic geometry result, with short bonds to the stronger ligand X, and intermediate and long bonds to the weaker ligand L. Such a geometry is observed, for instance, for the complex $Cu(H_2O)_2Cl_4^{2-}$. In general, it is apparent that, depending on the balance between Jahn–Teller coupling and ligand field asymmetry, a wide range of distortions may occur for complexes with degeneracy in the e_g^* orbitals.

Two vibrations, those of ε_g and τ_{2g} symmetry, are Jahn–Teller active for an octahedral complex with a degeneracy in the t_{2g} orbital set, such as occur for metal ions with d^1, d^2, low-spin d^4, d^5, and high-spin d^6 and d^7 configurations. The ε_g mode involves metal–ligand stretching (Fig. 7.1a). On the other hand, the three components of the τ_{2g} mode each involve metal–ligand bending vibrations. The relevant coupling coefficients can be estimated using AOM bonding parameters (Section 3.1). They suggest that interaction with the ε_g vibration is likely to be the more important; however, any such interaction is likely to be much smaller than that resulting from degeneracy in the e_g^* set. This is because the bonding interactions are of π rather than σ symmetry and π bonds are significantly weaker than are σ bonds.

As noted, an estimate of the relative magnitudes of the interactions may be obtained from the AOM bonding parameters, of which some typical values are given in Section 3.2.6. The linear-coupling coefficient of the ε_g vibration for complexes with a t_{2g} degeneracy is proportional to $4e_\pi$, compared with the value of $3e_\sigma$ when the degeneracy is in the e_g^* orbital set. For strong π donors, such as F^-, and OH^-, $e_\pi \approx e_\sigma/4$; for NH_3 and aliphatic amines the π-interaction is expected to be small. Moreover, for ligands such as CN^-, e_π is negative, corresponding to a bonding rather than an antibonding interaction. The distortion is thus expected to vary significantly with the nature of the ligands in a complex.

The Jahn–Teller stabilization energy E_{JT} is in the range of 1200 to 2500 cm^{-1} for complexes with an e_g^* orbital degeneracy. When the degeneracy is in the t_{2g}^* orbitals, however, E_{JT} is much smaller: ~ 400 to 800 cm^{-1} for strong π donors such as F^-, and close to 0 for weak or non π bonding ligands such as NH_3. Therefore, E_{JT} may be comparable to, or less than, the energy of the ε_g vibration which means that many complexes of this type do not undergo a simple geometric distortion. Instead, a marked softening of the ε_g vibration occurs relative to non-Jahn–Teller active metal ions. This often has a significant effect on the optical spectrum, because the large amplitude of the ε_g mode in the ground state leads to what is, in effect, a splitting of the components of the e_g^* orbitals in the excited state (Section 8.2.1). Also, because this Jahn–Teller coupling may be comparable to spin–orbit interactions, it can affect the magnetic properties by partially quenching the orbital angular momentum associated with the orbitally degeneracy. This aspect is considered further in Section 9.6.1.

7.1.2.2   *Tetrahedral Complexes.*

Because of the relatively small d orbital splitting (4/9 that of a octahedral complex with similar ligands and bond lengths), virtually all tetrahedral complexes are high spin. Thus the only d orbital configurations with orbital degeneracies in the antibonding t_2^* set are d^3, d^4, d^8, and d^9. For such complexes, three vibrations are Jahn–Teller active, bending modes of ε and τ_2 symmetry and a stretching mode of σ_2 symmetry. The electronic coupling constants V for each vibration may be estimated in terms of AOM bonding parameters. To first order, the distortion is given by the ratio V/f (Eq 7.7), where f is the force constant of the vibration. The ε vibration, the active form of which is shown in Figure 7.5a, has a low force constant, which means that coupling with this mode is expected to produce the largest distortion. The potential surface resulting from the coupling is shown in Figure 7.5b. The Jahn–Teller energy depends on the sign of the distortion. For the d^9 configuration of the Cu^{2+} ion, stabilization occurs if the angle θ increases with respect to the tetrahedral value, but the reverse is true for the d^8 configuration of the Ni^{2+} ion.

Simple calculations suggest that the angle θ is expected to increase from $109.5°$ to $\sim 130°$ on going from the $ZnCl_4^{2-}$ to the $CuCl_4^{2-}$ ion (3). A value close to the latter, $127°$, is indeed observed for Cs_2CuCl_4. No other clear-cut distortions attributable to factors other than lattice perturbations are observed. This suggests that coupling with the τ_2 modes is quenched by the interaction with the more active ε vibration. The crystal structures of many compounds containing the four-coordinate $CuCl_4^{2-}$ ion have been determined. Most have angles θ in the range of $125°$ to $135°$, but a number exhibit much larger distortions. In particular, several compounds contain planar $CuCl_4^{2-}$ groups ($\theta = 180°$). These all involve countercations that form hydrogen bonds to the chloride ions, and those may help stabilize the planar geometry. When Cu^{2+} is doped into ZnO, the high strength of the interconnecting bonds of the continuous lattice causes the force constant of the Jahn–Teller vibration to be rather large. EPR measurements suggest that the distortion away from a regular tetrahedral geometry is then quite small. In general, although most four-coordinate Cu^{2+} complexes adopt the D_{2d} geometry predicted by the Jahn–Teller theorem, the size of the distortion varies considerably from one compound to another. This occurs because the force constant of the Jahn–Teller active vibration is so low that the distortion is easily modified by factors such as lattice forces and the ligand stereochemistry.

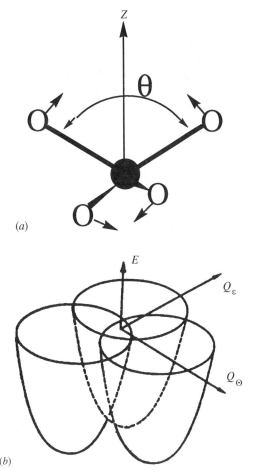

Figure 7.5. (*a*) Form of the active component of the Jahn–Teller active ε vibration of a tetrahedral complex with an orbital degeneracy in the t_2 orbital set. (*b*) Potential surface of a Jahn–Teller distorted tetrahedral complex.

Although the linear-coupling constant of the $e^4 \cdot t_2^{*4}$ configuration of a tetrahedral d^8 metal ion is similar to that of a d^9 metal ion, tetrahedral Ni^{2+} complexes are much less distorted than are similar Cu^{2+} complexes (3). For example, the angular distortion of the $NiCl_4^{2-}$ ion away from a regular tetrahedron appears to be rather small. Two factors appear to be responsible for the small distortions observed for tetrahedral Ni^{2+} complexes. First, the distortion corresponds to a contraction of the L-M-L angle, rather than the expansion that occurs for Cu^{2+} complexes. This contraction leads to an increase in ligand–ligand repulsions, which act to oppose the distortion. Second, and probably more important, the $^3T_1(F)$ ground state of the d^8 Ni^{2+} ion does not correspond exactly to the strong-field electron configuration $e^4 t_2^{*4}$ predicted simply on the basis of the relative energies of the d orbitals. Interelectronic repulsions act to mix in other electron configurations (Section 6.7), which reduces the linear Jahn–Teller coupling constant and hence the size of the distortion.

7.1.3 Bond Length Differences Between High- and Low-spin Complexes

Six-coordinate complexes of metal ions with d^4, d^5, d^6, and d^7 electron configurations may have ground states with two possible spin multiplicities (Sections 6.9 and 9.10). The same is true for d^8 metal ions, although here the low-spin state becomes the ground state only for highly distorted complexes and one pair of ligands is usually actually lost to yield a planar geometry. Conversion from the high- to the low-spin state always involves a change in the number of electrons and/or the creation or loss of orbital degeneracy in the antibonding e_g^* orbital set and hence a change in the metal–ligand bond lengths. The d orbital energy and concomitant structural changes associated with the spin changes are shown for the d^6, d^7, and d^8 configurations in Figure 7.6. Here, any π interactions with the t_{2g} orbital set have been ignored.

7.1.3.1 d^5 and d^6 Configurations. For the d^5 and d^6 configurations, a transition from the high- to the low-spin state involves the transfer of a pair of electrons from the σ antibonding e_g^* orbital set to the t_{2g} orbitals. In both cases, an orbital degeneracy occurs in the t_{2g} orbitals in one spin state. This, however—as

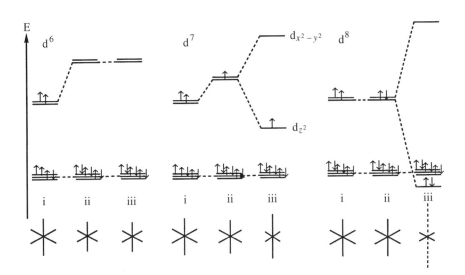

Figure 7.6. Change in structure and d orbital energies accompanying a high- to low-spin conversion for the d^6, d^7, and d^8 electron configurations. *i*, The high-spin state; *ii*, the change in overall M–L bond length; *iii*, any Jahn–Teller distortions. For the d^8 configuration, the low-spin state can become the ground state only in iii and the axial ligands are usually lost to give a planar geometry.

discussed in Section 7.1.2.1—does not cause a significant Jahn–Teller distortion, because these orbitals interact relatively weakly with the ligands. The dominant change in the geometry is, therefore, a significant shortening of the metal–ligand bonds in the low-spin state, which may be estimated using Eq. 7.5. Note that here it is assumed that the difference in interelectron repulsion energy of the two spin states, S in Eq. 7.3, is not significantly affected by the bond length change. In practice, the shorter bond lengths associated with the low-spin spin state mean that such a complex will be somewhat more covalent than the corresponding high-spin complex, which may lead to a reduction in S. In the case of Fe(II), comparison with experiment is possible because both spin states are observed for certain complexes.

For typical high-spin Fe^{2+} amine complexes, $r_0 \approx 2.18$ Å, $\Delta \approx 12000\,cm^{-1}$, and $f \approx 0.012\,pN\,Å^{-1}$ for the α_{1g} mode. Substituting these in Eq. 7.5 together with $n = 5$ and $m = 2$, yields the estimate $\delta r = 0.16$ Å for the decrease in the metal–ligand bond length expected when going to the low-spin complex. The observed bonds are indeed ~ 0.15 to 0.2 Å shorter in low-spin Fe^{2+} amines than in the high-spin forms. Thus Eq. 7.5 provides a realistic description of the balance between the electronic and structural forces for this metal ion. Moreover, this is also the difference in ionic radius estimated from the analysis of a range of compounds containing Fe^{2+} in the high- and low-spin states (7). Somewhat smaller bond length differences are observed for the high- and low-spin states of Fe^{3+} complexes, probably because the force constant f increases more than Δ when the metal is in a higher oxidation state.

The structural change accompanying the high- to low-spin transition for Fe^{2+} complexes has important biological consequences. Hemoglobin, the oxygen-carrying species in blood, consists of four heme groups, each containing an Fe^{2+} ion that can bind to one oxygen molecule. An usual feature of the system is that uptake of dioxygen by one heme group significantly improves the ability of the other heme groups to take up oxygen. In its deoxygenated form, the Fe^{2+} complex is high spin, but coordination of the oxygen molecule causes spin pairing to occur, the iron remaining Fe^{2+}. It is the decrease in ionic radius associated with the spin change that produces a change in the protein packing and enhances the ability of neighboring heme groups to take up oxygen.

7.1.3.2 d^4 and d^7 Configurations.

For both the d^4 and d^7 electron configurations, the high- and low-spin states differ by a single electron in the occupancy of the e_g^* orbitals. Moreover, in each case, the e_g^* orbitals contain a single electron in one of the spin states, so that these configurations are subject to a strong Jahn–Teller distortion. For complexes of metal ions with these electron configurations, the geometries of the two spin states are, therefore, expected to show a difference in overall bond length owing to the change in e_g^* occupancy, plus a displacement in the Jahn–Teller ε_g mode owing to the change in orbital degeneracy.

Experimental data involving d^4 metal ions in both spin states are sparse. Almost all Cr^{2+} complexes are high spin and exhibit the expected Jahn–Teller distortion, but few low-spin complexes other than with the cyanide ligand are known. Somewhat more data are available for d^7 metal ions. Most Co^{2+} complexes are high spin and exhibit the expected approximately regular octahedral geometry. The low-spin state, however, does occur for several complexes with ligands similar to the high-spin case. This allows at least a semiquantitative comparison of the geometries of the two spin states. Because the e_g^* occupancy in the high- and low-spin states differs by a single electron, Eq. 7.5 suggests that the difference in bond length should be about half that for the spin states of Fe^{2+}. In agreement with this, a difference in ionic radius of ~ 0.085 Å has been deduced from a range of crystal structures involving Co^{2+} in the two spin states (7). The low-spin state invariably exhibits the expected Jahn–Teller distortion, with two bonds being ~ 0.18 Å longer than the other four.

7.1.3.3 *d^8 Configuration.*

For the d^8 configuration, for a regular octahedral geometry, the low-spin state always corresponds to an excited state. Because the high- and low-spin complexes of d^8 metal ions, in particular Ni^{2+}, have such different shapes—octahedral and planar, respectively—it is easy to overlook the way in which the two geometries and their associated d configurations are related. In fact, the same principles apply to the geometries associated with the spin states of the Ni^{2+} ion as to those of other d configurations (4). The d orbital occupancies associated with the high-spin $^3A_{2g}$ and excited low-spin ^1E_g states of an octahedral Ni^{2+} complex are shown in Figure 7.6. The total number of electrons in the e_g^* orbitals remains unchanged when going from one to the other. On this basis, therefore, there is no driving force for a change in the *overall* metal–ligand bond length. The ^1E_g state, however, is orbitally degenerate and is hence subject to a Jahn–Teller distortion.

It is the Jahn–Teller stabilization energy that provides the driving force for the distortion, and distortion will occur when this is greater than the difference in inter electron repulsion energy, which favors the high-spin state. Here the orbital degeneracy reflects the fact that the paired electrons in the e_g^* orbitals may occupy either the $d_{x^2-y^2}$ or the d_{z^2} orbital. Upon distortion, the split components of the e_g^* set differ, therefore, by two electrons, and to a first approximation the first-order Jahn–Teller coupling constant V for the ^1E_g state of Ni^{2+} is twice as large as that of the ^2E_g state of Cu^{2+} ($m = 2$ rather than 1). Substituting this value into Eq. 7.13, together with $n = 5$ and the force constant and Δ splitting value appropriate to the $Ni(NH_3)_6^{2+}$ ion, yields the estimate $Q \cong 0.44$ Å for the Jahn–Teller distortion of the ^1E_g state of such a complex. For the elongated tetragonal distortion, this corresponds to a compression of the in-plane bond lengths of δx, $\delta y \approx 0.13$ Å, and an elongation of $\delta z \approx 0.25$ Å in the axial bond lengths (Eq. 7.11).

In practice, the in-plane bonds decrease by considerably more than this; the Ni-N bonds are ~ 0.18 Å shorter in low-spin ammine complexes compared to high-spin ones. Moreover, in low-spin Ni^{2+} complexes, the axial bonds are either very long or, more commonly, altogether absent, leading to the square planar geometry. Two factors probably act to stabilize the four-coordinate stereochemistry. First loss of the axial ligands:

$$ML_6 \rightarrow ML_4 + 2L \qquad (7.14)$$

increases the number of degrees of freedom of the system. This causes a rise in entropy, which lowers the energy of the system. Second, the interaction between the 4s and d_{z^2} orbitals that accompanies a tetragonal distortion (Section 3.2.2) depends on the size of the distortion. Because that reaches a maximum when the axial ligands are lost completely, this factor also favors the square planar geometry.

7.1.4 Variation of Bond Lengths on Crossing the Transition series

When comparing the bond lengths of different transition metal ions in the same oxidation state, we must consider two factors. First, the effective nuclear charge on the metal, which decides the radial extent of its orbitals; second, the influence of any antibonding d electrons, as discussed in the preceding sections of this chapter.

On crossing each transition series from left to right, the effective nuclear charge rises, causing a contraction of the metal orbitals, a decrease in the ionic and covalent radii, and, other things being equal, a shortening of metal–ligand bond lengths. This is nicely illustrated by the changes in the ionic radii of the lanthanide and actinide ions (Fig. 7.7). The f electrons interact only weakly with the ligands, so effects owing to changes in the f configuration are minimal, and the metal–ligand bond lengths and ionic radii decrease smoothly with increasing atomic number.

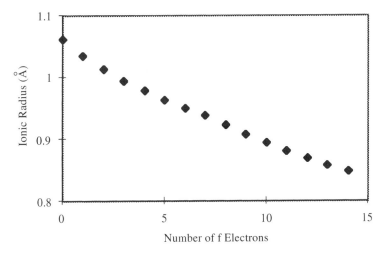

Figure 7.7. Change in ionic radius as a function of the number of f electrons for the trivalent lanthanide ions.

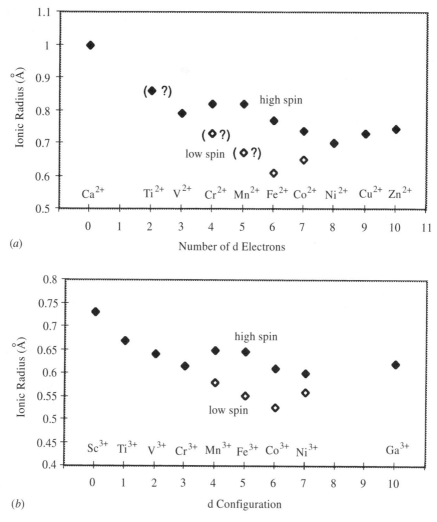

Figure 7.8. Experimental effective ionic radii of octahedrally coordinated transition metal ions in the divalent (*a*) and trivalent (*b*) state, as derived by Shannon and Prewitt (7). values designated? have a large uncertainty.

A more complicated pattern is observed for the ionic radii of the transition ions. Values derived from the structures of a wide range of compounds are shown in Figure 7.8 for divalent and trivalent metal ions. Superimposed on the overall contraction owing to the increase in effective nuclear charge is a slight increase whenever an electron occupies the antibonding e_g^* orbitals. This increase δr may be estimated using Eq. 7.5. It depends on the balance between the $e_g^* - t_{2g}$ energy separation Δ and the force constant f of the totally symmetric metal–ligand stretching vibration. For octahedrally coordinated divalent metal ions, the increase in bond length is expected to vary from ~ 0.04 Å per electron for ligands that are low in the spectrochemical series and have high force constants, such as H_2O, to ~ 0.08 Å for ligands such as NH_3 or α, α'-bipyridyl. Similar, although somewhat smaller changes are expected for trivalent metal ions, because although both Δ and f increase, the rise is somewhat greater for the latter. For any particular ligand, the ratio Δ/f is not expected to alter much on going from the beginning to the end of the transition series, because the modest decrease in Δ is approximately counterbalanced by the underlying decrease in metal–ligand bond length. A plot of the change in ionic radius for the divalent metal ions estimated assuming the bond lengths increase by the average value evaluated using Eq. 7.5, $\delta r = 0.06$ Å every time an electron occupies the e_g^* orbitals, is shown in Figure 7.9. Here, an underlying decrease in ionic radius analogous to that observed for the lanthanide ions has been assumed. The pattern is quite similar to that observed for the ionic radii (Fig. 7.8), suggesting that the irregularities in the curves are indeed the result of the influence of the electrons in the antibonding e_g^* orbitals. Note that the deviation from the smooth curve occurs at d^4 for the high-spin state, but d^7 for the low-spin. Effects owing to π bonding have been neglected. These would simply change the underlying slope of the line.

The M-O bond lengths of the hexaaqua complexes of the transition-metal ions are compared in Figure 7.10. Similar behavior is observed for other complexes. The d configuration of each metal ion is also indicated. The pattern is similar to that observed for the ionic radii (Fig. 7.8a) and estimated using Eq. 7.5 (Fig. 7.9). The complexes of the divalent metal ions are all high spin. The complex ions $Cr(H_2O)_6^{2+}$ and $Cu(H_2O)_6^{2+}$ exhibit highly distorted structures owing to a Jahn–Teller distortion (Section 7.1.2.1).

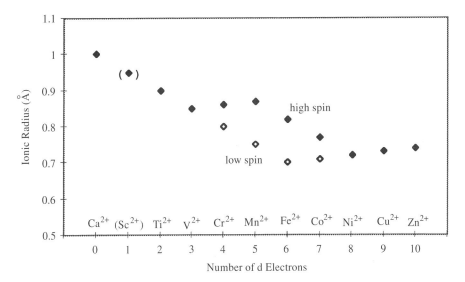

Figure 7.9. Ionic radii of octahedrally coordinated transition metal ions estimated using Eq. 7.5 (see text). An underlying decrease of 0.05 Å per atomic number owing to the rise in effective nuclear charge is also assumed. ◆, high-spin states; ◇, low-spin states.

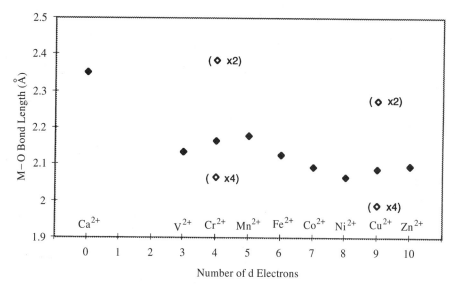

Figure 7.10. Correlation between the metal–oxygen bond lengths observed for the hexaaqua complexes of the divalent metal ions of the first transition series and their d configurations. ◆, average values of the Cr^{2+} and Cu^{2+} ions; ◇, long and short values of the Jahn–Teller distorted complexes.

For tetrahedral complexes, the influence of the occupancy of the antibonding t_2 orbitals is expected to be comparatively modest, because the d orbital splitting is only $\sim 4/9$ that in a corresponding octahedral complex. This is partially compensated for, however, by the smaller number of ligands, and Eq. 7.5 implies a displacement $\delta r \approx 2/3$ that of the corresponding octahedral complex when an electron occupies the t_2 rather than the e orbital set of a tetrahedral complex. As yet, little experimental data are available for structurally related tetrahedral complexes. However, because of the reversed ordering of the doubly and triply degenerate orbital sets, a plot of bond length against d configuration would be expected to change slope at the d^3 and d^8 configurations for the tetrahedral geometry. It occurs at d^4 and d^9 for octahedral complexes.

7.2 DEPENDENCE OF THE STABILITY OF A COMPLEX ON ITS d CONFIGURATION

When considering the relative stabilities of transition metal complexes, it is important to distinguish clearly between thermodynamic and kinetic stability. The first refers to the standard free energy changes of the reaction of a complex, and the second refers to the rates of those reactions. A thermodynamically unstable complex may be quite kinetically inert, whereas a stable one may nevertheless undergo rapid reactions. Historically, this played an important role in the development of transition-metal chemistry, because the isolation of thermodynamically unstable isomers played a crucial role in resolving the arguments over the nature of complexes. An example of a thermodynamically unstable, but kinetically inert complex is $Co(NH_3)_6^{3+}$ in slightly acidic aqueous solution. Here, the stability constant for the reaction:

$$Co(NH_3)_6^{3+} + 6H_2O \rightarrow Co(H_2O)_6^{3+} + 6NH_3 \tag{7.15}$$

is $\sim 10^{25}$, but in solution the complex remains unchanged for days. Conversely, the stability constant for the corresponding reaction for $Ni(CN)_4^{2-}$ is $\sim 10^{-22}$, but if isotopically labeled cyanide is added to an aqueous solution of the complex, it exchanges with that bound to the metal within a few seconds. This is a thermodynamically stable, but kinetically labile complex. Although a detailed account of these topics is beyond the scope of this book, both the thermodynamic and kinetic stability of transition-metal complexes are strongly influenced by the d configuration of the metal ion, and this aspect is now discussed briefly.

One of the successes of the electrostatic crystal field model (CFM) has been its ability to explain some of the observed trends in the thermodynamic and kinetic stability of transition metal complexes. Conventionally, this was done in terms of the crystal field stabilization energy (CFSE) of the metal ion. The terminology is unfortunate, because it implies that the d electrons exert a stabilizing influence; however, the stabilization is only with respect to the *baricenter*, or average energy of the d orbitals.

In fact, except for the rare situation when d electrons occupy t_{2g} orbitals interacting with π acceptor ligands, the d electrons always exert a *destabilizing* influence. Energetically the dominant effect, the interaction with the antibonding e_g^* (or t_2^*, in the case of tetrahedral complexes) orbitals, is always destabilizing as far as the d electrons are concerned, although of course, the ligand electrons are stabilized by the interaction. This being said, as far as comparing the energies of two different d configurations is concerned, the difference between the crystal field and a ligand field approach such as the AOM is largely a question of reference point. In the CFM this is the average energy of the d orbitals in the complex, whereas in the AOM it is the energy of the d orbitals before the complex is formed. This means that the conclusions drawn using crystal field stabilization energies to explain *relative* stabilities of complexes with the same geometry, therefore, apply equally well if covalent interactions, as in the AOM, are invoked.

7.2.1 Thermodynamic Effects

7.2.1.1 Bond Strengths. The way in which the metal–ligand bond energies vary on crossing the transition series may be illustrated by considering the standard hydration enthalpies of the divalent metal ions:

$$M^{2+}(g) + 2H^+(aqu.) + 2e(g) = M^{2+}(aqu.) + H_2(g) \tag{7.16}$$

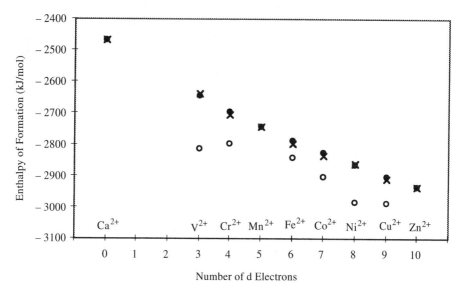

Figure 7.11. Standard enthalpy of formation of the hexaaqua complexes of the divalent first-row transition-metal ions (o). ✗, values obtained after correcting for ligand field effects alone; ●, values after allowing for ligand field, nephelauxetic and spin–orbit effects and bond-length variations. See text for details.

The enthalpy of formation of the complexes $M(H_2O)_6^{2+}$, ΔH_{hyd}, from the gaseous ions is shown by the open circles in Figure 7.11 for those first-row transition-metal ions for which data are available. Overall, the enthalpy becomes increasingly negative upon crossing the series, but superimposed on this are two curves with minima at V^{2+} and Ni^{2+}. The bond strengths of the complexes formed by the metal ions with an empty, half-filled and filled d shell, Ca^{2+}, Mn^{2+}, and Zn^{2+}, lie on a smooth line. Conventionally, this has been taken as the reference point for the discussion of the double-humped form of the curve. The influence of the d configuration of each metal ion is then related to the energy of the baricenter of the d orbitals by means of the CFSE. This is easily obtained from the d orbital splitting observed for each complex (shown in Tables 8.3 and 8.4), and the resulting energy changes $\delta E_{d-split}$ are listed in Table 7.3. When the enthalpies corrected for $\delta E_{Ed-split}$ are plotted, the pattern shown by the crosses in Figure 7.11 is obtained. They form a regular curve, which includes the reference ions Ca^{2+}, Mn^{2+}, and Zn^{2+}. This suggests that the predominant cause of the irregularities in the raw data is indeed related to the energy differences between the d electrons in the complexes. This observation has, in fact, often been taken as strong evidence supporting the CFM and, by inference, subsequent ligand field models.

Other factors also influence the way in which bond strengths change on crossing the transition series, however. Perhaps the most obvious is the variation in metal–ligand bond lengths, which follows a similar, although inverted, pattern

TABLE 7.3 Data Used to Estimate the Residual Heat of Formation of the Hexaaqua Complexes of the Divalent Transition Metal Ions.[a]

Metal	ΔH_{hydr}	$\delta E_{d-split}$	δe_{relax}	δE_{er}	δE_{so}	ΔH_{res}
Ca	−2468	0	0	0	0	−2468
V	−2814	−177	24	−18	2	−2645
Cr	−2799	−95	4	−16	2	−2694
Mn	−2743	0	0	0	0	−2743
Fe	−2843	−45	2	−16	3	−2787
Co	−2904	−68	5	−19	6	−2828
Ni	−2986	−123	10	−18	12	−2867
Cu	−2989	−80	1	−16	10	−2904
Zn	−2936	0	0	0	0	−2936

[a] Data presented in kilojoules / mole.

to the bond enthalpies (Section 7.1.4; Fig. 7.10). The deviations in the M-O bond lengths from the smooth curve interpolated between Ca^{2+}, Mn^{2+}, and Zn^{2+} may be used to estimate the influence of this factor. The bond-length deviations cause the energy to rise according to the relationship:

$$U = fQ^2/2 \qquad (7.17)$$

where f is the force constant of the totally symmetric stretching vibration Q of the $M(H_2O)_6^{2+}$ complex ion. If the bonds "relaxed" to lengths appropriate to the values extrapolated from those of the d^0, d^5, and d^{10} metal ions, the energy changes δE_{relax} listed in Table 7.3 would result. These changes act in an opposite sense to $\delta E_{d-split}$ because the "relaxation" is considered to lower the energy of the complex.

A more subtle factor influencing the relative bond strengths results from a change in the interelectron interaction energy as a result of complex formation. A detailed account of this is beyond the scope of this book, but is discussed by Johnson and Nelson (8,9). It basically results from the fact that the energy contribution owing to exchange interactions between pairs of electrons with parallel spins does not vary smoothly as a function of the number of d electrons. The nephelauxetic effect (Section 8.4.2) causes interelectron repulsion in a complex to be smaller than that in the free ion. The expected differences in interelectron repulsion energy δE_{er} between the free ions and their hexaaqua complexes are indicated in Table 7.3, and it may be seen that they follow a similar pattern to $\delta E_{d-split}$. A final small contribution, δE_{so}, comes from changes in the spin−orbit interactions which accompany complex formation.

Although the d orbital occupancy is the dominant cause of the irregular variation of the bond enthalpies, the other factors are by no means negligible. Their effect is moderated by the fact that the influences of interelectron repulsion and the variation in metal−ligand bond lengths oppose one another and

TABLE 7.4   Data Used to Estimate the Residual Heat of Formation of the Hexafluoro-Complexes of the Trivalent Transition-Metal Ions.[a]

Metal	ΔH_{6F}	$\delta E_{d-split}$	δE_{relax}	δE_{er}	δE_{so}	ΔH_{res}
Sc	-8001	0	0	0	0	-8001
Ti	-8337	-84	2	-35	1	-8221
V	-8464	-140	5	-45	3	-8287
Cr	-8637	-231	14	-37	5	-8388
Mn	-8640	-144	2	-37	4	-8465
Fe	-8534	0	0	0	0	-8534
Co	-8730	-67	2	-57	5	-8613
Ni	-8903	-134	5	-110	14	-8678
Cu	-9034	-202	14	-156	16	-8706
Ga	-8825	0	0	0	0	-8825

[a]Data presented in kilojoules/mole.

approximately balance. The net result after subtracting each of the above contributions, $\delta E_{d-split}$, δE_{relax}, δE_{er}, and δE_{so} from the experimental bond enthalpies δH_{hyd} is indicated by the solid circles in Figure 7.11. These show that when all the factors are taken into account, the resultant enthalpies follow a smooth curve connecting the metal ions with an empty, half-filled and full d shell.

Essentially similar arguments apply to the variation in the bond enthalpies of complexes formed by the metal ions in the trivalent state. The hexafluoro complexes MF_6^{3-} may be taken as an example. Table 7.4 shows the values of the standard enthalpy changes of the reaction:

$$3K^+(g) + M^{3+}(g) + 6F^-(g) = K_3MF_6(s) \qquad (7.18)$$

in which the solid compound K_3MF_6 is formed from gaseous ions. The values for the series Sc^{3+} to Ga^{3+} are shown in Figure 7.12. A similar "double-humped" behavior to that observed for the hexaaqua complexes of the divalent metal ions is observed, although as required, the curves are offset by one element. In this case, subtraction of the relative effects of the d orbital splitting, $\delta E_{d-split}$, as listed in Table 7.4, from the enthalpies does not completely remove the double-humped behavior, as shown by the crosses in Figure 7.12. When the influence of the irregularities in the metal–ligand bond lengths, interelectron repulsion, and spin–orbit coupling are included, a smooth curve does result (Fig. 7.12). The reason why ligand field effects alone are insufficient to account for the irregular variation in the bond lengths for the trivalent hexafluorides, although they do account satisfactorily for the corresponding trends in the divalent hexaaqua complexes, appears to be the much larger influence of the nephelauxetic effects in the former complexes. Because of their greater covalency, the interelectron repulsion parameters undergo a much larger reduction for the complexes of the trivalent metal ions. This means that this factor, represented by δE_{er}, has a much

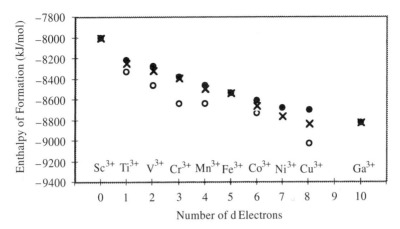

Figure 7.12 Standard enthalpies of formation of the solid hexafluoro complexes of the trivalent first-row transition-metal ions, from Eq. 7.18 (**o**). **✗**, values obtained after correcting for ligand field effects alone; **●**, values after also including nephelauxetic and spin–orbit effects.

greater influence than is the case for the divalent ions, as may be seen by comparing the values in Tables 7.3 and 7.4.

7.2.1.2 The Irving-Williams Series. For any particular ligand, a comparison of stability constants shows a rather similar pattern to that of the heats of hydration. It is often remarked that some measures of the biological activity of metal ions of the transition series show similar trends; both phenomena are generally referred to as the Irving–Williams series (10,11). It is important to remember that stability constants are not *absolute*. Rather, they are a measure of *relative* stability. Normally, because most measurements are made by adding ligands to an aqueous solution of the metal ion, the reference state is the hexaaqua complex. Moreover, unlike the overall enthalpy of hydration, it is possible to compare the relative stabilities of complexes that differ by a single ligand, rather than a complete set of water molecule ligands.

Most data are available for the metal ions in the second half of the first-transition series. There, the complexes formed by amine ligands, for instance, are considerably more stable than those formed by water. Typical plots of the negative logarithms of the formation constants for the addition of the first, second, and third ligand, K_1, K_2, and K_3, and of the overall formation constant of the M (en)$_3^{2+}$ complex, $\beta = K_1 K_2 K_3$ (not to be confused with the nephelauxetic ratio) are shown in Figure 7.13 (en-1,2-diaminoethane, ethylenediamine). The trend in the behavior of the overall stability constant on crossing the series is quite similar to that observed for the bond enthalpies (see preceding section). This is to be expected because factors such as ligand field splitting Δ and the nephelauxetic effect are greater for the amine than for aqua complexes. Any trends caused by these are, therefore, enhanced and influence the relative

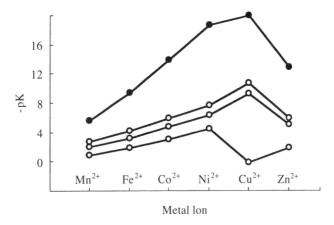

Figure 7.13. Stepwise (o) and overall (●) stability constants of the ethylenediamine (en) complexes of several transition-metal ions in aqueous solution.

energies of the complexes and hence the stability constants. Thus the tendency for β to rise on crossing the series correlates with the overall increase in magnitude of the bond energy. The deviation in this trend observed for the last two members of the series can be explained largely in terms of the energy change occurring when electrons are placed in the antibonding e_g^* orbitals, with smaller contributions from the other factors discussed in the preceding section.

Inspection of the individual stability constants shows that for every metal ion except Cu^{2+} they behave in a similar manner as do the overall stability constants. K_1 and K_2 maximize at this metal ion, whereas K_3 has a very small value. This is owing to the Jahn–Teller effect (Section 7.1.2.1), which causes this metal ion to form four short, strong metal–ligand bonds to two ethylenediamine (en) ligands; the addition of the third ethylenediamine is less facile, because it involves two long, weak axial bonds.

7.2.1.3   Site Preferences in Oxide Lattices.   Certain transition-metal oxides contain two types of lattice sites: one involving octahedral and the other tetrahedral coordination of the metal. The site adopted by different metal ions appears to depend on their d configurations. The electrostatic crystal field theory has traditionally been used to explain most of the observed structures by comparing the CFSE associated with the two possible coordination arrangements. As already mentioned, however, the CFSE is measured relative to a reference point at which the negative charges of the ligands are spread evenly over a sphere of radius equal to the metal–ligand bond length. In fact, the electrons are *repelled* by the formal ligand charges. When considering the relative stabilities of the two coordination geometries, the term *stabilization energy* must be taken in context. Generally, the different reference points used in various bonding models do not influence comparisons between different d configurations; however, care must be taken when the absolute energies of different stereochemistries are compared (1,12).

To illustrate this, the CFSE of the d^1 configuration in an octahedral crystal field $-4Dq_{oct}$ may be compared with that in the corresponding tetrahedral field $-6Dq_{tet}$. Applying the relationship $Dq_{oct} = 9Dq_{tet}/4$ leads to the conclusion that the octahedral geometry is stabilized by $4Dq_{oct}/3$ compared to the tetrahedral one. This result may be compared with that given by the AOM (Chapter 3), which describes the effects of different ligand fields in terms of a weak covalent interaction, with the added advantage that the energies of both stereochemistries may be expressed in terms of the single pair of metal–ligand bonding parameters e_σ and e_π. Here identical bond lengths are assumed for the octahedral and tetrahedral sites, and the π bonding is assumed to be cylindrical about the metal–ligand bond axis. For the d^1 configuration, the energy is $4e_\pi$ in an octahedral ligand field and $8e_\pi/3$ in a tetrahedral one. Remembering that for a π donor such as the O^{2-} ion the bonding parameters represent an antibonding interaction, the octahedral coordination is, therefore, *less* stable than the tetrahedral by $4e_\pi/3$ as far as the d electrons are concerned.

This argument, however, ignores the stabilization of the ligand electrons owing to the interaction with the metal d orbitals. To a good approximation they are stabilized by the same amount that the d orbitals are destabilized. As the ligand orbitals are filled, this leads to a stabilization of $12e_\sigma + 24e_\pi$ for octahedral coordination and $8e_\sigma + 16e_\pi$ for tetrahedral complexes. This means that for the d^1 case, the stabilization of the ligand electrons less the destabilization of the d electron, the overall molecular orbital stabilization energy (MOSE), is $12e_\sigma + 20e_\pi$ for octahedral coordination and $8e_\sigma + 40e_\pi/3$ for the tetrahedral case. The octahedral coordination is thus far more stable than the tetrahedral, by $4e_\sigma + 20e_\pi/3$, as far as this aspect is concerned.

The MOSE and CFSE—in units of Dq_{oct}—are tabulated for each d configuration for the tetrahedral and octahedral stereochemistries in Table 7.5. For the CFSE, data are given for both high- and low-spin situations, although it must be remembered that the spin-pairing energy has a large destabilizing effect in the latter case. The energy difference between the two coordination geometries is also given. Because the d electrons are antibonding, the MOSE becomes increasingly less negative (i.e., less stabilizing) on crossing the transition series for both stereochemistries. It must be stressed that this makes only a minor contribution to the total bond strength. That shows the opposite trend, tending to increase in magnitude because of the increase in the interaction between the metal 4s and 4p and ligand orbitals. The relative stability of the octahedral compared to the tetrahedral coordination geometry varies significantly from one d configuration to another. Because the dominant effect is once again the promotion energy required to place an electron in the upper e_g^* or, for the tetrahedral geometry, t_2^* set, similar trends are shown for MOSE and CFSE. In general, they suggest that octahedral coordination is particularly stable for the d^3 and d^8, and low-spin d^5 and d^6 configurations.

In considering the site occupancies of transition-metal oxides, most data are available for the spinels of general formula $A^{II}B^{III}_2O_4$, where A^{II} and B^{III} represent divalent and trivalent metals. The oxide ions adopt a cubic close-

TABLE 7.5 CFSE and MOSE for the d^n Configurations in Octahedral and Tetrahedral Coordination[a]

n	CFSE (Dq_{oct})[b]					MOSE (angular overlap bonding parameters)				
	tet	oct (hs)	$E_{oct}-E_{tet}$	oct (ls)	$E_{oct}-E_{tet}$	tet	oct (hs)	$E_{oct}-E_{tet}$	oct (ls)	$E_{oct}-E_{tet}$
0	0	0	0	—	—	$-8e_\sigma-16e_\pi$	$-12e_\sigma-24e_\pi$	$-4e_\sigma-8e_\pi$	—	—
1	$-2\frac{2}{3}$	-4	$-1\frac{1}{3}$	—	—	$-8e_\sigma-13\frac{1}{3}e_\pi$	$-12e_\sigma-20e_\pi$	$-4e_\sigma-6\frac{2}{3}e_\pi$	—	—
2	$-5\frac{1}{3}$	-8	$-2\frac{2}{3}$	—	—	$-8e_\sigma-10\frac{2}{3}e_\pi$	$-12e_\sigma-16e_\pi$	$-4e_\sigma-5\frac{1}{3}e_\pi$	—	—
3	$-3\frac{5}{9}$	-12	$-8\frac{4}{9}$	—	—	$-6\frac{2}{3}e_\sigma-9\frac{7}{9}e_\pi$	$-12e_\sigma-12e_\pi$	$-5\frac{1}{3}e_\sigma-2\frac{2}{9}e_\pi$	—	—
4	$-1\frac{7}{9}$	-6	$-4\frac{2}{9}$	-16	$-14\frac{2}{9}$	$-5\frac{1}{3}e_\sigma-8\frac{8}{9}e_\pi$	$-9e_\sigma-12e_\pi$	$-3\frac{2}{3}e_\sigma-3\frac{1}{9}e_\pi$	$-12e_\sigma-8e_\pi$	$-6\frac{2}{3}e_\sigma+\frac{8}{9}e_\pi$
5	0	0	0	-20	-20	$-4e_\sigma-8e_\pi$	$-6e_\sigma-12e_\pi$	$-2e_\sigma-4e_\pi$	$-12e_\sigma-4e_\pi$	$-8e_\sigma+4e_\pi$
6	$-2\frac{2}{3}$	-4	$-1\frac{1}{3}$	-24	$-21\frac{1}{3}$	$-4e_\sigma-5\frac{1}{3}e_\pi$	$-6e_\sigma-8e_\pi$	$-2e_\sigma-2\frac{2}{3}e_\pi$	$-12e_\sigma$	$-8e_\sigma+5\frac{1}{3}e_\pi$
7	$-5\frac{1}{3}$	-8	$-2\frac{2}{3}$	-18	$-12\frac{2}{3}$	$-4e_\sigma-2\frac{2}{3}e_\pi$	$-6e_\sigma-4e_\pi$	$-2e_\sigma-1\frac{1}{3}e_\pi$	$-9e_\sigma$	$-5e_\sigma+2\frac{2}{3}e_\pi$
8	$-3\frac{5}{9}$	-12	$-8\frac{4}{9}$	—	—	$-2\frac{2}{3}e_\sigma-1\frac{7}{9}e_\pi$	$-6e_\sigma$	$-3\frac{1}{3}e_\sigma+1\frac{7}{9}e_\pi$	—	—
9	$-1\frac{7}{9}$	-6	$-4\frac{2}{9}$	—	—	$-\frac{4}{3}e_\sigma-\frac{8}{9}e_\pi$	$-3e_\sigma$	$-1\frac{2}{3}e_\sigma+\frac{8}{9}e_\pi$	—	—
10	0	0	0	—	—	0	0	0	—	—

[a] tet, tetrahedral; oct, octahedral; hs, high spin; ls, low spin.

[b] Estimated assuming $Dq_{tet} = \frac{4}{9}Dq_{oct}$.

packed-type arrangement, with four octahedral and eight tetrahedral holes per formula unit. The metal ions occupy some of these holes, and several factors might be expected to influence the type of site they adopt. A high charge on the metal is better counterbalanced by six, rather than four O^{2-} ions. The small ionic radius that normally accompanies a higher positive charge, however, means that the metal fits better into the tetrahedral site. If these factors approximately balance, then the bonding influence of the d configuration should decide the site distribution of the metal ions. In so-called normal spinels, the divalent metal ions A^{II} occupy one-eighth of the tetrahedral sites and the trivalent ions B^{III} occupy one-half of the octahedral sites. In inverse spinels, the lattice positions of the divalent metal ions and half the trivalent metal ions are interchanged, so that the former occupy octahedral sites and half the latter tetrahedral sites. Some spinels lie between these two extremes, and B^{III} is disordered between the tetrahedral and octahedral sites. The type of spinel can be characterized by the parameter λ, which gives the fraction of B^{III} which is in a tetrahedral site. A value $\lambda = 0$ corresponds to a normal spinel, $\lambda = 0.5$ to an inverse spinel. The values of λ found for some typical spinels are given in Table 7.6.

For $B^{III} = Al^{3+}$ the only spinel in which a significant fraction of the divalent cations occupy octahedral sites is for $A^{II} = Ni^{2+}$, which is consistent with the particular stability of the d^8 configuration in an octahedral environment (Table 7.5). All the spinels with $B^{III} = Cr^{3+}$ are normal, as expected given the high relative stability of the d^3 configuration in octahedral coordination. For Co_3O_4, the high stabilization of low-spin Co^{3+} (d^6) in an octahedral ligand field, combined with the relative stability of Co^{2+} (d^7) in a tetrahedral field strongly favors the normal lattice. For Fe_3O_4, on the other hand, the high-spin d^5 configuration of the Fe^{3+} ion contributes no CFSE and only a small MOSE in either stereochemistry, and it seems that the relatively small gain in energy of the high-spin d^6 configuration of the Fe^{2+} ion in octahedral coordination is enough to tip the balance in favor of the inverse structure. The cation distributions in nearly all transition metal spinels do in fact conform to the CFSE and MOSE predictions. Although this naturally lends credence to the usefulness of ligand field theory in explaining the structures of transition-metal compounds, it must again be stressed that the d electrons play only a minor role as far as bond

TABLE 7.6   Fractional Occupancy λ of the Tetrahedral Sites by the Ions B in Spinels $A^{II}B_2^{III}O_4$.[a]

B^{3+} \ A^{2+}	Mg^{2+}	Mn^{2+}	Fe^{2+}	Co^{2+}	Ni^{2+}	Cu^{2+}	Zn^{2+}
Al^{3+}	0	0	0	0	0.38	—	0
Cr^{3+}	0	0	0	0	0	0	0
Fe^{3+}	0.45	0.1	0.5	0.5	0.5	0.5	0
Mn^{3+}	—	0	—	—	—	—	0
Co^{3+}	—	—	—	0	—	—	—

[a] Values of 0 and 0.5 correspond to normal and inverse spinels, respectively;—indicates that the type is unknown.

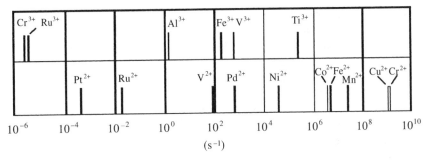

Figure 7.14. Rate constants for water exchange for various divalent and trivalent metal ions in aqueous solution. [From Docommum, Y. and Merbach, A. E. in "Inorganic High Pressure Chemistry", R. van Eldrik, Ed., 1986, Elsevier, Amsterdam, Netherlands. Used with permission.]

energies go. They play a deciding role only when other factors are in balance, as seems to be the case for the spinels.

7.2.2   Kinetic Effects

When kinetic stability is considered, the important factor is the activation energy, the relative energy of the reaction intermediate compared with the starting materials, as shown in Fig. 7.14 (13). The complexes of different transition-metal ions vary dramatically in their kinetic stability, and this aspect of their chemistry has been the subject of detailed and protracted studies. Despite this, it has so far proved impossible to derive simple mechanisms for the kinetic behavior of complexes in general. Certain features, however, in particular the kinetic inertness of the complexes formed by metal ions with particular d configurations, are readily apparent. These have traditionally been interpreted in terms of CFSE, the energy with respect to the baricenter of the d orbitals. As already discussed, such an approach must be treated with caution, because it relies on comparisons made using an artificial reference point. Moreover, the d electron bonding interactions make a rather small contribution to the total energy of a complex. A comparison, however, of the activation energies of different metal ions may be realistic enough to allow broad conclusions to be drawn.

Most experimental data are available for substitution reactions of octahedral complexes and planar complexes of the Pt^{2+} ion. For the former, reaction could occur either via loss of a ligand to form a five-coordinate transition state with a square-based pyramidal geometry or by gain of a ligand to produce a seven-coordinate pentagonal bipyramidal transition state. The changes in CFSE should give an idea of the relative activation energy to be expected for these two processes and these are listed for different d configurations in Table 7.7. Here, a negative value indicates a loss of stabilization energy and a high activation energy. Of course, some reactions do not fall neatly into either of the two classes. One ligand may leave as the second enters the coordination sphere in what is termed a "concerted" process. Moreover, it must be stressed again that it is only

TABLE 7.7   Change in CFSE for an Octahedral Complex on Forming a Five-Coordinate (square-based pyramidal) or Seven-Coordinate (pentagonal bipyramidal) Transition State[a]**. (From "Mechanisms of Inorganic Reactions", 2nd Edition, F. Basolo and R. G. Pearson, Wiley, 1958. Used with permission)**

d config.	High Spin		Low Spin	
	Five Coordinate	Seven Coordinate	Five Coordinate	Seven Coordinate
d^0	0	0	—	—
d^1	0.57	1.28	—	—
d^2	1.14	2.56	—	—
d^3	−2.00	−4.26	—	—
d^4	3.14	−1.07	−1.43	−2.98
d^5	0	0	−0.86	−1.70
d^6	0.57	1.28	−4.00	−8.52
d^7	1.14	2.56	1.14	−5.34
d^8	−2.00	−4.26	—	—
d^9	3.14	−1.07	—	—
d^{10}	0	0	—	—

[a] Units are in Dq; negative values correspond to a loss of stabilization energy, i.e., destabilization.

the *relative* values for the different d configurations that are important—the *absolute* values have no significance in this context because interactions with the d orbitals have relatively little effect on the overall bond energy. As noted in the preceeding section, the d^3 and d^8, and low-spin d^6 configurations are particularly stable in octahedral coordination, so it is not surprizing that these are predicted to have especially high activation energies. This correlates well with experimental observations, as may be seen from the typical rate constants for the exchange of water in hexaaqua complexes indicated in Figure 7.14. Here the exchange rates of the metal ions with d^9 (Cu^{2+}) and high-spin d^4 (Cr^{2+}) configurations are enhanced by the presence of Jahn–Teller distortions (Section 7.1.2.1). It may be seen that for first-row ions the complexes of Cr^{3+} (d^3) are particularly inert, and V^{2+} (d^3) and Ni^{2+} (d^8) form the most kinetically stable complexes of the divalent metal ions. It thus appears that the general kinetic behavior of transition-metal complexes does correlate quite well with the expectations of ligand field theory.

REFERENCES

1. Burdett, J. K., *'Molecular Shapes'*, Wiley-Interscience: New York, 1980.
2. Bersuker, I. B., *The Jahn–Teller Effect and Vibronic Interactions in Modern Chemistry*, Plenum Press, New York, 1984.
3. Reinen, D., and Atanasov, M., *Mag. Reson. Rev.*, 1991, **5**, 167.
4. Deeth, R. J., and Hitchman, M. A., *Inorg. Chem.*, 1986, **25**, 1225.

5. Warren, K. D. *Struct. Bond.*, 1984, **57**, 119.

6. Ceulemans, A., Beyens, D., and Vanquickenborne, L. G., *J. Amer. Chem. Soc.*, 1984, **106**, 5824.

7. Shannon, R. D., and Prewitt, C. T., *Acta Crystallogr.*, 1969, **B25**, 925.

8. Johnson, D. A., and Nelson, P. G., *Inorg. Chem.*, 1995, **34**, 3253 and 5666.

9. Johnson, D. A., and Nelson, P. G., *J. Chem. Soc., Dalton Trans.* 1995, 3483.

10. Williams, R. J. P., *Discuss. Faraday Soc.*, 1958, 123.

11. Hill, H. A. O., and Williams, R. J. P., *Coord. Chem. Rev.*, 1993, **122**, 1.

12. Reinen, D., *Struct. Bond.* 1970, **7**, 114.

13. Wilkins, R. G., *Kinetics and Mechanism of Reactions of Transition Metal Complexes*, VCG, Weinheim, 1991.

CHAPTER 8

THE ELECTRONIC SPECTRA OF COMPLEXES

The electronic spectrum provides probably the most direct way of studying the energy levels in a metal complex (1). A typical spectrum consists of a number of peaks, or bands, each of which has a characteristic energy, intensity, width, and shape; this chapter considers each of these aspects.

8.1 IMPORTANT FEATURES OF ELECTRONIC SPECTRA

8.1.1 Band Intensities

The intensity of light transmitted by a sample of molar concentration C and path length l, I_T is related to the incident intensity I_0 by the expression for the transmittance T:

$$T = I_T/I_0 = 10^{-\varepsilon Cl} \tag{8.1}$$

where ε is the molar extinction coefficient. Often, a more useful measure of the light absorption is the absorbance A sometimes referred to as the optical density, which is given by:

$$A = \log_{10}(I_0/I_T) = \varepsilon Cl \tag{8.2}$$

Equation 8.2 expresses the Beer–Lambert law, which states that the absorbance of a sample is proportional to C and to l. The light absorption of a band is generally characterized by the extinction coefficient ε, at the wavelength of the band maximum, for a sample of 1 M concentration when using a path length of 1 cm. This provides a convenient way of comparing band intensities, because the absorbances at band maxima are obtained easily from the spectra. It must be recognized, however, that ε is related only indirectly to the quantum mechanical description of light absorption. The band intensity is more properly measured by the *oscillator strength f*, the name of which arises because comparison is made with a standard quantum mechanical oscillator system. The oscillator strength is measured by the area of the band in a plot of extinction coefficient versus energy. If the band stretches from ν_1 to ν_2 (measured in cm^{-1})

$$f = 4.32 \cdot 10^{-9} \int_{\nu_1}^{\nu_2} \varepsilon \, d\nu \qquad (8.3)$$

Electronic transitions may occur through electric dipole, magnetic dipole, or electric quadrupole mechanisms. For the 'd-d' and charge-transfer transitions of transition-metal complexes, the electric dipole mechanism predominates, although occasionally the magnetic dipole mechanism does make a significant contribution to band intensities. The probability that a photon is absorbed to cause a transition from the state specified by ψ_1 to that specified by ψ_2 is proportional to the square of the transition moment integral Q:

$$Q = \langle \psi_1 | \mathbf{r} | \psi_2 \rangle \qquad (8.4)$$

where $\mathbf{r}$ is the electric dipole moment operator, which has the properties of a vector. The oscillator strength is related to the transition moment integral by:

$$f = (8\pi^2 mc/he^2) G\nu Q^2 \qquad (8.5)$$

Here h is Planck's constant, m and e are the mass and charge of the electron, c is the velocity of light, and ν is the energy of the transition in cm^{-1}. The parameter G takes into account the degeneracies of the states. For states of degeneracy d, $G = 1/d$, and the oscillator strength is the sum over all possible pairs of upper and lower states.

It is not generally the practice experimentally to measure the area of a band as defined above, and as already mentioned, band intensities are usually quantified and compared in terms of ε at the band maximum. This is useful as long as bands have similar widths, as is often the case. Then ε is related approximately to the oscillator strength by the expression:

$$f \propto \varepsilon \delta_{\frac{1}{2}} \qquad (8.6)$$

where $\delta_{\frac{1}{2}}$, (usually in cm^{-1}), is the so-called half-width, the width of the band between the points where ε has half its maximum value. The factors influencing band widths are discussed in Section 8.1.3. Eq. 8.6 also assumes a Gaussian line shape for the band, which may not always be justified (Section 8.1.3.3).

The intensities of electronic transitions, depending as they do on the wavefunctions of the ground and excited states (Eq. 8.4), provide important information about the electronic structure of a transition metal complex. We now consider the various factors influencing this aspect.

8.1.1.1   Selection Rules.   Important deductions concerning the magnitude of Q can be made from the form of the ground and excited state wavefunctions. A wavefunction may be broken down, approximately, into component product functions in the following manner:

$$\psi = \psi_{\text{orbital}} \cdot \psi_{\text{spin}} \cdot \psi_{\text{vibrational}} \cdot \psi_{\text{rotational}} \cdot \psi_{\text{translational}} \tag{8.7}$$

Except for the gas phase, the last two terms in this expansion are unlikely to change for a molecule in a time comparable to the lifetimes of excited electronic states, so that they integrate to unity in the expression for Q. From the remaining part of the expansion it may be shown that Q is zero unless:

$$S_{\psi_1} = S_{\psi_2} \tag{8.8}$$

This relationship, the *spin selection rule*, is satisfied only if the two wavefunctions have the same spin quantum number S. It provides a powerful limitation on the intensities of electronic transitions. Within the framework of the Russell–Saunders coupling scheme, in which the spin and orbital momenta are considered separately (Section 5.5), transitions between states with identical spin quantum numbers—*spin-allowed* transitions—have extinction coefficients typically 10^2 to 10^3 times larger than those having different spin quantum numbers; the latter are termed *spin-forbidden* transitions.

We now consider the orbital parts of the wavefunctions. Because the electric dipole operator $\mathbf{r}$ is antisymmetric to inversion through the center of the system, integration of the function:

$$\langle \psi_1(\text{orbital})|\mathbf{r}|\psi_2(\text{orbital})\rangle \tag{8.9}$$

is nonzero only if ψ_1 and ψ_2 have different parities; i.e., one of them must be symmetric (g) and the other antisymmetric (u) to inversion. This *parity* selection rule, often called the *Laporte* rule, means that Q is zero for the 'd-d' transitions of an octahedral or planar complex, in which both wavefunctions are g, unless vibrations are also considered.

The above integral is most conveniently investigated with the aid of group theory. If ψ_1 and ψ_2 transform as the irreducible representations Γ_1 and Γ_2 of the group to which the molecule belongs, and $\mathbf{r}$ transforms as Γ_r, of that group, Q is zero unless the direct product:

$$\Gamma_1 \times \Gamma_r \times \Gamma_2 \tag{8.10}$$

contains the totally symmetric irreducible representation of the group $A_{1(g)}$. In the groups O_h and T_d, the radius vector $\mathbf{r}$ transforms in the same manner as the set of Cartesian coordinates. For O_h and T_d, $\mathbf{r}$ transforms as T_{1u} and T_2, respectively.

In the absence of vibrations, every term derived from the d configuration of an octahedral complex transforms as a symmetric representation. The transitions between terms are, therefore, Laporte forbidden because the direct product of the representations of the wavefunctions with that of $\mathbf{r}$, T_{1u}, cannot contain A_{1g}.

An important result in connection with the Laporte selection rule concerns the transitions between the e^n and t_2^n orbitals of tetrahedral complexes. The direct product:

$$E \times T_2 \times T_2 = 2T_1 + 2T_2 + 2E + A_2 + A_1 \tag{8.11}$$

contains A_1, so transitions of this type are formally Laporte allowed. In considering the matrix elements of the type $\langle \psi_e^n | \mathbf{r} | \psi_{t_2}^n \rangle$, however, it must be remembered that this must be zero if just the d orbitals are considered, because they are centrosymmetric functions. Extending the treatment to include the metal 4s and 4p orbitals, it may be noted that, although the e^n set is composed essentially of metal d orbitals, the t_2^n set may contain both d and p atomic orbitals. The d orbitals have g character, whereas the p orbitals have u properties, leading to a nonzero matrix element.

Extending the picture further, to a scheme in which the e^n and t_2^n sets are molecular orbitals that have some ligand character (Section 3.1.1), shows that nonzero matrix elements also occur involving cross-terms between metal d and ligand p functions. In fact, it is thought that this latter contribution provides most of the absorption intensity for tetrahedral complexes. It is important to remember that, although the molecular orbital bonding scheme of octahedral complexes also involves contributions from ligand π orbitals to the e_g and t_{2g} sets, this does not induce intensity into the electronic transitions between these orbitals. This is because the symmetry properties of the orbitals make the sum of the cross-terms involving the metal d and ligand p functions zero. Thus we conclude: *The ligand field transitions that are forbidden in octahedral complexes may be partially allowed in tetrahedral complexes, because of d-p mixing.* For both spin-allowed and spin-forbidden transitions, the 'd-d' bands of tetrahedral complexes are usually 10 to 100 times more intense than those of octahedral complexes. Although the intensities are high, it should be noted that they depend largely on the ligand contribution to the e and t_2 orbital sets, typically $\sim 30\%$, and the extinction coefficients are still considerably lower than those of typical 'charge

TABLE 8.1. Order of Magnitude of the Extinction Coefficients ε and Oscillator Strengths f Typically Observed for the Different Types of Electronic Transitions of Transition-Metal Complexes

Electronic Transitions	ε	f
'd-d', spin forbidden, Laporte forbidden (O_h)	0.1	10^{-7}
'd-d', spin forbidden, Laporte allowed (T_d)	1	10^{-6}
'd-d', spin allowed, Laporte forbidden (O_h)	10	10^{-5}
'd-d', spin allowed, Laporte allowed (T_d)	100	10^{-4}
Charge transfer, spin and Laporte allowed	1000	10^{-3}

transfer' transitions. There, transitions are between an orbital that is largely p, and one that is largely d in character, giving rise to intense bands. The typical intensities of different kinds of electronic transition are listed in Table 8.1.

If the spin and Laporte selection rules were obeyed rigorously, the number of absorption bands observed for transition-metal complexes would be rather small. In fact, although they provide a useful guide to intensities, neither of the rules holds at all strictly, because there are mechanisms that afford some relaxation.

8.1.1.2 Spin–Orbit Coupling. The separation between spin and orbital momenta that forms the basis of the Russell–Saunders coupling scheme (Section 5.5) is only an approximation, and spin–orbit coupling provides the most important mechanism by which the spin selection rule is relaxed. As expected, the intensity of spin-forbidden bands relative to spin-allowed bands increases as the spin–orbit coupling constants rise on crossing each transition series, and on going from the first- to the third-transition series. The fact that, because of spin–orbit coupling, both the ground and the excited state wavefunctions no longer have well-defined spin quantum numbers provides a fundamental mechanism by which the spin selection rule is broken. For metal ions with high-spin d^5 electron configurations, for which every excited d configuration state has a different value of S from the ground state, this is the only way in which this selection rule is relaxed, and here the 'd-d' transitions are always very weak.

The operator involved $\lambda \mathbf{L} \cdot \mathbf{S}$ can be expanded and rearranged:

$$\begin{aligned} \mathbf{L} \cdot \mathbf{S} &= \mathbf{L}_x \mathbf{S}_x + \mathbf{L}_y \mathbf{S}_y + \mathbf{L}_z \mathbf{S}_z \\ &= \mathbf{L}_z \mathbf{S}_z + \frac{1}{2}(\mathbf{L}_+ \mathbf{S}_+ + \mathbf{L}_- \mathbf{S}_-) \end{aligned} \tag{8.12}$$

where $\mathbf{L}_\pm$ and $\mathbf{S}_\pm$ are the "ladder" angular momentum operators (Appendix 8). These latter alter the values of M_L and M_S, respectively, by one unit. Consequently, perturbation by spin–orbit coupling can mix into the ground term small quantities of excited terms differing from the ground term by $\Delta S = 1$ or

$\Delta L = 1$. The amount of mixing should be proportional to the value of the spin–orbit coupling constant λ and inversely proportional to the separation between the ground and excited term energies. Typically, for a first-transition series element, $\lambda \sim 100\,\mathrm{cm}^{-1}$ and $\Delta E(\psi_1 \rightarrow \psi_2) > 10,000\,\mathrm{cm}^{-1}$ so that admixture coefficients for terms with multiplicity different from that of the pure ground term are expected to be $\sim 10^{-2}$.

Of more importance is the spin–orbit interaction between excited states, because these can be relatively close in energy. If an excited state ψ_{2a} of the same spin multiplicity as the ground state is separated in energy by $\Delta E_{a,b}$ from an excited state with different spin multiplicity ψ_{2b} then, to a first aproximation, spin–orbit coupling causes a small admixture of the first wavefunction into that of the latter, which becomes:

$$\psi'_{2b} = \psi_{2b} + c\psi_{2a} \qquad (8.13)$$

where the mixing coefficient c is given by:

$$c = \langle \psi_{2b} | \lambda \mathbf{L} \cdot \mathbf{S} | \psi_{2a} \rangle / \Delta E_{a,b} \qquad (8.14)$$

The intensity of the formally spin-forbidden state is:

$$\langle \psi_{1a} | \mathbf{r} | \psi'_{2b} \rangle^2 = \langle \psi_{1a} | \mathbf{r} | \psi_{2b} \rangle^2 + c^2 \langle \psi_{1a} | \mathbf{r} | \psi_{2a} \rangle^2 \qquad (8.15)$$

or c^2 times that of the spin-allowed transition. Because the admixture of the two excited state wavefunctions is a two-way process, the intensity of the spin-allowed transition is reduced by the same amount. For this reason, the spin-forbidden transition is said to *borrow* or *steal* intensity from the spin–allowed transition.

The strong dependence of the intensity of a spin-forbidden transition on the energy separation to the nearest spin-allowed transition is well illustrated in the electronic spectra of octahedral Ni^{2+} complexes. The spin-forbidden transition $^3A_{2g} \rightarrow {}^1E_g$, which to a good approximation simply involves an electron spin-flip, is largely independent of the ligand field splitting parameter Δ. The transition $^3A_{2g} \rightarrow {}^3T_{2g}$, on the other hand, occurs at exactly Δ. When the spectra of a range of complexes involving ligands with different Δ values are compared (Fig. 8.1), it is seen that when Δ is small, the peak owing to the $^3A_{2g} \rightarrow {}^1E_g$ transition occurs as a weak shoulder on the high-energy side of that owing to the $^3A_{2g} \rightarrow {}^3T_{2g}$ transition. As Δ increases, the spin-allowed transition approaches the spin-forbidden one, and the latter increases in intensity. For high values of Δ, the $^3A_{2g} \rightarrow {}^1E_g$ transition is observed as a shoulder on the low-energy side of the spin-allowed transition. When the spin-forbidden and spin-allowed excited states have almost identical energies, the spins of the two states are extensively mixed. Then two peaks of about equal intensity are observed on either side of a

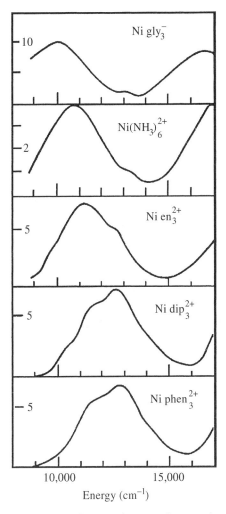

Figure 8.1. The bands owing to the $^3A_{2g} \rightarrow {}^3T_{2g}$ and $^3A_{2g} \rightarrow {}^1E_g$ transitions of several octahedral Ni^{2+} complexes. [From *The Theory of Transition Metal Ions* by J. S. Griffith, Cambridge University Press, Cambridge, U.K., 1964. Used with permission].

sharp dip, which occurs at the energy that the two states would have in the absence of spin–orbit coupling. The mechanism by which this band shape is generated is called Fano antiresonance.

8.1.1.3   Influence of Vibronic Coupling on Band Intensities. The Laporte selection rule is based on the neglect of vibrations. Vibrations modulate the ligand field by changing the donor atom positions. This modulation can cause a mixing of the electronic wavefunctions. Taking $\psi_{1(g)}$ and $\psi_{2(g)}$ to describe states of g parity, then u-type vibrations cause the admixture of a small proportion of u-type electronic wavefunctions into each of these states. Generally, these u states are charge transfer in origin and are much closer in energy to the excited states than to the ground state, so that coupling to the former predominates in this *vibronic* (vibrational-electronic) intensity-inducing mechanism. Each excited state then becomes:

$$\psi_2' = \psi_{2(g)} + \sum_{\text{u vibrations}} C_i \psi_{i(u)}' \qquad (8.16)$$

where the C_i are the vibronic coupling coefficients. For the transition moment integral $\langle \psi_{1(g)} | \mathbf{r} | \psi_2' \rangle$, expansion of the excited-state wavefunction now gives nonzero matrix elements of the form $C_i \langle \psi_{1g} / \mathbf{r} | \psi_{i(u)}' \rangle$.

In principle, excited states of odd parity are available for all possible symmetry types, and even simple centrosymmetric complexes always have at least one u vibration available to induce intensity for every 'd-d' transition. The vibronic coupling mechanism requires that the vibrational quantum number of the intensity-inducing mode changes by one unit in conjunction with the electronic transition. This means that at low temperature, when the complex is in the ground vibrational state, the electronic transition is shifted to higher energy by one quantum of this vibration. Moreover, upper vibrational levels of the ground state are thermally populated at higher temperatures, producing greater ligand displacements. The intensity produced by vibronic coupling, therefore, usually increases substantially when a complex is warmed from near to 0 K to room temperature. The influence of temperature on optical spectra is considered more fully in Section 8.1.4. Typically, octahedral complexes exhibit bands with intensities corresponding to ε in the range 1 to $50 \, \text{mol}^{-1} \, \text{l} \, \text{cm}^{-1}$.

It should also be mentioned that many complexes that formally belong to a noncentrosymmetric point group have only a small asymmetric contribution to the ligand field. Depending on the size of this contribution, the intensity lies between that of a truly centrosymmetric complex and one with a highly asymmetric ligand field, such as a distorted tetrahedral or square-based pyramidal complex. When the geometric distortion is so small that it is comparable with vibrational amplitudes, the band intensities are not much greater than those in typical centrosymmetric complexes. This is the case, for instance, in tris(ethylenediamine) complexes, in which the bands are not much more intense than in corresponding hexaammine complexes. When asymmetry is the result of a difference in the ligand type on either side of the metal, the deviation depends on how "different" the ligands are. The bands in the spectrum of a complex such as the *cis*-$Co(NH_3)_4Cl_2^+$ ion are two to three times as intense as those in the corresponding *trans* complex. Here, to compare oscillator strengths, the band areas must be used, rather than just extinction coefficients, because different band splittings occur for the two isomers.

Irrespective of whether the mechanism is 'static' or 'dynamic' in origin, ligand orbital participation plays an important part in the 'd-d' intensity-inducing mechanism. That explains the general observation that band intensities tend to increase as the metal–ligand bond becomes more covalent. Thus aqua complexes have relatively low band intensities compared to ammine complexes, but chloro and bromo complexes typically have relatively intense spectra. The intensities of electronic transitions provide one of the important means of characterizing transition-metal complexes, and typical extinction coefficients and oscillator strengths observed for octahedral and tetrahedral complexes are summarized in Table 8.1.

8.1.1.4 Polarized Spectra of Low-Symmetry Complexes. In complexes with noncubic symmetry, the twofold or threefold orbital degeneracy of states is partly or completely removed (2). The selection rules for the different components then depend on the orientation of the electric dipole vector $\mathbf{r}$ with respect to the Cartesian axes. A polarizing device may be used to orient $\mathbf{r}$ with respect to a complex in a single crystal, and the way in which this affects intensities provides a powerful method of assigning spectral bands.

8.1.1.4.1   PARITY-ALLOWED TRANSITIONS. For noncentrosymmetric complexes, the polarization properties of the electronic transitions may be derived easily using group theory. Occasionally, the spin–orbit components of electronic states are resolved, and then the double group must be used; however, this will not be considered here. We consider the spectrum of the complex $CuCl_4^{2-}$ as an example. Jahn–Teller coupling causes the complex to be distorted away from a regular tetrahedral geometry (Section 7.1.2.2), producing a complex belonging to the point group D_{2d} (Fig. 8.2). Such a complex has a $^2B_2$ ground state, and excited states of ^2E, $^2B_1$, and $^2A_1$ symmetry. To determine whether the transition moment integral is nonzero for a particular direction of the electric vector, it is necessary simply to consider the direct product of its irreducible representation in this direction with those of the ground and excited state wavefunctions. If the product contains the totally symmetric representation, the transition is allowed. In our example, the electric dipole components transform as the translations (xy) and z, which means that a transition is allowed in, for instance, z polarization, if

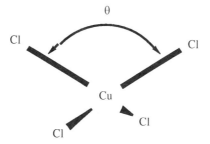

Figure 8.2. Geometry of a $CuCl_4^{2-}$ ion of D_{2d} symmetry.

the product of the ground and excited state-irreducible representations equals that of z. In this way the selection rules are easily derived as:

$$\text{Transition :} \quad {}^2B_2 \rightarrow {}^2E \quad {}^2B_1 \quad {}^2A_1 \tag{8.17}$$

$$\text{Polarization :} \quad xy \quad - \quad z \tag{8.18}$$

This means that when the electric vector is in the xy plane of the complex the only allowed transition is ${}^2B_2 \rightarrow {}^2E$, whereas when it is along z, only ${}^2B_2 \rightarrow {}^2A_1$ is allowed. The transition ${}^2B_2 \rightarrow {}^2B_1$ is parity forbidden for all orientations of the electric vector. It might, however, be observed weakly through vibronic coupling and/or 'borrowing' intensity from the allowed transitions via spin–orbit coupling.

The electronic spectrum of a typical compound containing the $CuCl_4^{2-}$ ion is shown in Figure 8.3, with the electric vector first approximately in the (xy) plane and then along the z axis of the complex. The spectrum is highly dichroic; the band at $\sim 9,000\,cm^{-1}$ has high intensity in (xy) polarization, and the one at $\sim 12,000\,cm^{-1}$ is intense in z polarization. This agrees well with the selection rules if the peaks are assigned to transitions to the ${}^2E$ and ${}^2A_1$ excited states, respectively. The Laporte forbidden band owing to the ${}^2B_2 \rightarrow {}^2B_1$ transition is

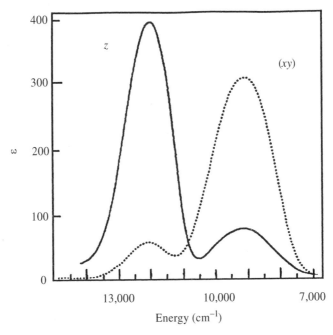

Figure 8.3. The spectrum of a distorted tetrahedral $CuCl_4^{2-}$ ion (D_{2d} point group) with the electric vector of polarized light parallel to the z axis and in the (xy) plane. [From McDonald, R. G., Riley, M. J., and Hitchman, M. A. *Inorg. Chem.* 1988, **27**, 894. Used with permission.]

not detected. Note that, because the complex has a geometry that is far from centrosymmetric, the extinction coefficients of the parity-allowed bands are high, being in the range typically observed for tetrahedral complexes.

8.1.1.4.2   VIBRONICALLY ALLOWED TRANSITIONS. In the case of centrosymmetric complexes, the treatment is basically similar to that for parity-allowed transitions, but with the direct product restriction extended to include those of the u vibrations inducing intensity. A transition is then vibronically allowed in **r** polarization only if the direct product of the irreducible representations of the ground and excited electronic states with that of **r** equals that of one of the u vibrational modes of the complex. The procedure is again best illustrated by an example. If the Jahn–Teller distortion of the $CuCl_4^{2-}$ ion occurs to the maximum extent, which is the case in certain crystal lattices, there results a centrosymmetric planar complex belonging to the point group D_{4h}. The vibronic selection rules are conveniently presented by a table that shows which u vibrations allow each electronic transition in each polarization; they are shown for planar $CuCl_4^{2-}$ in Table 8.2. This complex has u vibrations of α_{2u}, β_{2u}, and ε_u symmetry, so that for only one transition, $^2B_{1g} \rightarrow {}^2B_{2g}$ in z polarization, is there no u vibration of the correct symmetry to induce intensity (the direct product $\Gamma(B_{1g}) \times \Gamma(z) \times \Gamma(B_{2g}) = B_{1u}$).

The electronic spectrum of a planar $CuCl_4^{2-}$ complex is shown in Figure 8.4, with the electric vector approximately along z and in the (xy) plane. The extinction coefficients of the bands are rather low and decrease markedly on cooling, as expected for parity-forbidden transitions. In agreement with the vibronic selection rules, just one band is absent in z polarization, and this may be assigned to the transition $^2B_{1g} \rightarrow {}^2B_{2g}$. This assignment agrees with the d orbital energy sequence expected for a planar complex (Section 3.2.2). The selection rules do not provide information that allows the other bands to be assigned simply from their polarizations.

8.1.2   Band Energies

8.1.2.1   Use of Tanabe–Sugano Diagrams to Derive Δ and B.

For transition-metal ions with a d^1 or d^9 electron configuration, the electronic spectrum simply

TABLE 8.2. Vibrations Inducing Intensity into the Electronic Transitions of a Planar Cu(II) Complex

	Polarization	
Transition	(xy)	z
$B_{1g} \rightarrow B_{2g}$	ε_u	—
$B_{1g} \rightarrow E_g$	α_{2u}, β_{2u}	ε_u
$B_{1g} \rightarrow A_g$	ε_u	β_{2u}

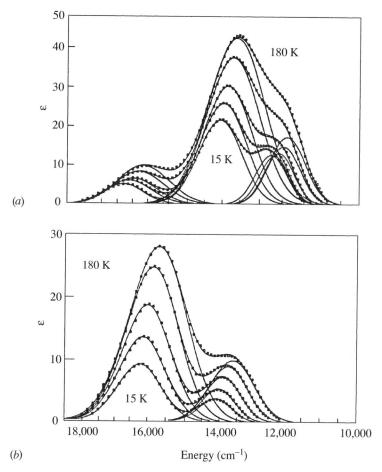

Figure 8.4. The spectrum of a planar $CuCl_4^{2-}$ ion (D_{4h} point group) at different temperatures between 180 K and 15 K with the electric vector of polarized light in the (xy) plane (a), and parallel to the z axis (b). [From McDonald, R. G., and Hitchman, M. A. *Inorg. Chem.*, 1986, **25**, 3273. Used with permission].

reflects the energy differences between the d orbitals in the complex. For many-electron metal ions, however, the band energies depend not only on the d orbital energies but also on the effects of interelectron repulsion. For first-row transition-metal ions, these latter are conventionally treated using the Russell–Saunders coupling scheme. The energy levels of complexes of near octahedral or tetrahedral symmetry are represented by Tanabe–Sugano plots, which present the term energies as a function of the ligand field splitting parameter Δ (Section 6.8). These diagrams can be used at different levels of sophistication. A way in which this can be done is best illustrated by an example; we consider the spectrum of the NiF_6^{4-} ion, which is formally present in $KNiF_3$.

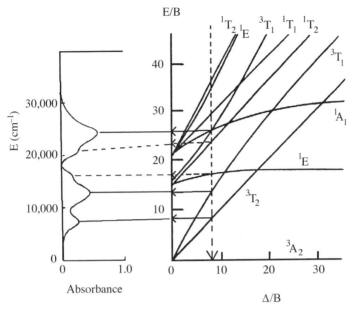

Figure 8.5. Matching of the electronic spectrum of the NiF_6^{4-} ion to the d^8 Tanabe – Sugano diagram.

To simply assign the spectrum, it is assumed that B does not differ too much from that used in plotting the Tanabe–Sugano diagram. The intensities indicate whether each band corresponds to a spin-allowed or spin-forbidden transition, and the fit to the calculated transitions on the vertical scale of the diagram is optimized. To illustrate the way in which each band correlates with a particular term, the spectrum of the NiF_6^{4-} ion is shown plotted vertically next to the d^8 Tanabe–Sugano diagram in Figure 8.5. The first spin-allowed transition to the $^3T_{2g}$ state occurs at precisely Δ, which thus has the value $7,400 \text{ cm}^{-1}$. Using this Δ value, a reasonably good fit for the other transition energies may be obtained if the peaks at 12,660 and $23,840 \text{ cm}^{-1}$ are assigned to transitions to the $^3T_{1g}(F)$ and $^3T_{1g}(P)$ states, respectively. The marked shoulders at $15,150 \text{ cm}^{-1}$ and $21,275 \text{ cm}^{-1}$ are owing to spin-forbidden transitions, which are relatively intense because of their close proximity to spin-allowed transitions.

When, as in this example, more than one spin-allowed transition is observed, the value of the Racah parameter B may be deduced by matching the ratio of the energies of two observed peaks with that predicted by the energy level diagram. If three bands are available, the question arises as to which pair should be used for the fitting process. The best result will be obtained when the ratio is most sensitive to the value of Δ. For a d^8 ion, this is the case for the $^3A_{2g} \rightarrow {}^3T_{2g}(F)$ and $^3A_{2g} \rightarrow {}^3T_{1g}(P)$ pair of transitions. For the NiF_6^{4-} ion, the ratio of the observed energies is 4.15, which occurs at $\Delta/B = 7.8$ on the Tanabe–Sugano plot. The corresponding intercept for the first transition occurs at

$7.8B = 7,400\,cm^{-1}$ on the vertical axis. Hence, $B = 950\,cm^{-1}$ for the Racah parameter in this complex. This represents a reduction of B to $\sim 90\%$ of the free ion value, implying that the complex is quite ionic in nature, which is expected with fluoride ions as the ligands. Because $\Delta = 7.8B$, the value $\Delta = 7400\,cm^{-1}$ for the complex (in fact the first transition occurs at exactly Δ in this example).

Because three spin-allowed transitions are observed, the self-consistency of the interpretation may be tested by comparing the experimental ratio of the first and second transitions. The experimental ratio, 1.67, is quite close to the value of 1.69 from the Tanabe–Sugano plot, showing that in this case the fit is indeed self-consistent.

The second Racah parameter, C, generally cannot be obtained by a similar method, because it influences the energies of only the spin-forbidden transitions, and these usually depend on C, B, and Δ via a set of linearly dependent equations. A ligand field computer program that calculates the excited state energies using Δ, B, and C as input is the best way of obtaining the value of C and of improving the estimates of the other two parameters (See references 23 and 28 of Chapter 3 for such computer programs).

For all d configurations, except high-spin d^5, the energy of the first spin-allowed transition occurs at exactly, or close to Δ, so a procedure similar to that given above may generally be used to estimate the parameters Δ and B. In doing this, it is important to check that the ratio of the transition energies is sensitive enough to Δ to allow an accurate fit to be obtained. For the high-spin d^6 configuration, only one spin-allowed transition is expected, so the Racah parameter B can be obtained only from complete analyses which include spin-forbidden transitions. The energies of all excited states of the high-spin d^5 configuration have a different spin from the ground state, and those in the visible region are rather insensitive to Δ, so that here both Δ and B can be obtained only approximately by using the Tanabe–Sugano diagram.

It must be remembered that Tanabe–Sugano diagrams should be used for complexes only of cubic or near-cubic symmetry. Whenever an orbitally degenerate ground state occurs for a regular octahedral or tetrahedral complex, the Jahn–Teller theorem requires that the complex distort to remove the degeneracy (Section 7.1.2). This ground state splitting causes orbitally degenerate excited states of E and T symmetry to split also. When the Jahn–Teller effect results from a σ interaction, the distortion is likely to be so large that the excited states bear little resemblance to those indicated on the Tanabe–Sugano diagram, which, therefore, should not be used to interpret the spectra. This will be the case for octahedral complexes with high-spin d^4 and low-spin d^7 configurations, for tetrahedral complexes with d^9, and to a lesser extent, d^8 configurations. When the Jahn–Teller coupling involves orbitals of π symmetry, the much smaller distortions cause only a slight splitting of the orbitally degenerate excited state terms, and the Tanabe–Sugano diagram may be used for interpreting the spectra. The former case holds for octahedral high-spin d^6 complexes, for which the transition is to the excited ^4E_g state, which is strongly influenced by a distortion along the Jahn–Teller mode. Here a splitting into two

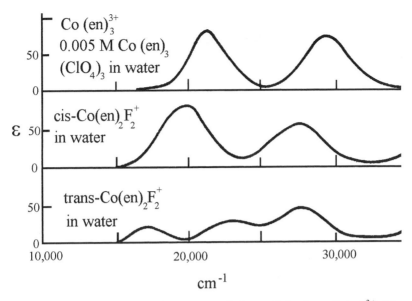

Figure 8.6. Electronic spectra of aqueous solutions of the ions Co(en)$_3^{3+}$ (a), *cis*-Co(en)$_2$F$_2^+$ (b), and *trans*-Co(en)$_2$F$_2^+$ (c).

components is generally observed, even when six identical ligands are bonded to the metal.

8.1.2.2 *Band Splittings and the Rule of Average Environment.*

When a complex involves two different ligands, band splittings can sometimes provide information not only on the symmetry of the complex but also on the relative σ and π bonding properties of the ligands.

The effects of the lowering in symmetry of the ligand field in a mixed ligand complex are best illustrated by an example. The visible spectra of the series Co(en)$_3^{3+}$, *cis*-Co(en)$_2$F$_2^+$, and *trans*-Co(en)$_2$F$_2^+$ are shown in Figure 8.6. The Co(en)$_3^{3+}$ complex has a ligand field of close to O_h symmetry. As expected from the Tanabe–Sugano diagram for d^6 (Fig. 6.14) the electronic spectrum consists of two symmetric bands owing to the transitions $^1A_{1g} \rightarrow {}^1T_{1g}$ and $^1A_{1g} \rightarrow {}^1T_{2g}$. The spectrum of *cis*-Co(en)$_2$F$_2^+$ is rather similar, but both peaks broaden and shift to lower energy by $\sim 2000\,\text{cm}^{-1}$. For *trans*-Co(en)$_2F_2^+$, the lower energy peak is clearly split into two components, and the band intensities are somewhat lower than those of *cis*-Co(en)$_2$F$_2^+$ and Co(en)$_3^{3+}$. This is consistent with the fact that the latter two complexes do not have an inversion center, so that the electronic transitions are not formally forbidden by the Laporte selection rule.

The basic cause of band splittings is the removal of the orbital degeneracy of excited states by the lowering in symmetry of the ligand field. This is illustrated by the energy level diagram in Figure 8.7, which shows the effect of an increasing tetragonal component of the ligand field on the $^1T_{1g}$ and $^1T_{2g}$ terms. The two lower energy bands of *trans*-Co(en)$_2$F$_2^+$ are the result of transitions to

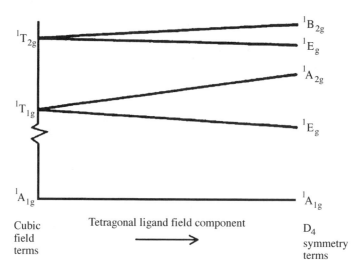

Figure 8.7. Splitting of the excited $^1T_{1g}$ and $^1T_{2g}$ terms of an octahedral Co^{3+} complex under the influence of a tetragonal ligand field.

the ^1E_g and $^1A_{2g}$ components of the $^1T_{1g}$ term. The $^1T_{2g}$ term undergoes a significantly smaller splitting, which is not clearly resolved even in the *trans*-complex. Within the framework of the AOM, the d orbital energies depend on the sum of the effects of the ligands along each axis, and it is the so-called *holohedrized symmetry* that decides the energy levels in a complex (Section 3.2.1). The tetragonal component in the ligand field of the *trans*- complex is, therefore, twice that in the *cis* one. In fact, no clear splitting of the bands can be resolved in the spectrum of *cis*-$Co(en)_2F_2^+$. Although it is not apparent in the solution spectra, the sign of the splitting differs for the *cis* and *trans* isomers, and Figure 8.7 shows the plot for the situation in which the axial ligand field is weaker than the in-plane field, as is the case in the *trans*-$Co(en)_2F_2^+$ ion. The magnitudes of the splittings are related to the difference in σ and π bonding of the axial and in-plane ligands. The band splittings of mixed–ligand complexes thus provide a powerful means of deriving metal–ligand bonding parameters. For a correct analysis, the bands must be properly assigned, and when possible, this is best achieved by measuring single crystal polarized spectra.

Taking the mean position of the split components of the $^1T_{2g}$ transition of *trans*-$Co(en)_2F_2^+$, the band positions are quite similar to those of *cis*-$Co(en)_2F_2^+$: 18,000 and 26,000 cm^{-1}. Following the procedure outlined in Section 8.1.2.1, this implies a ligand field splitting $\Delta \sim 21,000$ cm^{-1} in these complexes. The *rule of average environment* states that in a mixed ligand complex the ligand field splitting is approximately equal to the weighted average of the cubic field ligand fields associated with a complete set of each ligand. For instance, the Δ value of an octahedral complex of the form MA_nB_{6-n} is:

$$\Delta = [n\Delta(MA_6) + (6-n)\Delta(MB_6)]/6 \qquad (8.19)$$

The ligand field splittings of the $Co(en)_3^{3+}$ and CoF_6^{3-} ions are $\Delta = 23,000$ and $13,000\,cm^{-1}$, respectively. Applying the rule of average environment to the $Co(en)_2F_2^+$ complexes suggests:

$$\Delta = [4 \cdot 23,000 + 2 \cdot 13,000]/6 = 20,000\,cm^{-1} \qquad (8.20)$$

This agrees quite well with the observed value of $21,000\,cm^{-1}$, particularly in view of the fact that the CoF_6^{3-} ion is high spin and hence has longer Co-F bonds than do the low-spin $Co(en)_2F_2^+$ complexes.

8.1.3 Band Widths and Shapes

It might be expected that the transition to a term corresponding to a single energy level would give rise to an exceedingly narrow peak. In practice, the converse is true: The bands at room temperature and in solution are mostly rather broad, with half-widths $\delta_{\frac{1}{2}} > \sim 1000\,cm^{-1}$, and are rarely even moderately narrow ($\delta_{\frac{1}{2}} < 100\,cm^{-1}$). Several factors cause this broadening. These are discussed individually below, but it must be remembered that some or all of them may operate together in any particular situation.

8.1.3.1 *Use of Tanabe–Sugano Diagrams to Rationalize Band Widths.* A useful qualitative rationalization of band widths may be obtained by considering that the ligand field splitting parameter Δ does not take a sharply defined value in a complex. Rather, it fluctuates in conjunction with the metal–ligand vibrations. This causes the Δ value in a complex at any instant of time to take the form of a curve (Fig. 8.8). Because electronic transitions take place much more rapidly than does the motion of the ligand atoms during a vibration (the Franck–Condon principle), the band shape of the electronic transition reflects the vertical projection of the distribution of Δ values on the term energies. Ignoring other factors, we see the band width associated with any particular term, therefore, depends on its sensitivity to Δ; i.e., its slope in the Tanabe–Sugano diagram. Figure 8.8 refers to the high-spin d^5 configuration, and it may be seen that the band widths in the spectrum of the $Mn(H_2O)_6^{2+}$ ion (shown later in Fig. 8.15i) do indeed follow the expected pattern.

This interpretation suggests that a sharp spectral line should be found whenever the Tanabe–Sugano diagram shows a horizontal line for an excited term. This situation occurs only for spin-forbidden transitions, which involve 'spin-flips', because here the energy of the transition involves interelectron repulsion but not the ligand field splitting. It must be recognized that this interpretation, although useful in a practical sense, is not rigorous, being based on a classical view of the metal–ligand vibrations. The way in which the band shape, and hence the band width, depends on the potential surfaces of the ground and excited states is discussed more fully in Section 8.1.3.3.

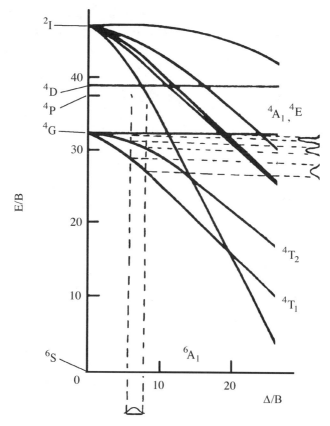

Figure 8.8. Dependence of the widths of 'd-d' transitions on the slopes of the term energies in a Tanabe–Sugano diagram.

8.1.3.2   *Influence of Spin–Orbit Coupling and a Low-Symmetry Ligand Field on Band Widths.* When orbital degeneracy occurs, each cubic field term may be split by spin–orbit coupling and low-symmetry components of the ligand field. For T terms split by spin–orbit coupling (Section 5.5), the energy difference between the highest and lowest states of a term (the overall multiplet width) is a simple multiple of the spin–orbit coupling constant λ (Fig. 5.4). This latter varies between about 100 and $1000\,\text{cm}^{-1}$ for the first-transition series, depending on the metal ion (Table 5.5). The peaks owing to transitions involving at least one T term are, therefore, expected to be distributed over a frequency range of some hundreds of cm^{-1}, and the majority of bands fall into this category.

Ideally, it should be possible to observe spin-orbit coupling as a fine structure to bands. For first-transition series complexes, however, it is rarely possible to resolve such structure in solution and at room temperature. An exception is tetrahedrally coordinated Co^{2+}, for which spin–orbit splittings are observed on the $^4A_2 \rightarrow {}^4T_1(P)$ transition (Section 8.2.2). With certain second- and third-transition series ions, for which the spin–orbit coupling constants are much larger, spin–orbit structure has been observed for solutions of their complexes. In the solid state, and particularly at low temperatures at which vibrational effects are minimized, spin–orbit structure may occasionally be resolved even for complexes of first-transition series ions.

Distortions from cubic symmetry, whether induced by the inequivalence of ligand atoms, crystal packing forces, or the Jahn–Teller effect, mean that it is no longer strictly permissible to label the wavefunctions of the system by the term symbols for cubic symmetry. In a great many cases, however, the low-symmetry ligand field component may be considered to be a perturbation on the main cubic ligand field potential. It is then convenient to retain the cubic nomenclature, because the band positions are at similar energies to those of cubic symmetry. In general, the bands broadened by the above effects always involve T or E terms. Sometimes it is possible to resolve the splittings owing to the departure from cubic symmetry and quantify the low-symmetry component as described in the previous section.

8.1.3.3 *Influence of Vibrational Structure on Band Shape.*

A quantitative description of the shape of an electronic spectral band must include the possibility that a change may occur not only in the electronic but also in the vibrational wavefunction of the complex. Bands owing to such vibronic transitions generally involve the excitation of multiple vibrational quanta. Group theoretical arguments require that this can only occur for a vibration that transforms as the totally symmetric irreducible representation of the point group of the molecule (or a Jahn–Teller active vibration, if the excited state is subject to such a distortion). The relative intensities of the different vibrational components of the band are decided by the overlaps between the ground and excited state vibrational wavefunctions. This in turn depends on the extent to which the excited state is displaced from the ground state in the vibrational mode in question.

A typical vibronic transition is illustrated in Figure 8.9. Here, the excited state is expanded in the totally symmetric metal–ligand 'breathing' vibration, as occurs when an electron is excited to a more antibonding orbital. At low temperature, only the lowest ($v = 0$) vibrational state of the ground electronic state is thermally occupied. The Franck–Condon principle requires that the electronic transition takes place rapidly with respect to nuclear motion; i.e., the nuclear geometry is that of the ground electronic state. The vibronic transitions are, therefore, represented by vertical arrows on the diagram. The band shape depends on the relative intensities of the transitions from the $v = 0$ level of the ground state to the various vibrational levels of the excited state. Typical intensity distributions observed for transitions involving different displacements of the excited state potential surface are shown in Figure 8.10. When the displacement is small, the largest overlap is between the $v = 0$ levels of the two states, and a highly asymmetric band results, with much of the intensity in the

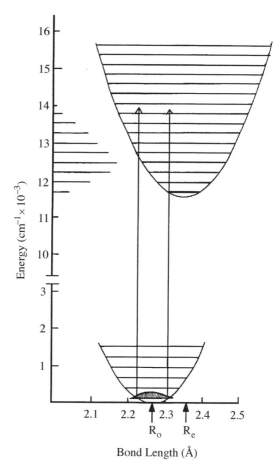

Figure 8.9. The way in which a change in the metal–ligand bond lengths generates a progression in the totally symmetric α_{1g} vibration.

first member. For a large displacement, the intensities first increase and then decrease as the vibrational quantum number of the excited state increases. This leads to the Gaussian line shape which characterizes the bands in many electronic spectra. By measuring the relative intensities of the components, it is possible to determine the change in geometry between the ground and excited electronic states.

Vibrational structure of this kind is resolved only occasionally, and then usually only at low temperature. An exception is the spectrum of the permanganate ion MnO_4^- (Fig. 8.11). When the displacement between the ground and excited states is large, bands usually appear smooth and quite symmetrical. The change in equilibrium bond length between two states depends on the difference in their antibonding energy (Section 7.1.1). A relatively large difference occurs for the spin-allowed transitions of octahedral complexes,

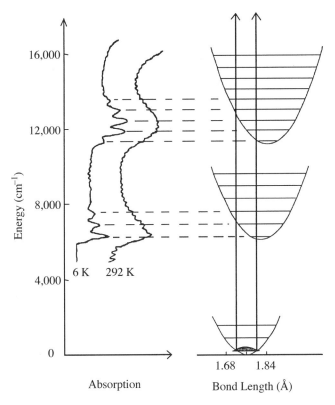

Figure 8.10. Indication of how the relative intensities of the members of a vibrational progression depend on the bond length change δr in the excited state. The spectra shown are for the excitation to the first and second excited states of the linear NiO_2^{3-} ion.

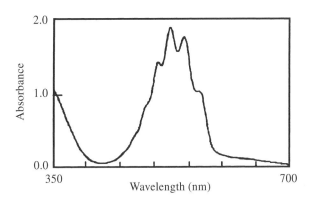

Figure 8.11. Electronic spectrum of a 10^{-3} M aqueous solution of the MnO_4^- ion.

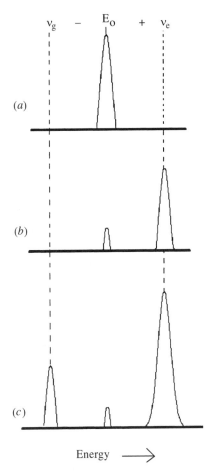

Figure 8.12. Relative energies of the electronic origin E_0 (a) the vibronic origin of a parity-forbidden transition at low temperature (b), and the vibronic origins of a parity-forbidden transition at high temperature (c). ν_g and ν_e are the energies of the intensity-inducing mode in the ground and excited states, respectively.

which typically exhibit symmetrical Gaussian band shapes with a halfwidth of ~ 1500 to $2000\,\text{cm}^{-1}$. For tetrahedral complexes, the ligand field splitting is much smaller, and the relatively small change in bond length accompanying the 'd-d' transition means that the bands may have a distinctly asymmetric shape, rising sharply on the low-energy side and with a pronounced tail at higher energy. As already mentioned, for some spin-forbidden transitions, there may be little change in the antibonding energy of the electrons. The excited state potential surface then lies almost directly over that of the ground state, and the band intensity is concentrated in the first member of the vibrational progression.

This discussion refers to a band built on a single origin — the electronic origin for a parity-allowed transition, as illustrated in Figure 8.12a, or the electronic origin plus one quantum of the u-exciting vibration for a parity-forbidden transition with just one u vibration active in producing intensity, as shown in Figure 8.12b. Very often, more than one origin occurs, which has the effect both of broadening the band and of complicating its underlying structure, often to the extent that this cannot be resolved. Moreover, at room temperature, the $v = 1$ level of an the inducing vibration is usually populated, so that 'hot bands' contribute to the band envelope (Section 8.1.4.1). Compared to the low-temperature spectrum, these are shifted to lower energy by approximately twice the energy of the inducing vibration (Fig. 8.12c), which leads to both a broadening of the band and a change in its shape.

8.1.4   Effect of Temperature on Electronic Bands

8.1.4.1   Parity-Forbidden Bands. The intensities of the parity-forbidden 'd-d' bands of centrosymmetric complexes arise through vibronic interactions (Section 8.1.1.3). Vibrational amplitudes increase as a function of the vibrational quantum number, so that the thermal population of upper vibrational levels of an intensity-inducing mode leads to increased intensity. Parity-forbidden transitions, therefore, usually show a substantial temperature dependence. For the ideal case of a single harmonic vibration of frequency ν, the intensity of a band at a temperature T, f_T, is related to that at absolute zero f_0 by the so-called coth rule given by the expression:

$$f_T = f_0 \coth(\nu/2kT) \tag{8.21}$$

Note that if ν is high enough, as may occur for light, tightly bound ligands, essentially only the ground level is occupied at room temperature and the coth term is close to unity.

Although derived for an idealized situation, Eq. 8.21 describes the intensities of the bands of many 'd-d' spectra quite well. Generally, the occupations assumed in deriving Eq. 8.21 are averages over the populations of several intensity-inducing modes. Of course, significant exceptions do occur. For example, in the presence of a weak Jahn–Teller effect, as may occur for complexes with orbitally triply degenerate ground states, a considerable distortion of the vibrational occupation scheme is likely. The implications of Eq. 8.21 are, however, sufficiently reliable that if the intensity of a band is invariant to temperature, then it is likely that the complex has no center of symmetry.

For simple complexes, the temperature dependence of the intensity of a vibronically allowed transition may be used to estimate the energy of the intensity-inducing vibration. As an example, for square planar copper (II) complexes, the $^2B_{1g} \rightarrow {}^2A_{1g}$ transition in z polarization is allowed through vibrations of β_{2u} symmetry. For monatomic ligands, there is just one vibration of this symmetry. The spectrum of planar $CuCl_4^{2-}$ in z polarization at various

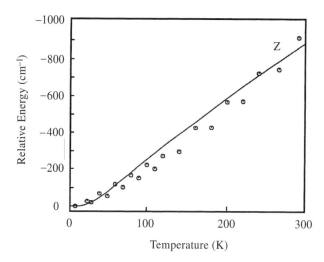

Figure 8.13. Temperature dependence of the intensity of the vibronically allowed $^2B_{1g} \rightarrow {}^2A_{1g}$ transition of a planar $CuCl_4^{2-}$ ion in z polarization. The line represents the behavior predicted by the coth rule when the intensity-inducing vibration has an energy of $64\,cm^{-1}$.

temperatures is shown in Figure 8.4. As expected, the bands show a significant increase in intensity as the temperature is raised. A plot of the intensity of the $^2B_{1g} \rightarrow {}^2A_{1g}$ transition versus temperature matches Eq. 8.21 quite well, assuming an energy of $64\,cm^{-1}$ for the β_{2u} mode (Fig. 8.13). The low energy of this vibration is related to the fact that it carries the complex into the distorted tetrahedral geometry generally observed for Jahn–Teller affected, four-coordinate copper (II) complexes (Section 7.1.2.2).

The vibronic selection rule $\delta v = \pm 1$ for parity-forbidden transitions (Section 8.1.1.1) means that as the $v = 1, 2 \ldots$ etc. levels of the intensity-inducing vibration become thermally populated, transitions from $v = 1, 2 \ldots$ of the ground state to $v = 0, 1 \ldots$ of the excited state occur. Because these develop as the temperature is raised, they are termed hot bands. As shown in Figure 8.12, bands built on these vibronic origins are about $2v$ lower in energy than those observed at low temperature, based on the $v = 0 \rightarrow 1$ vibronic transition. Hot bands may sometimes be resolved in high-resolution studies using single crystals. Their effect, however, is usually simply to broaden a vibronically allowed transition and cause the band maximum to shift to lower energy, as the complex is warmed from near 0 K to room temperature. The thermal population of upper levels of the α_{1g} vibration can also contribute to the broadening, as discussed in the following section for parity-allowed transitions. Because room temperature corresponds to $\sim 200\,cm^{-1}$, intensity-inducing modes of this energy or lower should cause a decrease in energy of the band maximum of up to twice this ($\sim 400\,cm^{-1}$), and experimental studies often show changes of about this magnitude.

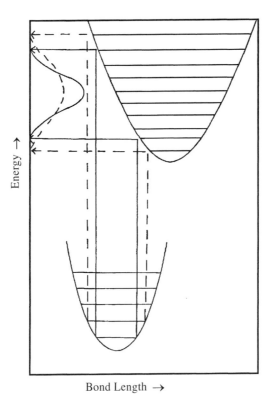

Bond Length $\rightarrow$

Figure 8.14. Mechanism by which the thermal population of upper vibrational states of the α_1 mode broadens electronic bands.

8.1.4.2 *Parity-Allowed Bands.*

Unlike parity-forbidden bands, the intensity of a parity-allowed transition is not expected to be significantly affected by temperature. The basic band shape is generally caused by the excitation of multiple quanta of modes of α_1 symmetry, as shown for the transitions from the $v = 0$ level in Figure 8.9. When the $v = 1, 2 \ldots$ levels of the α_1 mode are thermally populated, the larger amplitudes of these mean that the intensity of the band is distributed over a wider range of vibrational levels of the excited electronic state (Fig. 8.14). An increase in temperature is, therefore, expected to cause a broadening of parity-allowed transitions, although the effect is generally rather less pronounced than for parity-forbidden transitions. This is, first, because there is no contribution from hot bands based on the intensity-inducing vibration and, second, because the α_1 breathing vibration usually has a rather high energy. Because the band area remains essentially constant, the extinction coefficient at the maximum of a parity-allowed transition often actually *decreases* as the temperature is raised. Quantitatively, the half-width of a parity-allowed transition is expected to depend on temperature in a manner analogous to the intensity of a parity-forbidden transition, so an expression exactly analogous to Eq. 8.21, but involving width rather than intensity, may be used to estimate the energy of the α_1 mode producing the band envelope.

8.2   CHARACTERISTIC SPECTRA OF COMPLEXES OF FIRST-ROW TRANSITION IONS

One of the most striking features of the transition-metal ions, particularly those of the first series, is the wide variety of colors that are exhibited by their compounds. It is useful to remember that the colors of the visible spectrum correspond to energy in roughly the following manner:

red	yellow	green	blue	violet
14,000 to 16,000;	$\sim 18,000$;	$\sim 20,000$;	21,000 to 25,000;	$\sim 25,000 \, cm^{-1}$.

An absorption band in one part of the spectrum causes a solution to have the complementary color: A band in the red leads to a blue solution, and so on.

In the following sections, the important characteristics of the spectra of first-row transition-metal ions are illustrated by considering the spectra of their aqueous solutions and, when these have been well characterized, tetrahalo-complexes. The band energies refer to the centers of gravity, rather than to the absorption maxima. Extinction coefficients have been corrected for the solvent background absorption. Except when noted otherwise, the interpretation assumes that the bands are owing simply to transitions between unperturbed cubic field terms. In view of the preceding discussion, the results obtained in this way can be only approximate. Detailed analysis allowing for vibronic and spin–orbit coupling, and low-symmetry ligand field components would lead to slightly different values for the ligand field and interelectronic repulsion parameters.

8.2.1   Hexaaqua Complexes of First-Row Ions

8.2.1.1 d^1. Aqueous solutions of Ti^{3+} salts are purple, owing to absorption of both blue and red light. The spectrum of $Ti(H_2O)_6^{3+}$ (Fig. 8.15a) consists of a broad, weak band ($\varepsilon = 4$, $f = 8 \cdot 10^{-5}$) at $\sim 20,300 \, cm^{-1}$ as a result of the transition $^2T_{2g} \rightarrow {}^2E_g$. It shows signs of double structure, caused by the splitting of the ^2E_g term by some $2,000 \, cm^{-1}$ owing to a low-symmetry ligand field component of Jahn–Teller origin (Section 7.1.2.1). The low band intensity is consistent with the fact that the transition is formally parity forbidden. The spectrum thus gives $\Delta = 20,300 \, cm^{-1}$ for $Ti(H_2O)_6^{3+}$. The absorption in the violet region is owing to the tail of a charge transfer band in the ultraviolet part of the spectrum.

The vanadyl ion, $VO(H_2O)_5^{2+}$, is responsible for the bright blue color often obtained on dissolving vanadium (IV) compounds in water. The visible spectrum consists of two bands at 13,000 and 16,000 cm^{-1} ($\varepsilon = 16$ and 7.5, respectively). The bond length in the VO^{2+} unit is very short (1.69 Å), so that the ligand field is

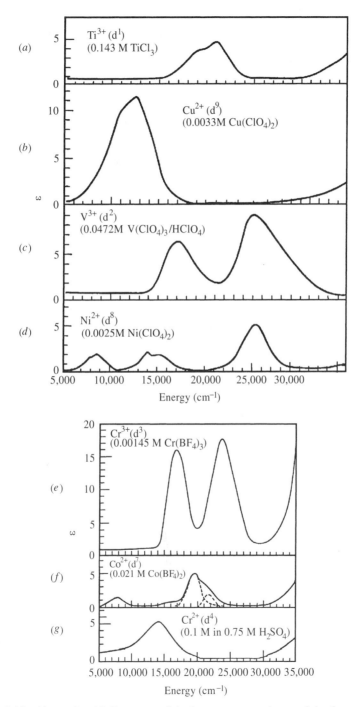

Figure 8.15. Absorption 'd-d' spectra of the hexaaqua complexes of the first-transition series measured in aqueous solution at room temperature.

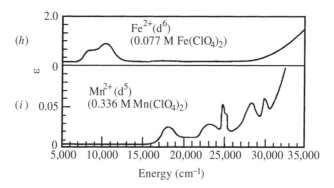

Figure 8.15 (*continued.*)

highly asymmetric. We assume square pyramidal symmetry C_{4v}. The lowest energy band is likely to be owing to the $^2B_2 \rightarrow {}^2E$ transition; the one at $16{,}000 \text{ cm}^{-1}$ is the result of $^2B_2 \rightarrow {}^2B_1$. In the noncentrosymmetric C_{4v} point group of the complex, the former transition is allowed in xy polarization, whereas the latter is Laporte forbidden. The first band is indeed somewhat more intense than the second and than the band of $Ti(H_2O)_6^{3+}$. The band is still relatively weak, however, possibly because the high charge on the metal causes the d orbitals to contract, reducing their overlap with the ligand orbitals, a requirement for significant 'd-d' intensity (Section 8.1.1). It seems likely that the third 'd-d' transition, $^2B_2 \rightarrow {}^2A_1$ is obscured under the intense charge transfer absorption commencing at $\sim 25{,}000 \text{ cm}^{-1}$. The high energies of the $^2B_2 \rightarrow {}^2E$ and $^2B_2 \rightarrow {}^2A_1$ transitions reflect the strong metal–ligand interaction in the VO^{2+} unit. To a first approximation, the $^2B_2 \rightarrow {}^2B_1$ transition occurs at Δ for the in-plane H_2O ligands. The energy, $16{,}000 \text{ cm}^{-1}$, is consistent with this if it is considered that the formal charge on the metal is substantially reduced by the strong interaction with the apical oxygen atom; i.e., the in-plane ligands experience the metal more as a VO^{2+} group than as a V^{4+} ion.

8.2.1.2 d^2. Aqueous solutions of V^{3+} salts are green. After a small correction for hydrolysis, the spectrum consists of broad weak bands as follows (Fig. 8.15c): $17{,}200 \text{ cm}^{-1}$ ($\varepsilon = 6; f = 1.1 \cdot 10^{-4}$), and $25{,}600 \text{ cm}^{-1}$ ($\varepsilon = 8; f = 2.4 \cdot 10^{-4}$). Very weak bands ($f \sim 10^{-7}$) also occur between $20{,}000$ and $30{,}000 \text{ cm}^{-1}$. The two stronger peaks are the result of spin-allowed transitions, whereas the very weak bands, which will not be discussed further here, are from spin-forbidden transitions to excited singlet terms. We examine the spectrum in terms of the appropriate Tanabe–Sugano diagram (Fig. 6.10). The $17{,}200 \text{ cm}^{-1}$ band is caused by the transition $^3T_{1g}(F) \rightarrow {}^3T_{2g}(F)$ and that at $25{,}600 \text{ cm}^{-1}$ by $^3T_{1g}(F) \rightarrow {}^3T_{1g}(P)$. By the procedure outlined in Section 8.1.2.1, a fit of the ratio:

$$[^3T_{1g}(F) \rightarrow {}^3T_{1g}(P)]/[^3T_{1g}(F) \rightarrow {}^3T_{2g}(F)] = 25{,}600/17{,}200 = 1.49 \quad (8.22)$$

to Figure 6.10 yields the estimate $\Delta/B \approx 0.28$. The points of intersection of the vertical line for $\Delta/B = 0.28$ and the lines for the $^3T_{2g}(F)$ and $^3T_{1g}(P)$ terms in Figure 6.10 correspond to the values of $E/B = 25.9$ and $E/B = 38.7$, respectively. From either of these results and the energy of the corresponding transition from the ground term, it is deduced that $B = 665\,\mathrm{cm}^{-1}$. Hence, for $V(H_2O)_6^{3+}$, $\Delta = 18,600\,\mathrm{cm}^{-1}$ and B is reduced to $\sim 75\%$ of the free ion value of $886\,\mathrm{cm}^{-1}$. The diagram suggests that the transition $^3T_{1g}(F) \rightarrow {}^3A_{2g}(F)$ should lie at $\sim 36,000\,\mathrm{cm}^{-1}$. A band corresponding to this transition has not been detected in aqueous solution. The failure is not unexpected, because the band should be very weak. Not only is it parity-forbidden, but to a first approximation it corresponds to the simultaneous excitation of two electrons. Moreover, charge transfer bands commence in about the same region in the spectrum. In the solid state, for V^{3+} in Al_2O_3, for which charge transfer occurs at higher energy, a weak band at about the right position has been observed. The oxygen ligand atoms in the Al_2O_3 structure produce about the same value of Δ as do the water molecules in $V(H_2O)_6^{3+}$.

8.2.1.3 d^3. In water, Cr^{3+} salts give light green solutions. The color is the result of absorption in the yellow and blue parts of the spectrum. Figure 8.15e shows bands at $17,000\,\mathrm{cm}^{-1}$ $(\varepsilon = 14, f = 1.9 \cdot 10^{-4})$, and $24,000\,\mathrm{cm}^{-1}$ $(\varepsilon = 15, f = 2.6 \cdot 10^{-4})$. There is also a rather weak band at $37,000\,\mathrm{cm}^{-1}$ occurring as a shoulder on the intense charge transfer band at higher energy. Weaker bands at about $15,000\,\mathrm{cm}^{-1}$ are the result of spin-forbidden transitions.

Following the procedure discussed in Section 8.1.2.1, the obvious assignment of the three stronger bands is to the transitions from the $^4A_{2g}$ ground term to the three excited quartet terms:

$$^4A_{2g}(F) \rightarrow {}^4T_{2g}(F)(= \Delta)\, 17,000\,\mathrm{cm}^{-1}$$
$$^4A_{2g}(F) \rightarrow {}^4T_{1g}(F) \qquad 24,000\,\mathrm{cm}^{-1}$$
$$^4A_{2g}(F) \rightarrow {}^4T_{1g}(P) \qquad 37,000\,\mathrm{cm}^{-1}$$

As described earlier, B is found by fitting the ratio of two band energies. Choosing the more pronounced, lower two bands, gives the estimate $\Delta/B \approx 24.5$ and yields the parameters: $B = 695\,\mathrm{cm}^{-1}$ and $\Delta = 17,000\,\mathrm{cm}^{-1}$.

For $Cr(H_2O)_6^{3+}$ B is, therefore, reduced to about 75% of the free ion value for Cr^{3+} $(933\,\mathrm{cm}^{-1})$. Using these values, it is predicted that the $^4T_{1g}$ (P) term lies at $37,000\,\mathrm{cm}^{-1}$, in agreement with experiment. Here the internal consistency of the assignment is good.

8.2.1.4 d^4. Aqueous solutions of Cr^{2+} are pale blue, and the spectrum (Fig. 8.15g) consists of a broad, weak band at $14,100\,\mathrm{cm}^{-1}$ $(\varepsilon = 4.2, f = 1 \cdot 10^{-4})$. The $Cr(H_2O)_6^{2+}$ ion has the spin-quintet ground term. The ^5E_g ground state expected for a regular octahedral complex undergoes a strong Jahn–Teller

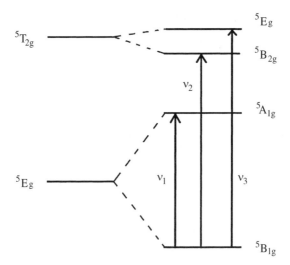

Figure 8.16. Energy levels and electronic transitions in a tetragonally elongated octahedral Cr^{2+} complex.

distortion, producing the energy splittings shown in Figure 8.16. The Tanabe–Sugano diagram, therefore, cannot be used to describe the energy levels in the complex. Studies of single crystals at low temperature have shown that the transition between the split components of the ^5E_g level, $^5B_{1g} \rightarrow {}^5A_{1g}$, occurs at $\sim 10,000\,cm^{-1}$, presumably contributing to the low-energy tail of the broad band in aqueous solution. The transition from the $^5B_{1g}$ ground state to the components of the $^5T_{2g}$ parent term, ν_2 and ν_3 in Figure 8.16 are unresolved and together give rise to the band at $\sim 14,000\,cm^{-1}$. This represents the Δ splitting corresponding to the four short Cr-O bond distances. Making the assumption that the Jahn–Teller effect produces a symmetrical splitting of the ^5E_g state, we can obtain the d orbital splitting of the parent octahedral complex from the energy of the $^5B_{1g} \rightarrow {}^5T_{2g}$ transition minus half the energy of the $^5B_{1g} \rightarrow {}^5A_{1g}$ transition. This yields the estimate $\Delta \approx (14,000-5,000) \approx 9,000\,cm^{-1}$ for the undistorted $Cr(H_2O)_6^{2+}$ ion, a value quite similar to the undistorted hexaaqua complexes of other divalent transition ions.

An interesting feature revealed by low-temperature spectra of crystals containing the $Cr(H_2O)_6^{2+}$ ion is the presence of weak peaks at $\sim 18,000\,cm^{-1}$ owing to the coupling of internal vibrations of the water molecules with the 'd-d' transitions. Similar structure has been observed in the low-temperature spectra of other hydrate complexes.

8.2.1.5 d^5. The aqueous solutions containing the Mn^{2+} ion are a very pale pink. The absorption spectrum shows a series of very weak peaks, some rather narrow, stretching through the green and blue portions of the visible spectrum into the ultraviolet (Fig. 8.15i). The $Mn(H_2O)_6^{2+}$ ion has a spin sextet ground term, and because the d^5 configuration has only one level of this multiplicity

($^6A_{1g}$), there can be no spin-allowed 'd-d' transitions. The spectrum is assigned using the diagram in Figure 6.13. The first pair of bands are assigned to transitions to the $^4T_{1g}(G)$, followed by the $^4T_{2g}(G)$ terms. The success of the prediction of the remaining bands confirms the correctness of the assignment.

The ratio of the energies of the first two transitions fits the Tanabe–Sugano diagram at $\Delta/B = 11$. This suggests that the first transition $^6A_{1g} \rightarrow {}^4T_{1g}(G)$ occurs at $E/B \approx 24.0$. Because it is observed at 18,600 cm^{-1}, it implies that for Mn(H$_2$O)$_6^{2+}$: $B = 770$ cm^{-1} and $\Delta = 8,500$ cm^{-1}. B is, therefore, reduced to about 90% of the free ion value for Mn^{2+} (859 cm^{-1}). A check of the internal consistency of the assignment is the observation that the transitions to the $^4E_g(G)$ and $^4A_{1g}(G)$ terms, which are predicted to be degenerate (Fig. 6.13), are separated by only about 250 cm^{-1} in the spectrum. Were these bands not considerably narrower than most observed for transition metal complexes, it would not be possible to resolve the splitting.

The sharpness of some of the peaks of the Mn(H$_2$O)$_6^{2+}$ spectrum, particularly the two mentioned in the previous paragraph, was discussed in Section 8.1.3.1 and 8.1.3.3. The energies of the $^4E_g(G)$ and $^4A_{1g}(G)$ terms alter little relative to the $^6A_{1g}$ ground term as Δ changes, so that the bands corresponding to transitions to these two terms are not appreciably broadened by vibronic coupling. Moreover, neither excited term is split by spin–orbit coupling.

The ferric ion is subject to hydrolysis in aqueous solution, and the yellow color commonly observed for aqueous solutions of ferric salts is not the result of ligand field bands of Fe(H$_2$O)$_6^{3+}$ but rather of charge transfer bands involving hydroxo-complexes. A complete analysis of the ligand field spectrum of the ion has not proved possible, but an estimate of $\Delta \sim 14,000$ cm^{-1} has been obtained.

8.2.1.6 d^6. In aqueous solution Fe^{2+} is pale green, which is the result of a weak absorption band in the red, at 10,000 cm^{-1} ($\varepsilon = 1.1, f = 1.6 \cdot 10^{-5}$) (Fig. 8.15h). The ground term is a spin quintet, so that the left side of the diagram in Figure 6.14 is involved. The band is assigned to the spin-allowed transition $^5T_{2g} \rightarrow {}^5E_g$, which corresponds to Δ. This, therefore, equals $\sim 10,000$ cm^{-1} for the Fe(H$_2$O)$_6^{2+}$ ion. The band shows a definite doublet structure, with a peak separation of about 2,000 cm^{-1}, very similar to that observed in the spectrum of the Ti(H$_2$O)$_6^{3+}$ ion. In both complexes, the splitting is owing to a Jahn–Teller distortion associated with the orbital degeneracy of a T_{2g} ground term, quintet or doublet respectively , (Section 7.1.2.1).

8.2.1.7 d^7. Aqueous solutions of Co^{2+} are pink. As Figure 8.15f shows, there is a weak peak at 8,100 cm^{-1} ($\varepsilon = 1.3, f = 1.4 \cdot 10^{-5}$), with a second band at $\sim 20,000$ cm^{-1}. The latter absorption is split into two components: one at 19,400 cm^{-1} ($\varepsilon = 4.8, f = 5.4 \cdot 10^{-5}$) and a weaker shoulder at 21,600 cm^{-1}. The Co(H$_2$O)$_6^{2+}$ ion has a spin-quartet ground term, so the spectrum must be fitted to the left side of the Tanabe–Sugano diagram (Fig. 6.15). There are three excited terms of the same multiplicity as the ground term. The first peak must correspond to the transition to the lowest excited quartet term; i.e.,

$^4T_{1g}(F) \rightarrow {}^4T_{2g}(F)$. A reasonable value of B can be obtained only if the peak at 19,400 cm^{-1} is assigned to the third, rather than the second excited term; i.e., $^4T_{1g}(F) \rightarrow {}^4T_{1g}(P)$ rather than $^4T_{1g}(F) \rightarrow {}^4A_{2g}(F)$. A fit of the observed ratio of the transition energies, 2.395, to Figure 6.15 thus yields the value B = 850 cm^{-1}, corresponding to 86% of the free ion value of 989 cm^{-1}. Substitution of this into the transition energies in Figure 6.15 yields the estimate $\Delta = 9200$ cm^{-1}, a value very similar to those derived for the hydrate complexes of other divalent transition ions. This fit predicts that the transition $^4T_{1g}(F) \rightarrow {}^4A_{2g}(F)$ should lie at $\sim 16,000$ cm^{-1}, but no peak is observed in this region. The transition, however, is expected to be weak, because to a first approximation it corresponds to a two-electron jump. High-resolution studies of single crystals of similar complexes at low temperature have in fact revealed a band assigned to this transition at approximately this energy. The shoulder at $\sim 21,600$ cm^{-1} is owing to a transition to one of the spin-doublet levels expected in this region, which gains intensity via spin–orbit coupling because of its proximity to the $^4T_{1g}(P)$ term.

8.2.1.8 d^8. Aqueous solutions of Ni^{2+} are a light green, which is caused by weak bands in the red and blue portions of the visible spectrum (Fig. 8.15d). The major band maxima occur at 8,700 ($\varepsilon = 1.6$, $f = 1.8 \cdot 10^{-5}$), 14,500 ($\varepsilon = 2.0$, $f = 3.0 \cdot 10^{-5}$), and 25,300 cm^{-1} ($\varepsilon = 4.6$, $f = 7.0 \cdot 10^{-5}$). These are associated with the spin-allowed transitions from the $^3A_{2g}$ ground term to the three excited triplet terms (Fig. 6.16). The first spin-allowed transition, to the $^3T_{2g}$ state, occurs at precisely Δ, which thus has the value 8,700 cm^{-1}. Using this value, a reasonably good fit for the other transition energies may be obtained if the peaks at 14,500 and 25,300 cm^{-1} are assigned to transitions to the $^3T_{1g}(F)$ and $^3T_{1g}(P)$ states, respectively. The ratio of the energy of the third to the first transition is 2.91, which occurs at $\Delta / B \approx 9.7$ on the Tanabe–Sugano plot. The corresponding intercept for the first transition occurs at 9.7B = 8,700 cm^{-1} on the vertical axis. Hence B = 900 cm^{-1} for the Racah parameter in the Ni(H$_2$O)$_6^{2+}$ ion, which represents a reduction to $\sim 86\%$ of the free ion value. Because three spin-allowed transitions are observed, the self-consistency of the interpretation may be tested by comparing the experimental ratio of the first and second transitions. The experimental ratio 1.67 is quite close to the value of 1.65 obtained from the Tanabe–Sugano diagram, showing that in this case also the fit is self-consistent.

The marked shoulder at $\sim 14,000$ cm^{-1} is the result of the spin-forbidden transition to the ^1E_g term. This is relatively intense because of its close proximity to a spin-allowed transition. The intensity borrowing mechanism is relatively effective because of the rather high spin–orbit coupling constant of the Ni^{2+} ion (600 cm^{-1} for the free ion).

8.2.1.9 d^9. In aqueous solution Cu^{2+} is a characteristicly pale blue, owing to the absorption of red light. The spectrum consists of a broad weak band centered at $\sim 12,000$ cm^{-1} ($\varepsilon = 11$, $f = 2.3 \cdot 10^{-4}$) (Fig. 8.15b). As for the d^4 Cr(H$_2$O)$_6^{2+}$ ion (Section 8.2.1.4), the ^2E_g ground state expected for a regular octahedral Cu(H$_2$O)$_6^{2+}$ complex is subject to a strong Jahn–Teller distortion. Polarized single crystal spectra obtained from compounds containing the deuterated complex (to avoid the interference of infrared overtones from the O-H vibrations), have shown that the transition between the split components of the $^2E_{2g}$ level, $^2B_g \rightarrow {}^2A_g$, occurs at $\sim 7000\,cm^{-1}$. The transitions from the $^2B_{2g}$ ground state to the $^2B_{1g}$ and ^2E_g components of the $^2T_{2g}$ parent term give rise to the band at $\sim 12{,}000\,cm^{-1}$, which corresponds to the Δ_{oct} value associated with the four short Cu-OH$_2$ bonds of the distorted complex. As discussed for the analogous Cr^{2+} complex, the ligand field splitting of the undistorted $Cu(H_2O)_6^{2+}$ complex may be estimated approximately as the energy of the $^2B_{2g} \rightarrow {}^2T_{2g}$ transition minus half the energy of the $^2B_{2g} \rightarrow {}^2A_g$ transition. Thus $\Delta \approx 8{,}500\,cm^{-1}$, which is quite similar to the values observed for the corresponding Ni^{2+} and Co^{2+} hexaaqua complexes.

8.2.2 Tetrahedral and Planar Complexes of First-Row Transition Ions

Because, unlike an octahedral complex, a tetrahedral complex does not have an inversion center, the 'd-d' spectra are formally Laporte allowed (Section 8.1). Consequently, the 'd-d' spectra of tetrahedral complexes are always considerably more intense than those of the octahedral complexes formed by similar ligands, typically by a factor between 10 and 50. Except for d^1 and d^9 metal ions, for which the energy levels are unaffected by interelectron repulsion, the transition energies of a tetrahedral complex of a metal ion with n d electrons are conventionally interpreted using the Tanabe–Sugano diagram for an octahedral complex of a metal ion with d^{10-n} electrons (Section 8.1.2.1). Moreover, the ligand-field splitting parameter Δ is typically about half the value of the corresponding octahedral complex agreeing with the theoretical relationship $\Delta_{tet} = -(4/9)\Delta_{oct}$. Typical spectra of the more stable tetrahedral complexes of first-row transition ions are now discussed briefly.

The spectra of the tetrahedral complexes $MnCl_4^{2-}$ and $MnBr_4^{2-}$ are quite similar to that of the $Mn(H_2O)_6^{2+}$ ion, except for the higher intensities (Fig. 8.17). This is because the energies of the transitions in the visible region are decided largely by interelectron repulsion effects and are relatively little affected by the strength of the ligand field. The peak positions of the $MnCl_4^{2-}$ ion correspond to a rather small ligand field splitting, $\Delta_{tet} \approx -3300\,cm^{-1}$, in agreement with the expected ratio of $\sim 4/9$ that of the corresponding octahedral complex ($\Delta_{oct} \approx 7{,}500\,cm^{-1}$ for the $MnCl_6^{4-}$ chromaphore in $MnCl_2$).

The transition $^5E \rightarrow {}^5T_2$ has been observed at $\sim 4000\,cm^{-1}$ in the electronic spectrum of the tetrahedral $FeCl_4^{2-}$ complex. From Figure 6.12, it may be seen that the transition occurs at precisely Δ (note that the diagram for octahedral d^4 is appropriate for a tetrahedral d^6 complex). In agreement with theory, the value is about half the value of $7600\,cm^{-1}$ observed for the corresponding octahedral complex.

The tetrahedral $CoCl_4^{2-}$ complex is characteristicly bright blue. It is particularly stable and is even formed in aqueous solution containing high concentrations of chloride ion. The spectrum, which is shown in Figure 8.18,

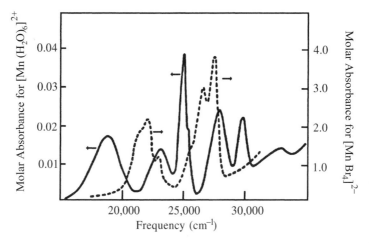

Figure 8.17. Comparison of the electronic spectrum of the tetrahedral $MnBr_4^{2-}$ ion (*dotted line*) with that of the octahedral $Mn(H_2O)_6^{2+}$ ion (*solid line*).

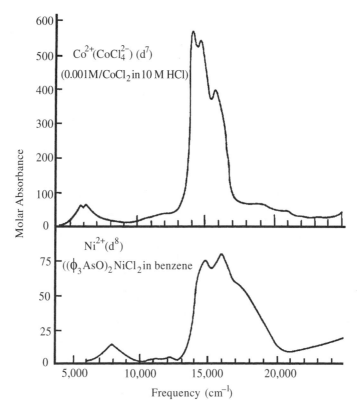

Figure 8.18. Electronic 'd-d' spectra of the tetrahedral complexes $CoCl_4^{2-}$ and $Ni[(C_6H_5)_3AsO]_2Cl_2$.

exhibits bands at $5,800\,cm^{-1}$ ($\varepsilon = 64$, $f=4.5 \cdot 10^{-4}$) and $15,000\,cm^{-1}$ ($\varepsilon = 600$, $f=4.8 \cdot 10^{-3}$), with further weak absorptions between 17,000 and $23,000\,cm^{-1}$ ($\varepsilon \approx 5$, $f \approx 10^{-4}$). The spectrum is interpreted using Figure 6.11, which suggests transitions occur from the $^4A_2$ ground term to the excited states $^4T_2$, $^4T_1(F)$, and $^4T_1(P)$ terms in order of increasing energy. If the band at $5,800\,cm^{-1}$ is assigned to the lowest energy transition, it is impossible to fit the ratio of the band energies to the energy level diagram. If, however, we assign this band to the $^4A_2 \rightarrow {}^4T_1(F)$ transition and the one at $15,000\,cm^{-1}$ to $^4A_2 \rightarrow {}^4T_1(P)$, we obtain an energy ratio that fits the one in the diagram when $\Delta/B = 0.44$. This gives the values $B = 730\,cm^{-1}$ and $\Delta = 3,200\,cm^{-1}$ for the tetrahedral $CoCl_4^{2-}$ ion. This assignment predicts that the $^4A_2 \rightarrow {}^4T_2$ transition should occur in the infrared region at $\sim 3,200\,cm^{-1}$, and a band has indeed been observed in this region in studies of the single crystal spectrum of this complex ion. The pronounced structure of the band at $15,000\,cm^{-1}$ is caused by spin–orbit splitting of the $^4T_1(P)$ term. The bands at $\sim 20,000\,cm^{-1}$ are the result of spin-forbidden transitions. The high intensity of these, comparable to that of spin-allowed transitions in octahedral complexes, is owing to the fact that the transitions are parity allowed. Because the intensity is borrowed via spin–orbit coupling from the neighboring spin-allowed $^4A_2 \rightarrow {}^4T_1(P)$ transition, it mirrors the high intensity of this band. The rather large reduction in B, 74%, compared with the free ion value ($986\,cm^{-1}$), suggests that the $CoCl_4^{2-}$ complex is rather covalent. The ligand field splitting is ~ 0.46 of the magnitude observed for octahedral coordination of chloride to Co^{2+}, in good agreement with the theoretical prediction.

The spectra of tetrahedral Ni^{2+} complexes are difficult to interpret quantitatively, because the $^3T_1(F)$ ground term is subject to a Jahn–Teller distortion, although it is likely that this is relatively small for the tetrahalide complexes (Section 7.1.2.2). The spectrum of a typical tetrahedral complex $Ni[(C_6H_5)_3AsO]_2Br_2$, is shown in Figure 8.18 and consists of bands at: $7,800\,cm^{-1}$ ($\varepsilon = 12$, $f=8 \cdot 10^{-5}$), $14,600\,cm^{-1}$ ($\varepsilon = 70$, $f=6 \cdot 10^{-4}$), and $16,400\,cm^{-1}$ ($\varepsilon = 71$, $f= 10^{-3}$). It seems likely that the two higher energy bands derive from the same excited term, split by spin–orbit coupling and the low-symmetry components of the ligand field associated with the Jahn–Teller distortion. Three spin-allowed transitions are expected for a tetrahedral Ni^{2+} complex (Fig. 6.10). Assuming that, as for $CoCl_4^{2-}$, the first band occurs too low in energy to be observed suggests the assignment $^3T_1(F) \rightarrow {}^3A_2$ at $7,800\,cm^{-1}$, and the multicomponent band centered at $\sim 16,000\,cm^{-1}$ is owing to transitions to the split components of $^3T_1(P)$. The observed energy ratio fits that of the Tanabe–Sugano diagram at $\Delta/B = 0.50$, which yields values of $B = 850\,cm^{-1}$ and $\Delta = 4,200\,cm^{-1}$ for the tetrahedral complex. The Racah parameter is reduced to about 82% of the free ion value. As expected, the ligand field splitting is quite small, about half that for the $Ni(H_2O)_6^{2+}$ complex.

Planar Ni^{2+} complexes are almost invariably low-spin, and are generally yellow or red, because the 'd-d' transitions occur in the green and blue part of the spectrum. Often just a single broad, asymmetric band is observed in the region from 20,000 to 24,000 cm^{-1}, with an extinction coefficient, if the complex is centrosymmetric, of $\varepsilon \sim 50$. This is because of the superposition of the bands associated with excitation of an electron from each of the filled d orbitals to the empty, antibonding $d_{x^2-y^2}$ orbital. To a first approximation, the energy corresponds to Δ_{oct} (the ligand field the ligands would produce in the corresponding octahedral complex with identical bond lengths). The value is about twice that for octahedral Ni^{2+} complexes with similar ligands. This reflects the much shorter bond lengths in planar compared with octahedral Ni^{2+} complexes. The relatively high extinction coefficients are probably associated with the rather covalent nature of these complexes.

Four-coordinate Cu^{2+} complexes are subject to strong Jahn–Teller coupling and almost always adopt geometries distorted substantially away from a regular tetrahedron. The most common stereochemistry is a compressed tetrahedron, and the single crystal polarized spectrum of a compound containing such a complex is shown in Figure 8.3. The spectrum of the same complex, but in a lattice that stabilizes a planar geometry, is shown in Figure 8.4 (see Section 8.1.1.4). As expected, the spectrum of the noncentrosymmetric distorted tetrahedral complex is considerably more intense than that of the planar complex, and the bands occur at considerably lower energy.

8.3   TYPICAL SPECTRA OF SECOND- AND THIRD-ROW TRANSITION ELEMENTS

On going from the first- to the second- to the third-transition series, the interelectron repulsion parameters decrease, and the spin–orbit coupling constants increase progressively. This means that the description of terms using separate spin and orbital angular momentum quantum numbers, the basis of the Russell-Saunders scheme, becomes progressively poorer. As far as the interpretation of electronic spectra is concerned, the terms become closer in energy and are split to a greater extent by spin–orbit coupling. The distinction between spin-allowed and spin-forbidden transitions becomes less pronounced. Moreover, the ligand field splitting increases substantially when going from each series to the next, so that the 'd-d' transitions are often obscured under charge transfer transitions. This being said, it is still sometimes possible to use the appropriate Tanabe–Sugano diagram to interpret the spectra of second-row element complexes, although it must be remembered that these diagrams can give only an approximate guide to the energy levels for the heavier transition elements.

As an example, the d^3 MoCl$_6^{3-}$ ion in aqueous solution shows spin-allowed bands at 19,000 and 24,000 cm^{-1}; using the procedure described for Cr^{3+}, Figure 6.11 yields the parameters $B = 480$ cm^{-1} and $\Delta = 19,000$ cm^{-1}. Similarly, the Rh(NH$_3$)$_6^{3+}$ ion has spin-allowed bands at 32,700 and 39,100 cm^{-1}; using Figure 6.14 as described for the spectra of Co^{3+} complexes gives the parameters $B = \sim 400$ cm^{-1} and $\Delta = 34,000$ cm^{-1}. In both cases, the

Racah parameter is considerably lower and the ligand field splitting parameter Δ higher than is observed for the corresponding complex of the first row metal ion.

Whereas low-spin, planar complexes of Ni^{2+} are observed only for high-field ligands, almost all the complexes of the second- and third-row ions Pd^{2+} and Pt^{2+} are of this type. They generally exhibit several moderately intense bands in the region from 25,000 to 35,000 cm^{-1} owing to spin-allowed 'd-d' transitions; weaker though still pronounced peaks owing to the corresponding spin-forbidden transitions occur some 10,000 cm^{-1} lower in energy.

8.4   THE SPECTROCHEMICAL AND NEPHELAUXETIC SERIES

8.4.1   The Spectrochemical Series

When the ligand field splittings observed for a variety of ligands coordinated to different metal ions are compared, various trends can be observed. Table 8.3 lists the Δ values obtained from the spectra of a range of octahedral and tetrahedral complexes. It is found that when the ligand field splitting is compared for a range of different ligands, these may be arranged in the *spectrochemical series*, in

TABLE 8.3. Ligand Field Splitting Parameter Δ ($cm^{-1} \cdot 10^{-3}$) Deduced from the Electronic Spectra of Octahedral and Tetrahedral Complexes of Different Ligands Bonded to Various Metal Ions.

Number of d Electrons	Metal Ion	6F	6Cl	6Br	6H$_2$O	3Ox	6NH$_3$	3en	6CN	4Cl
1	Ti^{3+}	—	13	—	20	—	17?	—	—	—
2	V^{3+}	16	13	—	19	18	18?	—	—	—
3	Cr^{3+}	—	13	—	17	17	22	22	26	—
	Re^{4+}	32	33	—	—	—	—	—	—	—
4	Cr^{2+}	—	13[a]	—	14[a]	—	—	18[a]	—	—
	Mn^{3+}	22[a]	20[a]	—	21[a]	—	—	—	30[b]	—
5	Mn^{2+}	8	8	—	8	—	—	10	—	2
	Fe^{3+}	14	—	—	14	14	—	—	35[b]	5
6	Fe^{2+}	—	—	—	10	—	—	—	33[b]	4
	Co^{3+}	13[c]	—	—	19	18	24	24	34	—
	Rh^{3+}	—	20	19	27	26	34	35	—	—
	Ir^{3+}	—	25	23	—	—	—	41	—	—
	Pt^{4+}	33	29	25	—	—	—	—	—	—
7	Co^{2+}	—	—	—	10	11	11	11	—	3.2
8	Ni^{2+}	—	7	6	9	—	11	12	—	3.5
9	Cu^{2+}	—	—	—	13[a]	—	15[a]	16[a]	—	—

[a] Value refers to the in-plane ligands of a tetragonally elongated octahedral complex.
[b] Low-spin complex (the other values refer to high-spin complexes for this metal ion).
[c] High-spin complex (the other values refer to low-spin complexes for this metal ion).

which the order is essentially independent of the metal ion. In this comparison, the oxidation state, coordination geometry, and spin state of the metal must remain constant, because all strongly influence the ligand field splitting.

The order of increasing Δ values for some important ligands (omitting charges on ions) is:

$$I < Br < NCS^* < F < OH < RCOO < Ox < ONO^* < H_2O < SCN^*$$
$$< glycine \sim edta < py \sim NH_3 < en < bipy < o - phenanthrolene$$
$$< N^*O_2 < CN,$$

where the asterisk specifies a donor atom. Here, a factor of two or three separates the Δ values of the lowest and highest field ligands. It is important to note that in this context, a high-field ligand is simply one that causes a relatively large splitting of the d orbitals. Because the metal–ligand bond strength is dominated by the interaction of the ligand orbitals with the metal s and p orbitals (Section 3.1.1), high-field ligands may not necessarily involve metal–ligand bonds that are particularly strong in a thermodynamic sense. The spectrochemical series was derived empirically; however, one factor that does influence the position of a ligand in the spectrochemical series is its relative σ and π bonding capacities. As discussed in Section 3.2.1 for an octahedral complex with cylindrical π bonding about the metal–ligand bond axis:

$$\Delta = 3e_\sigma - 4e_\pi \tag{8.23}$$

A ligand is, therefore, high in the spectrochemical series when it is a strong σ donor and/or a weak π donor or π acceptor. In agreement with this, saturated amines such as NH_3, for which π interactions are expected to be negligible, lie at the center of the series. Ligands with filled orbitals of π symmetry, which are likely to be strong π donors, are low in the series. Ligands high in the series almost invariably have low-lying vacant orbitals of π symmetry, so that they may be expected to show considerable π acceptor character. The influence that the π bonding capacity of a ligand has on its position in the spectrochemical series is well illustrated by the linkage isomerism exhibited by the nitrite ion NO_2^- as a ligand. For this ligand, the negative charge resides in a nonbonding orbital of π symmetry localized on the two oxygen atoms. When it bonds via oxygen (nitrito-coordination) it is expected to be a strong π donor, and indeed the nitrito group is a rather weak ligand lying close to acetate in the spectrochemical series. The empty antibonding orbitals of the nitrite π system are positioned so that they overlap well with the t_{2g} orbitals of the metal when the ligand bonds via nitrogen (nitro-coordination) so that there it exhibits π acceptor character. In agreement with this, the nitro group lies high in the spectrochemical series and is slightly above conjugated aromatic amines such as o-phenanthrolene.

From the band energies of low-symmetry complexes, it has become possible to measure the σ and π bonding parameters of ligands independently. These are

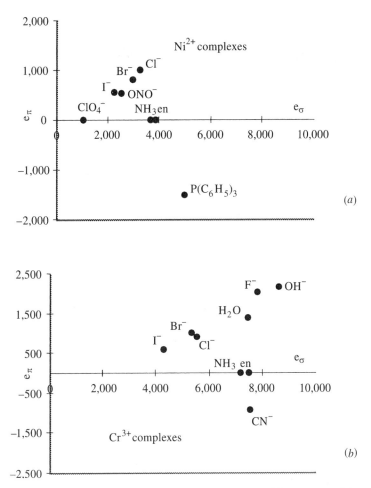

Figure 8.19. Two-dimensional spectrochemical series for Ni^{2+} (*a*) and Cr^{3+} (*b*) ions.

usually measured in terms of the angular overlap bonding parameters e_σ and e_π (Section 3.2.6) and they may be used to construct a two-dimensional spectrochemical series. This is shown for a limited number of ligands for the Ni^{2+} and Cr^{3+} ions in Figure 8.19. When data are available for both metal ions, the ligand characteristics are generally similar. An exception is pyridine, which appears to be a weak π acceptor toward Cr^{3+} but a weak π donor when bonded to Ni^{2+}. It is quite possible that the π bonding capacity of a ligand may change from one metal to another, because it is expected to be influenced by the relative energy of the metal and ligand orbitals, which in turn depends on factors such as the oxidation state of the metal and its atomic number. In some cases, a comparison between the one- and two-dimensional spectrochemical series provides interesting insights into the metal–ligand bonding. Thus fluoride and hydroxide ions are actually *stronger* σ donors toward the Cr^{3+} ion than are ammonia and ethylenediammine, although they lie considerably lower in the one-dimensional spectrochemical series. The low Δ value of these ligands occurs because they are particularly strong π donors toward the Cr^{3+} ion. This is probably related to the partly filled t_{2g} shell of this metal ion, which means that the interaction with the filled ligand orbitals of π symmetry has a stabilizing effect. Further expansion of the two-dimensional spectrochemical series to other metal ions and a greater range of ligands should greatly enhance the general understanding of metal–ligand bonding characteristics.

Several trends concerning the dependence of ligand field splitting parameters on the nature of the metal ion are also apparent from the data in Table 8.3. The value of Δ increases substantially for low-spin compared to high-spin complexes, owing to the shorter metal–ligand bond lengths. An increase in the value of Δ also accompanies an increase in oxidation state, although here it is sometimes enhanced by the pronounced shortening accompanying a change from high to low spin. When other factors are kept constant, there is a similar increase in Δ on going from the first- to the second- and to the third-transition series for any particular d configuration. It is hard to discern any consistent trend in the variation of Δ on crossing each transition series, other factors being kept constant. The Δ values of the complexes of the early metal ion V^{2+} are some 20% higher than those of the later metal ions Co^{2+} and Ni^{2+}, but those of the intermediate metal Mn^{2+} show the lowest Δ values of any transition metal ion.

There are many gaps in Table 8.3, and to make an approximate prediction of the value of Δ for any combination of ligand and metal ion, the following relationship may be used:

$$\Delta = fg(cm^{-1} \cdot 10^{-5}) \tag{8.24}$$

where f and g are the empirical constants listed in Table 8.4. It must be remembered that this table may not work well if the metal ion is in an uncommon spin state; for instance, the Δ value of a high-spin Co^{3+} complex is significantly overestimated using this equation.

8.4.2 The Nephelauxetic Series

Interelectron repulsion is invariably reduced in a complex compared to the free ion. This has two causes: First, the lower effective charge on the metal ion means that the d orbitals are expanded in a complex compared to the free ion, reducing interelectron repulsion by the so-called *nephelauxetic* (cloud-expanding) effect. Second, the interelectron repulsion is lowered by delocalization of the electrons into ligand orbitals. Both effects are related to the covalency of the complex. One way of treating the reduction is to lower the energy separation of the terms compared to the free ion case, applying a so-called Trees correction (3). The most widely used method of treating interelectron repulsion, certainly by inorganic chemists, is the parametrization scheme developed by Racah. The

TABLE 8.4. Empirical Parameters for Estimating Δ and B for Any Combination of Ligand and Metal Ion[a]

Metal Ion	g	k	Ligand	f	h
Co^{2+}	9.3	0.24	3acac	1.2	—
Co^{3+}	19.0	0.35	6Br	0.76	2.3
Cr^{2+}	14.1	—	$6CH_3CO_2$	0.96	—
Cr^{3+}	17.0	0.21	4Cl	−0.36	—
Cu^{2+}	12.0	—	6Cl	0.80	2.0
Fe^{2+}	10.0	—	6CN	1.7	2.0
Fe^{3+}	14.0	0.24	6NCS	1.03	—
Ir^{3+}	32	0.3	2den	1.28	—
Mn^{2+}	8.5	0.07	edta	1.20	—
Mn^{3+}	21	—	ddtp	0.86	2.8
Mn^{4+}	23	0.5	3bipy	1.43	—
Mo^{3+}	24	0.15	3en	1.28	1.5
Ni^{2+}	8.9	0.12	6F	0.9	0.8
Pt^{4+}	36	0.5	3glycine	1.21	—
Re^{4+}	35	0.2	$6H_2O$	1.00	1.0
Rh^{3+}	27.0	0.3	$6NH_3$	1.25	1.4
Ti^{3+}	20.3	—	$6NO_2$	1.5	—
V^{2+}	12.3	0.08	6OH	0.94	—
V^{3+}	18.6	—	3Ox	0.98	1.5
			3o-phen	1.43	—
			6py	1.25	—
			6urea	0.91	1.25

[a] The following relationships were used: $\Delta = fg$ $(cm^{-1} \cdot 10^{-5})$ and $B_{complex} = B_{free\ ion}(1-hk)$. The units of g and h are $cm^{-1} \cdot 10^{-3}$, and k and f are dimensionless. (Charges on ligand ions have been omitted).

nephelauxetic effect for the Racah parameter B, for which most experimental data are available, may be measured by the parameter β:

$$\beta = B_{complex}/B_{free\ ion} \qquad (8.25)$$

Typical values of β are listed in Table 8.5 for various ligands and metal ions. Ligand atoms may be arranged in a series of decreasing β, the nephelauxetic series, which is approximately independent of the metal ion. This series for some common ligands (charges on ions are ignored) is $F > H_2O > NH_3 > en \sim Ox > SCN^* > Cl \sim CN > Br > I$.

The series represents the transition from ionic, in which β is close to unity, towards covalent, in which β is quite small, in the nature of the metal–ligand bonding. It is noteworthy that many ligands occupy quite different positions in the nephelauxetic and spectrochemical series. For instance, Cl and CN have similar positions in the former, but occupy opposite positions in the latter. This

TABLE 8.5. Racah Parameters B and the Nephelauxetic Ratio β for some Typical Complexes

Complex	d configuration	Symmetry	B (cm^{-1})	β
VF_6^{3-}	d^2	O_h	627	0.71
VCl_6^{3-}	d^2	O_h	536	0.61
VCl_4^{-}	d^2	T_d	505	0.57
VCl_2	d^3	O_h	615	0.81
VBr_2	d^3	O_h	530	0.70
VI_2	d^3	O_h	510	0.67
$V(NH_3)_6^{2+}$	d^3	O_h	660	0.87
CrF_6^{3-}	d^3	O_h	896	0.96
$CrCl_6^{3-}$	d^3	O_h	512	0.55
$Cr(NH_3)_6^{3+}$	d^3	O_h	657	0.70
MnF_6^{2-}	d^4	O_h	650	0.60
$MnCl_4^{2-}$	d^5	T_d	650	0.76
$MnBr_4^{2-}$	d^5	T_d	630	0.73
FeF_6^{3-}	d^5	O_h	835	0.84
$FeBr_4^{-}$	d^5	T_d	470	0.46
$Fe(CN)_6^{4-}$	d^6	O_h	490	0.55
CoF_6^{3-}	d^6	O_h	787	0.73
$Co(NH_3)_6^{3+}$	d^6	O_h	615	0.57
$Co(CN)_6^{3-}$	d^6	O_h	400	0.37
$Co(NH_3)_6^{2+}$	d^7	O_h	885	0.89
$Co(NH_3)_4^{2+}$	d^7	T_d	710	0.72
$CoCl_4^{2-}$	d^7	T_d	710	0.72
$CoBr_4^{2-}$	d^7	T_d	695	0.70
CoI_4^{2-}	d^7	T_d	665	0.67
$KNiF_3$	d^8	O_h	843	0.81
$CsNiCl_3$	d^8	O_h	838	0.80
$CsNiBr_3$	d^8	O_h	777	0.75
$Ni(NH_3)_6^{2+}$	d^8	O_h	881	0.85

implies that although chloro and cyano complexes may have a similar features of covalency, the nature of the metal bonding of the two ligands is quite different, as discussed in the preceding section.

In an octahedral complex, it is expected that the covalent interaction involving the σ antibonding e_g^* orbitals will be stronger than that affecting the t_{2g} orbitals, which are involved in relatively weak π interactions. Because the occupancy of these two types of orbitals differs for the various terms, it is sometimes possible to derive separate nephelauxetic parameters for the occupancy of the e_g^* and t_{2g} orbital sets. The parameters β_{33} and β_{35}, so-named for historical reasons, have been defined and are associated with the interelectron repulsion between a pair of electrons in two t_{2g} orbitals and in one t_{2g} and one e_g^* orbital, respectively. Typical values for several fluoro-complexes are listed in Table 8.6, which shows that, as expected, the β_{35} values are somewhat lower than the β_{33} values.

TABLE 8.6. The Nephelauxetic Ratios β_{33} and β_{35} for Some Fluoro-Complexes

Complex	β_{33}	β_{35}
CrF_6^{3-}	0.93	0.89
MnF_6^{2-}	0.77	0.56
NiF_6^{2-}	0.36	—
OsF_6	0.52	—
IrF_6	0.43	—

The Racah parameter B has been measured for only a limited number of metal ions and ligands. The empirical relationship:

$$B_{complex} = B_{free\ ion}(1 - hk) \tag{8.26}$$

has been devised to provide estimates when data are not available. Here h and k are constants that define the nephelauxetic effects of each ligand and metal ion, respectively. Values of h and k are given for a few complexes in Table 8.4.

8.5   CHARGE TRANSFER SPECTRA

Charge transfer spectra, as the name implies, reflect a redistribution of the electron density in the complex (3). This can occur in one of two ways: The transition is either from a filled ligand orbital to a partly filled or empty d orbital on the metal or from a filled or partly filled metal d orbital to a vacant ligand orbital. These are termed ligand $\rightarrow$ metal $(L \rightarrow M)$ and metal $\rightarrow$ ligand $(M \rightarrow L)$ charge transfer transitions, respectively. Because relative to the ground state the metal is reduced or oxidized in the excited electronic state, they are often referred to as redox spectra. The relative ease with which a metal or ligand participates in such a transition is said to depend on its *optical electronegativity*. Because the ligand orbitals are generally of the correct symmetry and type to satisfy the parity selection rule, and the spin selection rule can always be obeyed, charge transfer transitions are usually intense $(\varepsilon \sim 10^3 - 10^4)$; however, this not a characteristic of *all* charge transfer transitions. The band at $\sim 20,000\ cm^{-1}$, which gives Ni^{2+} nitro complexes their characteristic red appearance, is now thought to be a low-energy $M \rightarrow L$ charge transfer transition, even though it has an intensity no higher than that of a typical 'd-d' transition. Charge transfer transitions are usually broad, because the equilibrium metal–ligand bond length is much longer in the excited state than in the ground state, so that the transition involves the excitation of multiple quanta of the α_1 "breathing" vibration in conjunction with the electronic transition (Section 8.1.3.3). The spectrum of the permanganate ion MnO_4^- is unusual in that this vibrational structure is readily apparent in the spectrum of an aqueous solution at room temperature (Fig. 8.11).

To a first approximation, charge transfer spectra are conveniently interpreted using the simple molecular orbital energy diagram developed earlier (Section 3.1.1). Because the redox process occurs in an opposite sense for the two types of transition, different criteria decide when each is likely to occur at low energy. For $L \rightarrow M$ transitions, the energy depends on the separation between the filled bonding and, when appropriate, nonbonding ligand orbitals and the partly filled metal d orbitals (Fig. 3.2). Low-energy transitions of this type are, therefore, expected when the metal is in a high oxidation state and/or the ligand is relatively electropositive. In addition, as there is a tendency for the ionization energy to increase when crossing each transition series, there is a concomitant decrease in the energy of $L \rightarrow M$ charge transfer transitions. Many transition metal complexes show this type of transition in the near-ultraviolet part of the spectrum. Quite frequently, the band tails into the visible region and influences the color of the complex. For instance, the green of $CuCl_2(H_2O)_2$ is caused by the absorption of red light by the 'd-d' transitions combined with absorption of blue light by the tail of the intense $L \rightarrow M$ charge transfer from the chloride ion. Sometimes the transitions are low enough in energy to be centered in the visible region, producing intensely colored complexes, such as the MnO_4^- and $Fe(H_2O)_4(NCS)_2^+$ ions, both of which have common use as visual indicators in redox titrations.

Low-energy $M \rightarrow L$ transitions require the metal orbitals to be high in energy and are, therefore, associated with a low oxidation state of the metal and tend to be more pronounced for early transition ions. They are less common than $L \rightarrow M$ transitions, because they also require that the ligand has low energy vacant orbitals to accept the electron. This condition is satisfied for complexes of large aromatic amines, such as bipyridyl and o-phenanthrolene. $Fe(bpy)_3^{2+}$ is a typical example of a complex that has a strong absorption band in the visible range owing this type of transition. This complex also is commonly used an indicator in redox titrations.

For $L \rightarrow M$ charge transfer transitions sufficient experimental data are available to note some consistent trends. Metal ions may be arranged in a series according to their ease of reduction:

$$Rh^{4+} > Ru^{4+} > Cu^{2+} > Os^{4+} > Fe^{3+} > Ru^{3+} > Pd^{4+} > Re^{4+} \sim Os^{3+}$$
$$\sim Pd^{2+} \sim Pt^{4+} \sim Rh^{3+} > Pt^{2+} > Ti^{4+} \sim Ir^{3+}.$$

For halides, the ease of oxidation follows the expected trend:

$$I > Br > Cl > F.$$

The extent to which the spectra conform to the simple molecular orbital energy diagram may be tested by comparing the band energies observed for

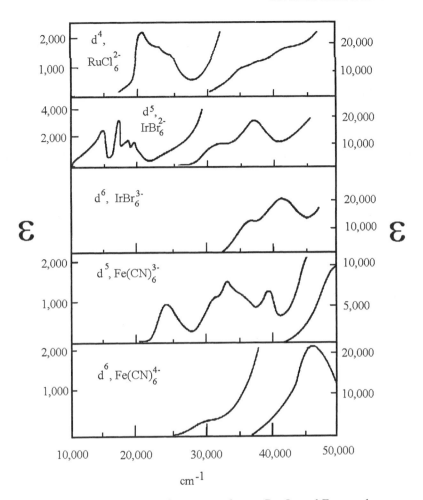

Figure 8.20. Charge transfer spectra of some Ru, Ir, and Fe complexes.

different complexes. The spectra of the ions $RuCl_6^{2-}$, $IrBr_6^{2-}$, and $IrBr_6^{3-}$ are shown in Figure 8.20. Each of the first two complexes shows two sets of bands, a weaker set at $\sim 20{,}000\,\mathrm{cm}^{-1}$ ($\varepsilon \sim 2000$) and a more intense set at $\sim 40{,}000\,\mathrm{cm}^{-1}$ ($\varepsilon \sim 20{,}000$). These are assigned as transitions from the filled t_{1u} and t_{2u} sets of ligand orbitals to the partly filled t_{2g} orbitals and the empty e_g^* orbitals on the metals, respectively. As expected, the bands occur at slightly lower energy for the less electronegative bromide ligand. In the spectrum of the $IrBr_6^{3-}$ complex, the lower energy set of bands is absent, and this is consistent with the above interpretation, because there the d^6 electron configuration of the metal requires complete filling of the metal t_{2g} orbitals. The only $L \rightarrow M$ transitions that may occur are, therefore, those to the empty e_g^* orbitals. As expected from the lower charge on the metal, the bands are at slightly higher energy than those of the $IrBr_6^{2-}$ ion. If this picture is correct, the energy difference between the two sets of bands at $\sim 20,000 \, cm^{-1}$ should be approximately equal to the ligand field splitting Δ. Although Δ cannot be measured directly, because the 'd-d' transitions presumably are obscured under the charge transfer bands, the energy difference is certainly of the right order of magnitude (Table 8.3).

Although the above way of interpreting charge transfer spectra gives a reasonably self-consistent picture of this type of transition, it must be remembered that it contains many simplifications. Interelectron repulsion effects are ignored in treating the transitions as electron jumps between orbitals. Moreover, for highly covalent complexes, for instance those involving the metal in a high formal oxidation state such as the MnO_4^- ion or with ligands such as dimaleothionitrile, the concept of charge transfer accompanying the electronic transitions is highly questionable, as the spectra can really be interpreted satisfactorily only in terms of delocalized molecular orbitals.

8.6   LUMINESCENCE SPECTRA

Although the electronic transitions of transition metal complexes are generally observed by measuring their absorption spectra, the same transition moment integral applies equally well to the process by which a photon is emitted rather than absorbed (4). Thus exactly the same selection rules apply to both processes. Because excited electronic states are generally much higher above the ground state than normal thermal energies and because nonradiative relaxation rates are usually rapid, the fraction of complexes in the excited state is usually too small to be detected by conventional absorption spectroscopy. The high sensitivity of modern equipment, however, means that it is now possible to measure luminescence spectra for many transition ions; and the importance of this in the development of lasers has led to considerable interest in the area.

Luminescence is of two kinds: Fluorescence and phosphorescence. In the former, emission occurs from the same excited state as that involved in the absorption process. This normally involves a spin-allowed transition, and because this is a partly allowed process, the lifetime in the excited state is relatively short, typically 10^{-6} to $10^{-10} \, s$. Because the transitions generally involve excitation between the t_{2g} and e_g^* orbitals of an octahedral complex, the two states have different metal–ligand bond lengths, and the absorption band consists of a progression in the α_{1g} mode (Section 8.1.3.3). When fluorescence occurs, the complex first relaxes to the lowest vibrational state of the excited electronic state and then emits light by dropping to various vibrational levels of the ground electronic state. This process is presented in Figure 8.21, which shows that the fluorescence spectrum is expected to be a mirror image, reflected to lower energy, of the absorption spectrum. To minimize thermal relaxation processes and maximize spectral resolution, luminescence spectra are normally measured at low temperature.

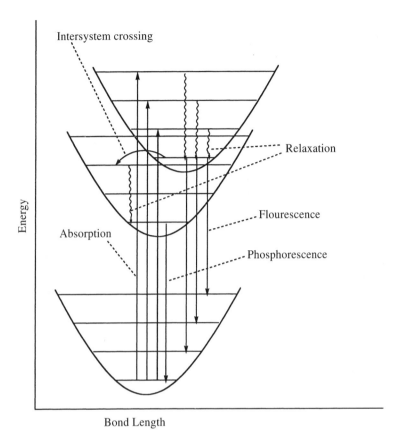

Figure 8.21. Comparison of the processes involved in absorption, fluorescence, and phosphorescence. *Wavy lines*, relaxation pathways; *vertical straight lines*, transitions involving the absorption or emission of optical photons.

In phosphorescence, unlike fluorescence, the complex crosses from the excited state reached by light absorption to a lower excited state that has a different spin quantum number from the ground state (intersystem crossing). Emission occurs when the complex drops from this second excited state to the ground state. Because the transition moment between these states of different multiplicity is small, the complex may be trapped in the excited state for a relatively long time, typically 10^{-6} to 10^{-3} s and sometimes as long as seconds (Fig. 8.21). Phosphorescent emission occurs at considerably lower energy than the absorption. For transitions corresponding to spin-flips there is no significant change in antibonding energy of the electrons, so that the excited state potential surface has a similar bond length to the ground state and the band may be sharp (Section 8.1.3.3).

A particularly important example of phosphorescence is provided by ruby, in which Cr^{3+} substitutes for $\sim 2\%$ of the Al^{3+} ions in corundum Al_2O_3. The

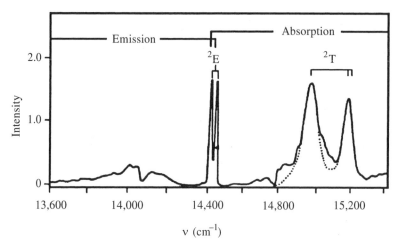

Figure 8.22. Absorption and emission spectra associated with the lower spin-forbidden transitions of Cr^{3+} in Al_2O_3 (ruby). [From *Inorganic Electronic Spectroscopy* (second edition), by A. B. P. Lever, Elsevier, Amsterdam, The Netherlands, 1984. Used with permission].

absorption and emission spectra in the low-energy part of the visible region are shown in Figure 8.22. The aluminium / chromium ion is in a trigonally distorted octahedral ligand field, and the absorption bands are caused by the spin-forbidden $^4A_{2g} \rightarrow {}^2T_{2g}$ and ^2E_g transitions of the Cr^{3+} ion. Emission occurs from the lower-energy ^2E_g state. The emission from and absorption to this state occur at the same energy, because the ground and excited states differ only in the spin component of the wavefunctions, so that the electronic transition exhibits no vibrational progression.

The phosphorescence exhibited by ruby was exploited in the development of the first laser, a devise making use of the light amplification by stimulated emission of radiation (5). For a substance to act as a laser, a 'population inversion' must be created between the ground and excited states. This is achieved by shining an intense burst of light in a region where the compound has a highly allowed transition: In the case of ruby, a parity-allowed L → M charge transfer state in the near-ultraviolet part of the spectrum. This 'pumps' most of the complexes into the excited state, from which they relax until they are trapped in the ^2E_g state. When a photon is emitted by one complex unit and strikes a neighboring complex unit that is in an excited state, it 'stimulates' this to also emit a photon. This pair of photons can stimulate two other complex units to emit, the process thus generating a cascade of photons by what is essentially an optical chain reaction (the population inversion is crucial here, otherwise the emitted photons would simply be reabsorbed). The process is shown in Figure 8.23. Because the transition energy is sharply defined, the radiation produced by the phosphorescence is highly monochromatic. Moreover, the radiation produced by stimulated emission is *coherent* (the wave properties of the photons

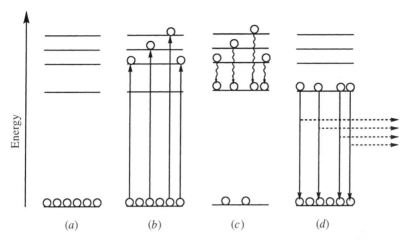

Figure 8.23. Processes involved in the operation of a laser. **a** Molecules are all in the ground state. **b**, Majority of molecules are pumped into excited states via allowed transitions. **c**, Molecules relax to produce a population inversion in an excited state that differs in spin from the ground state. **d**, A molecule phosphoresces, and the photon it emits stimulates other molecules to phosphoresce.

are in phase with one another). These features, together with the intense nature of the radiation, make lasers a powerful tool in the development of modern technology. The ruby laser emits in the red part of the spectrum, and there has been considerable interest in preparing transition-metal compounds that act as lasers in other parts of the electromagnetic spectrum.

REFERENCES

1. Lever, A. B. P., *Inorganic Electronic Spectroscopy*, 2nd ed., Elsevier, Amsterdam, 1984.
2. Hitchman, M. A., and Riley, M. J., in Solomon, E. I., and Lever, A. B. P., eds. *Inorganic Electronic Structure and Spectroscopy*, Vol. 1, Wiley, New York, 1999, chapter 4.
3. Lever, A. B. P., and Dodsworth, E. S. in Ref. 2, chapter 4.
4. Brunhold, T. C., and Güdel, H. U., in Ref. 2, chapter 5.
5. Krausz, E., and Riesen, H., in Ref. 2, chapter 6.

CHAPTER 9

MAGNETIC PROPERTIES OF COMPLEX IONS

9.1 THE THEORY OF MAGNETIC SUSCEPTIBILITY

9.1.1 General

The magnetic field inside a substance differs from the free-space value of the applied field (1,2). The difference can be put in the form shown in Figure 9.1:

$$H = H_0 + \Delta H \tag{9.1}$$

where H_0 is the free-space magnetic field and ΔH is the field produced by the magnetic polarization of the substance. More usually, working in the SI system of units, the relationship is given as:

$$\mathbf{B} = H_0 + I \tag{9.2}$$

where $\mathbf{B}$ is the magnetic induction (identical to H in Eq. 9.1) and $\mathbf{I}$ is the intensity of magnetization. I is the magnetic moment, per unit volume, that is necessary to account for ΔH. Dividing the last expression by H_0, we have:

$$\mathbf{B}/H_0 = 1 + \chi \tag{9.3}$$

$\mathbf{B}/H_0$ is referred to as the magnetic permeability of the substance. χ, which is a measure of how easy it is to magnetically polarize the substance, is called the magnetic susceptibility (per unit volume). More usually, the quantity measured

228

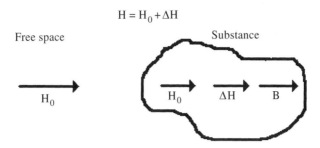

Figure 9.1. The relationship between the magnetic induction (**B**) and the applied magnetic field (H$_0$).

is the specific magnetic susceptibility χ_m, which is related via the density of the substance:

$$\chi_m = \chi/\rho \qquad (9.4)$$

The units for χ_m are cubic meters. (In the cgs system of units, $\kappa(\text{cgs}) = 10^6 \cdot \chi(\text{SI})/4\pi$ and $\chi(\text{cgs}) = 10^3 \chi_m(\text{SI})/4\pi$) and the units are cubic centimeters).

The magnetic properties of substances can be developed with the supposition that each atom of a molecule possesses the attributes of a magnetic dipole; i.e., it acts as if it were an atomic scale bar magnet. This magnetic dipole may be intrinsic or it may be induced by the applied field. In the presence of the applied field, these dipoles are quantised so that they take on one of a limited number of directions relative to that of the field. In this picture, the process of magnetic polarization involves the tendency of the magnetic dipoles to achieve the arrangement of lowest energy, which is alignment with the applied magnetic field. Using that model of magnetic polarization, it is possible to obtain an alternative, but completely equivalent, definition of the magnetic susceptibility. The magnetic field lowers the energy per unit mass W of the substance, and it was derived by Van Vleck (1) that:

$$\chi_m = -(1/\mathbf{B})(\partial W/\partial \mathbf{B}) \qquad (9.5)$$

This definition of χ_m is the more useful for the discussion that follows. The tendency for the dipoles to align along the direction of the applied magnetic field is counteracted by the randomizing effect of the thermal energy. The thermal energy available to the system kT is, except at very low temperatures, much larger than the alignment energy, so that the degree of alignment achieved is usually small.

Interaction between the atomic dipoles in a system, however, can greatly increase the magnetic alignment energy by making it the property of a large number of units rather than of a single one. Under such circumstances, the alignment energy may be comparable with or greater than the thermal energy. Substances in which there are no appreciable interactions between adjacent atomic dipoles are said to be *magnetically dilute*. In *magnetically concentrated* substances, interactions between adjacent atomic dipoles must be considered. Magnetically concentrated substances may show complicated magnetic polarization effects, for example the well-known phenomenon of ferromagnetism or the less well known but more common antiferromagnetism. Except for a limited class, we shall not deal with magnetically concentrated substances, and the following development of magnetic properties applies for magnetically dilute conditions, unless otherwise specified.

Two types of magnetic behavior are found experimentally in magnetically dilute substances. In each type, for moderate magnetic fields the magnetic susceptibility is independent of the strength of the applied magnetic field. In *diamagnetic* substances χ_m is negative; in *paramagnetic* substances it is positive.

Classically, diamagnetism arises from the motions of electrons in the applied magnetic field on account of their electric charge. In a crude manner, it may be pictured as arising from the circulation of electric currents as the electrons move from one atomic orbital to another. The interaction between the current and the magnetic field causes the plane of the circulation of the current to precess (the Lamor precession). Diamagnetism is a small effect, but because all materials contain electrons, it is always present. It is of some importance, because the presence of a large number of diamagnetic atoms can make the diamagnetism an appreciable contribution to the susceptibility of a molecule that contains a paramagnetic atom. In dealing with paramagnetism, it is usually necessary to correct for the presence of the diamagnetic atoms in the molecule. Any paramagnetic susceptibilities mentioned here are to be taken as corrected for the presence of the diamagnetic portions of the molecule. The diamagnetic susceptibility of an atom can be calculated to be proportional to the average squared radius of the orbitals housing its electrons:

$$\chi_A = -(N_A e^2/6\,m\,c^2) \sum_i \overline{r_i^2}$$
$$= -3.56 \times 10^8 \sum_i \overline{r_i^2} \tag{9.6}$$

where χ_A is the susceptibility per kilogram atom, viz. $\chi_m \cdot$ (atomic mass) for the element. The summation over i includes all the electrons of the atom. The results of calculations performed with the aid of this expression do not agree particularly well with the experimental data for elements. Thus it is customary to treat the diamagnetic susceptibilities of atoms as well as molecules on an empirical basis. It is assumed that an additive law applies to atomic and molecular diamagnetic susceptibilities, with contributions coming from the atoms and from the interactions between them—the chemical bonds. Then the diamagnetic susceptibility for a molecule is put in the form

$$\chi_M = \sum_i \chi_{Ai} + \sum_j \chi_{Bj} \tag{9.7}$$

Here, χ_{Ai} is the atomic susceptibility associated with atom i, and χ_{Bj} is that associated with the jth bond in the molecule. Lists of such susceptibilities are available for atoms in organic molecules (Pascal's constants) and for metal ions and inorganic radicals (3). For instance, a carbon atom in an organic molecule is usually taken to contribute $-71 \cdot 10^{-12} \, \text{m}^3 \, \text{mol}^{-1}$ to the susceptibility of the molecule. Diamagnetism is independent of temperature.

Paramagnetism arises from the angular momenta of electrons. The angular momentum may be orbital or spin in origin. In fact, few cases arise in which a system possesses orbital angular momentum in the absence of spin. For that reason, paramagnetism is most usually associated with the presence of unpaired electrons. It is found in the following places in the Periodic Table:

s electrons. Alkali metal vapors.

p electrons. A few molecules of main group elements, such as NO, O_2 and ClO_2, and organic free radicals.

d electrons. The three transition series.

f electrons. The lanthanide (rare earth) and actinide (transuranic) series.

We are, of course, concerned mainly with the d electron series here. In coordination compounds there are sufficient diamagnetic atoms surrounding the central metal ions, in most instances, to reduce interaction between the latter to small values. Coordination complexes are, then, usually magnetically dilute, at least as far as the present purposes are concerned. At low temperature even a small amount of magnetic interaction between neighboring dipoles is likely to be important. Most paramagnetic substances show features of magnetically concentrated behavior to some extent at temperatures near absolute zero. The characteristics of different kinds of magnetic behavior are summarized in Table 9.1.

For chemical purposes it is most convenient to deal with susceptibilities per mole, the *atomic* and *molecular* susceptibilities, of which the former was introduced earlier in connection with diamagnetism:

$$\chi_M = \chi_m \cdot (\text{molecular mass}) = \chi \cdot (\text{molecular volume}) \tag{9.8}$$

the units are cubic meters per mole. For paramagnetic substances, χ_A is obtained from χ_M by correction for the diamagnetic portion of the molecule that contains the paramagnetic atom:

$$\chi_A = \chi_M - \chi_{\text{diam}} \tag{9.9}$$

TABLE 9.1. Types of Magnetic Behaviour and Their Characteristics

Type of Susceptibility	Sign	Approximate Magnitude of χ_A ($m^3\ mol^{-1}$)	Dependence on H_0	Origin
Diamagnetism	− ive	10^{-9}	Independent	Motion of charge
Paramagnetism	+ ive	0–10^{-7}	Independent	Angular momenta
Ferromagnetism	+ ive	0–10^{-2}	Dependent	↑↑ dipole coupling
Antiferromagnetism	+ ive	0–10^{-6}	May depend	↑↓ dipole coupling

(In the cgs system, $\chi_M(cgs) = 10^6 \chi_M(SI)/4\pi$, and the units are cubic centimeters per mole.)

9.1.2   Paramagnetism

In terms of the magnetic dipole model, the magnetic dipole moment in the z direction for an atom possessing angular momentum is see p. 286:

$$\mu_z = (\mathbf{L}_z + g_e\mathbf{S}_z)\mu_B \simeq (\mathbf{L}_z + 2\mathbf{S}_z)\mu_B \qquad (9.10)$$

where μ_B is the Bohr magneton, $0.92732 \cdot 10^{-23}\ J\,T^{-1}$. Here, the unit of magnetic field T is the Tessla; $1T = 10^4\ G$. It is derived in Appendix 7 that an assemblage of noninteracting dipoles of moment μ_z, quantized to point either along or against a magnetic field in the z direction and each possessing thermal energy kT, gives rise to a susceptibility of the form:

$$\chi_m = C/T \qquad (9.11)$$

This relationship is the well-known Curie law for paramagnetic substances, and C is the Curie constant.

For chemical purposes, the magnetic properties of substances are most frequently discussed in terms of a quantity called the *effective magnetic moment* μ_{eff} rather than of the susceptibility itself. This quantity possesses the properties its name implies and is related to but not identical with μ_z, mentioned above. It is defined as:

$$\mu_{eff} = (3kT/N_A\mu_B^2)^{\frac{1}{2}}(\chi_A T)^{\frac{1}{2}}$$
$$= 797.5(\chi_A T)^{\frac{1}{2}} \qquad (9.12)$$
$$(= 2.828(\chi_A T)^{\frac{1}{2}}\ \text{in the cgs system})$$

where N_A is Avogadro's number. If the Curie law is obeyed, μ_{eff} is independent of temperature.

In fact, comparatively few systems obey the Curie law accurately. There are many reasons for departure from the ideal behavior required for the Curie law, some of which will be discussed in due course. Often it is possible to account for the variation of magnetic susceptibility with temperature much better, at least over a large portion of the range, with a simple modification of the Curie law, the Curie–Weiss law:

$$\chi_m = C/(T - \theta) \qquad (9.13)$$

where θ is the Weiss constant. In magnetically concentrated systems, θ has a fundamental significance and can be related to the magnetic exchange present. In magnetically dilute substances, its role is essentially empirical. When the origin of θ is known to be in magnetic exchange, it is reasonable to calculate the effective magnetic moment as:

$$\mu_{\text{eff}} = (3k/N_A\mu_B^2)^{\frac{1}{2}}[\chi_A/(T - \theta)]^{\frac{1}{2}} \qquad (9.14)$$

but this procedure may give misleading results for magnetically dilute systems.

9.1.3 Quantum Mechanical Treatment of Paramagnetic Susceptibilities

The magnetic dipole model for the magnetic polarization of substances has severe limitations for the discussion of magnetochemistry in detail. The general theory for the paramagnetic susceptibility of an atomic system is now developed on a quantum mechanical basis (1,2).

The energy that corresponds to a wavefunction ψ_i in the presence of a magnetic field H can be expanded as a power series in the field:

$$W_i = W_i^0 + W_i^I B + W_i^{II} B^2 + \dots \qquad (9.15)$$

The perturbation of the energy levels of an atom by the magnetic field is known as the Zeeman effect. The quantities W_i^I and W_i^{II} are, respectively, the first- and second-order Zeeman coefficients for the state specified by the wavefunction ψ_i. W_i^0 is, of course, the unperturbed energy of the wavefunction.

We assume a Boltzmann, distribution among the energy levels of the atoms and apply Eq. 9.5. There results, as set out in by Van Vleck (1) and Gerloch (2), the *Van Vleck equation*:

$$\chi_A = N_A \sum_i [(W_i^I)^2/kT - 2W_i^{II}]\exp\left(-W_i^0/kT\right)/\sum_i \exp\left(-W_i^0/kT\right) \quad (9.16)$$

The coefficients of the Zeeman effect are related to the wavefunction by the application of the magnetic moment operator. This is, for the z direction:

$$\mu_z = (\mathbf{L}_z + 2\mathbf{S}_z)\mu_B \qquad (9.17)$$

with equivalent expressions for the x and y directions. The relationship is, for example:

$$W_{iz}^{I} = \langle \psi_i | \mathbf{L}_z + 2\mathbf{S}_z | \psi_i \rangle \qquad (9.18)$$

$$W_{iz}^{II} = \sum_{j} [\langle \psi_i | \mathbf{L}_z + 2\mathbf{S}_z | \psi_j \rangle \mu_\beta]^2 / [W_i^0 - W_j^0] \qquad i \neq j \qquad (9.19)$$

Because the wavefunctions are available as eigenfunctions of $\mathbf{L}_z$ and $\mathbf{S}_z$, the evaluation of the first- and second-order Zeeman coefficients for this direction is facile. For instance, if the wavefunction is specified by $M_L = 3$ and $M_S = 1$:

$$\langle \{L, 3\}[S, 1] | \mathbf{L}_z + 2\mathbf{S}_z | \{L, 3\}[S, 1] \rangle = 3 + 2 \cdot 1$$
$$= 5 \qquad (9.20)$$

The form of W^{II} needs some further consideration. The summation over the index j includes all wavefunctions of the atom other than ψ_i. If it should be found for a wavefunction that is degenerate with ψ_i (so that $W_i^0 = W_j^0$) that:

$$\langle \psi_i | \mathbf{L}_z + 2\mathbf{S}_z + | \psi_j \rangle \neq 0 \qquad (9.21)$$

then the second-order Zeeman effect seems to be infinite. That, obviously, is not physically possible. The escape from this dilemma is pointed out in Section 9.4.

The application of Eq. 9.16 may be considered under four headings.

1. There are energy levels only $\langle\langle kT$ above the ground level(s). In this case the only wavefunctions to be taken into account are those of a set that is degenerate in the absence of a magnetic field (or nondegenerate). Then all the coefficients W_i^0 are infinite (if i does not belong to the set) or zero (if i belongs to the set). These lead, respectively, to exponentials of zero or unity. Furthermore, all the coefficients W_i^{II} must be zero in view of the preceding argument. The first-order Zeeman effect alone contributes to the susceptibility. It is readily seen that the susceptibility is then of the form:

$$\chi_A = C/T \qquad (9.22)$$

which is the Curie law defined above. Indeed, the requirement of energy levels $\langle\langle kT$ is that which is necessary for the magnetic dipole alignment model of polarization to be valid. It is rarely met strictly in practice.

2. There are energy levels only $\rangle\rangle kT$ above the ground level(s), which has no first-order Zeeman effect. If this requirement is met, then we are dealing with a ground level, or group of levels, for which all members have the coefficient $W_i^I = 0$. Of course, we assume that the wavefunctions of the set are chosen to eliminate the second-order Zeeman effect among themselves. As in application 1, the exponentials in Eq. 9.16 approach unity (i pertains to the ground level(s)) or zero (i does not belong to the ground level(s)). W_i^0 is zero for the ground level(s). Therefore the only contribution to the susceptibility comes from the second-order Zeeman effect between the ground level(s) and the higher levels which involves the W_i^{II} values that belong to the ground level(s).

It arises from the sum of expressions of the form, for example, for the z direction:

$$-2[\langle\psi_i|\mathbf{L}_z + 2\mathbf{S}_z|\psi_j\rangle\mu_B]^2/(W_i^0 - W_j^0) \tag{9.23}$$

where i belongs to the ground level(s) and j to the higher levels. The numerator is a positive quantity. Three points should be noted about this second order Zeeman effect susceptibility. (a) the susceptibility is independent of temperature, because the expression now does not contain T; (b) it is small, because the separation $(W_i^0 - W_j^0)$ is large compared to kT, unless the temperature is low; and (c) it is positive, because $(W_i^0 - W_j^0)$ is a negative quantity.

The second-order Zeeman effect contribution to the susceptibility of the ground level(s) is conveniently put in the form:

$$\chi_A = N_A\alpha \tag{9.24}$$

where

$$\alpha = -(2/n)\sum_{i=1}^{n} W_i^{II} \tag{9.25}$$

the summation taking place over the n wavefunctions of the n-fold degenerate ground level(s).

The situation represented by the requirement given for heading 2 is not uncommon among ions in transition-metal complexes. The term temperature independent paramagnetism (TIP) or Van Vleck high-frequency paramagnetism has been given to such susceptibilities. It is best thought of simply as the second-order Zeeman effect contribution to the paramagnetism.

3. There are energy levels both $\langle\langle$ and $\rangle\rangle kT$ (but not $\sim kT$) relative to the ground level(s). In these circumstances, as in the previous two applications, the exponentials in Eq. 9.16 approach zero or unity. This simply produces a contribution to the susceptibility from both the first-order Zeeman effect of the ground level(s) and their second-order Zeeman effect with higher levels. The result is a combination of the Curie law and Eq. 9.24:

$$\chi_A = C/T + N_A\alpha \tag{9.26}$$

This formula for the susceptibility is often referred to as the Langevin–Debye equation and has quite wide applicability (1). The second term in Eq. 9.26 is usually fairly small compared to the first term; in other words, the second-order Zeeman effect is relatively minor, and the Curie law holds approximately.

4. There are energy levels $\sim kT$ (and probably also $\langle\langle$ and $\rangle\rangle$ kT) relative to the ground level(s). In this case, the susceptibility is a complicated function of temperature, and each system must be treated individually; however, the behavior of the second-order Zeeman effect when this involves energy levels separated by an amount comparable to kT is of special interest. We trace the contribution it makes to the susceptibility as the separation of these energy levels varies from $\rangle\rangle$ kT to $\langle\langle$ kT.

We examine the second-order Zeeman effect between the n-fold degenerate ground levels and a group of levels m-fold degenerate at energy δW^0. Considering only that system, the second-order Zeeman effect for the ground levels is, say:

$$\sum_{i=1}^{n} W_i^{II} = -a\mu_B^2 \tag{9.27}$$

whereas the effect for the higher levels must be the negative of this:

$$\sum_{i=1}^{m} W_i^{II} = a\mu_B^2 \tag{9.28}$$

The relationship follows because of the two equations:

$$\langle \psi_i | L_z + 2S_z | \psi_j \rangle = \langle \psi_j | L_z + 2S_z | \psi_i \rangle \tag{9.29}$$

$$(W_i^0 - W_j^0) = -(W_j^0 - W_i^0) \tag{9.30}$$

Applying Eq. 9.16, for example, to the system of the two groups of levels, or more simply taking a Boltzmann distribution for their occupations, with susceptibility contributions $2N_A a\mu_B^2$ and $-2N_A a\mu_B^2$ we obtain:

$$\chi_A = 2aN_A \mu_B^2 [1 - \exp(-\delta W^0/kT)]/[n + m \cdot \exp(-\delta W^0/kT)] \tag{9.31}$$

This expression reduces to Eq. 9.24 for $\delta W^0 \rangle\rangle kT$, which is the condition specified in application 2. It also produces the Curie law, with:

$$C = 2aN_A\mu_B^2 \delta W^0/k(n+m) \tag{9.32}$$

for $\delta W^0 \langle\langle kT$, which is the condition for the degeneracy of the ground levels imposed in Section 9.1. Thus we see that the second-order Zeeman effect transforms smoothly from the temperature-independent paramagnetism of levels far apart to the Curie law of degenerate levels. Note that if the numbers of the ground and excited levels are the same ($m = n$) Eq. 9.31 simplifies to:

$$\chi_A = (2aN_A\mu_B^2/n)\tanh(\delta\,W^0/2kT) \tag{9.33}$$

9.2 THE MAGNETIC PROPERTIES OF FREE IONS

9.2.1 The First-Order Zeeman Effect

For the system composed of a term whose degeneracy is lifted by spin–orbit coupling to give rise to states (Section 5.5), the remaining degeneracy of the states is lifted on the application of a magnetic field. For the present, attention is confined to the ground state, it being assumed that the other states lie at $\rangle\rangle kT$.

The first-order Zeeman effect for a state specified by J gives levels at energies (Fig. 9.2):

$$W_{M_J} = M_J g_J \mu_B \mathbf{B} \tag{9.34}$$

It is assumed that the zero of energy is that of the unperturbed state. g_J, the so-called spectroscopic or Landé splitting factor, is a number peculiar to the state J and lying often in the region of 2.0. The separation between any adjacent pair of M_J levels is $g_J\mu_B\mathbf{B}$. The application of Eq. 9.16 (the conditions are those of Section 9.1.2 application 1) gives:

$$\begin{aligned}
\chi_A &= N_A\mu_B^2 \sum_{M_J=-J}^{J} M_J^2 g_J^2 / kT(2J+1) \\
&= (N_A\mu_B^2/3kT)g_J^2 J(J+1)
\end{aligned} \tag{9.35}$$

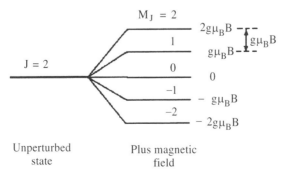

Figure 9.2. The splitting of the members of a state by a magnetic field. It is assumed that the unperturbed state is at zero energy. J $= 2$ is used as an example.

Hence:

$$\mu_{\mathrm{eff}}^2 = g_J[J(J+1)]^{\frac{1}{2}} \tag{9.36}$$

Eq. 9.36 gives the result for the magnetic moment of a system that consists of a ground state only. It applies quite well to most of the ions of the lanthanide series, in which the approximation that the ground state alone is concerned in determining the magnetic properties is reasonable. In those ions, spin–orbit coupling is usually so large that states other than the ground state are thermally inaccessible and, as far as the f electrons are concerned, the ions are essentially 'free'.

The splitting factor g_J is a function of the amount of orbital and spin angular momenta that the state J possesses. If a state arises from a term specified by L and S, there follows:

$$g_J = 1 + [S(S+1) - L(L+1) + J(J+1)]/2J(J+1) \tag{9.37}$$

For a system in which there is no orbital angular momentum (L = 0), then J = S and $g = g_e \approx 2.00$. This is a general result; when spin angular momentum alone is the basis of the magnetic behavior of an ion, g = 2.00. Also, from Eq. 9.36:

$$\mu_{\mathrm{eff}} = 2[S(S+1)]^{1/2} \tag{9.38}$$

If n is the number of unpaired electrons of an ion, then S = n / 2. Equation 9.38 can thus be rewritten:

$$\mu_{\mathrm{eff}} = [n(n+2)]^{1/2} \tag{9.39}$$

The eqs. 9.38 and 9.39 are known as the *spin-only* formulae for the magnetic moment, because they involve no contribution from the orbital angular momentum, only from the spin. A list of spin-only magnetic moments for various numbers of unpaired electrons, together with the corresponding atomic magnetic susceptibilities at 300 K, is given in Table 9.2.

When the spin–orbit coupling constant λ is on the order of kT, more than one state is thermally accessible. Then the susceptibility owing to the first-order Zeeman effect for each state contributes to the susceptibility of the system in proportion to its Boltzmann population. We return to this point later, after the second-order Zeeman effect has been considered in further detail.

9.2.2   The Second-Order Zeeman Effect

If a state for which J = 0 lies lowest and that with J = 1 lies at $\rangle\rangle kT$ above it, then the conditions of application 2 in Section 9.1.3 are met. The susceptibility arises from the second-order Zeeman effect alone. The expression for the coefficient of

TABLE 9.2. Spin-only Magnetic Moments and Atomic Magnetic Susceptibilities at 300 K for Various Numbers n of Unpaired Electrons[a].

n	$2S+1$	μ_{eff}^{S-O}	χ_A
1	2	1.73	15.71
2	3	2.83	41.88
3	4	3.87	78.54
4	5	4.90	125.66
5	6	5.92	183.47
6	7	6.93	251.33
7	8	7.94	329.87

[a] The moments are in units of μ_B. The susceptibilities are in units of 10^{-9} m^3 mol^{-1}.

the second-order Zeeman effect for a state, owing to the state with J greater by unity, is:

$$F_{J,J+1} = \sum_{M_j=-J}^{J} W_{MJ}$$

$$= -(J+L+S+2)(-J+L+S)(J-L+S+1)$$
$$\times (J+L-S+1)\mu_B^2/12(J+1)^2\lambda \qquad (9.40)$$

The coefficients between states for which J differs by more than unity are zero.

The case just described does occur among the lanthanide elements. For Eu^{3+} (f^6) the ground state is $^7F_0$, with $^7F_1$ some 100's cm^{-1} higher. (In this context, note that kT at 300 K corresponds to 210 cm^{-1}.) Instead of the value of zero predicted by Eq. 9.36, the magnetic moments of compounds of trivalent europium at ambient temperature are found to be about 3.6 μ_B. The susceptibility at lower temperatures is, however, independent of temperature, so that the magnetic moment is *not* independent of temperature. At the lower temperatures, it is proportional to T$^{\frac{1}{2}}$.

9.2.3 States $\sim kT$

When there are states separated from the ground state by an energy on the order of kT then their first-order Zeeman effects contribute to the total susceptibility according to the Boltzmann weights of the thermal distributions among them. In addition, there are the contributions from the second-order Zeeman effects between adjacent states. As a result, the magnetic moment is a complicated function of temperature. The Eu^{3+} ion comes into this category at higher temperatures, at which the thermal population of the J = 1 state is appreciable. Sm^{3+} (f^5) is also relevant here. For it, there is a first-order Zeeman effect contribution for the $^6H_{2\frac{1}{2}}$ ground state together with a second-order Zeeman

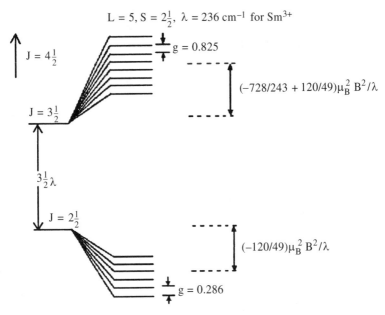

$$L = 5, S = 2\tfrac{1}{2}, \lambda = 236 \text{ cm}^{-1} \text{ for Sm}^{3+}$$

$J = 4\tfrac{1}{2}$

$g = 0.825$

$(-728/243 + 120/49)\mu_B^2 B^2/\lambda$

$J = 3\tfrac{1}{2}$

$3\tfrac{1}{2}\lambda$

$J = 2\tfrac{1}{2}$

$(-120/49)\mu_B^2 B^2/\lambda$

$g = 0.286$

Figure 9.3. The application of Eq. 9.16, or more obviously Eq. 9.41, to the first- and second-order Zeeman effects for the $J = 3\tfrac{1}{2}$ and $J = 2\tfrac{1}{2}$ states of the ^{6}H term of f^4. For this term, states exist up to $J = 7\tfrac{1}{2}$. The entry $-(728/243)\,\mu_B^2\, \mathbf{B}^2/\lambda$ arises from the second-order Zeeman effect between the $J = 3\tfrac{1}{2}$ and $J = 4\tfrac{1}{2}$ states.

effect from the ^{6}H$_{3\frac{1}{2}}$ state, which lies some 100s cm^{-1} higher. At higher temperatures, there is also population of the ^{6}H$_{3\frac{1}{2}}$ state, and hence a contribution also from its first- and second-order Zeeman effects. At ambient temperature, the magnetic moment for Sm^{3+} compounds is about 1.54 μ_B. The magnetic moment predicted by Eq. 9.36 is 0.84 μ_B. The first- and second-order Zeeman effect contributions for the $J = 2\tfrac{1}{2}$ and $J = 3\tfrac{1}{2}$ states of Sm^{3+} are given in Figure 9.3.

Applied to a set of states arising from a term, Eq. 9.16 can, in the light of the preceding discussion, be put in the form:

$$\chi_A = N_A \mu_B^2 \sum_{J=|L-S|}^{|L+S|} [g_J^2 J(J+1)(2J+1)/3kT - 2F_{J,J+1}$$

$$+ 2F_{J-1,J}]\exp\left(-W_J/kT\right) \Big/ \sum_{J=|L-S|}^{|L+S|} (2J+1)\exp\left(-W_J/kT\right) \quad (9.41)$$

Like g_J, W_J is a function of J, L, and S:

$$W_J = \frac{1}{2}[J(J+1) - |L-S|(|L-S|+1)]\lambda \quad (9.42)$$

For the Sm^{3+} ion, as set out in Figure 9.3:

$$\chi_A = N_A\mu_B^2[0.286^2 \cdot 2\tfrac{1}{2} \cdot 3\tfrac{1}{2} \cdot 6 + 240\,kT/49 + \{0.825^2 \cdot 3\tfrac{1}{2} \cdot 4\tfrac{1}{2} \cdot 8$$
$$+ \frac{[(1456/243 - 240/49)]kT/\lambda\}\exp\left(-3\tfrac{1}{2}\lambda/kT\right) + \cdots]}{6 + 8\exp\left(-3\tfrac{1}{2}\lambda/kT\right) + \cdots} \tag{9.43}$$

$$\mu_{eff} = 1.38\mu_B \quad \text{(at 300 K)} \tag{9.44}$$

9.2.4 States $\langle\langle$ kT

Although there are no transition-metal systems for which the spin–orbit coupling is so small that the approximation that states of a term lie $\langle\langle kT$ above the ground state is valid, it is illuminating to deduce the magnetic properties of such a case. If, as required by this condition, for the term concerned all the factors W_i^0 in Eq. 9.16 are zero, these equations reduce, after the collection of terms and conversion to μ_{eff}, to:

$$\mu_{eff} = [4\,S(S+1) + L(L+1)]^{\frac{1}{2}}\mu_B \tag{9.45}$$

If L is dropped from this expression, it becomes identical to Eq. 9.38, the formula for the spin-only magnetic moment, a result which, of course, is expected. Eq. 9.45 has been used to estimate the amount of orbital contribution to the magnetic moment. It is of little value in that regard, however, because it is based on behavior only at infinite temperature; results at finite temperatures can lead to quite different estimates. The magnetic moment may deviate from the spin-only value by much more than is given by Eq. 9.45. The maximum deviation occurs when states are separated by $\sim kT$ (see below).

9.3 QUENCHING OF ORBITAL ANGULAR MOMENTUM BY LIGAND FIELDS

It is not difficult to see qualitatively that there is likely to be a loss of orbital angular momentum on the incorporation of a free ion into a complex. Orbital angular momentum about an axis is associated with the ability to rotate an atomic or molecular orbital about the axis to give an *identical* and *degenerate* orbital. For instance, in the free ion, the d_{xz} orbital can be rotated around the Z axis by $2\pi/4$ to give the d_{yz} orbital (and vice versa), and the d_{xy} orbital can be rotated about the same axis by $2\pi/8$ to give the $d_{x^2-y^2}$ orbital. In the presence of a cubic ligand field, however, the d_{xy} and $d_{x^2-y^2}$ orbitals are no longer degenerate, and no orbital angular momentum arises between them. Not all the orbital angular momentum is quenched by the ligand field because, for example, the d_{xz} and d_{yz} orbitals remain degenerate. Orbital angular momentum remains to some extent within the $t_{2(g)}$ orbital set because, just as rotation about the Z axis turns d_{yz} into d_{xz}, so rotation about the X or Y axes turns d_{xy} into d_{xz} or d_{yz}, respectively. Inspection shows that no rotation can turn the d_{z^2} orbital into the $d_{x^2-y^2}$ orbital, if only because they differ in shape. There is, therefore, no orbital angular momentum associated with the $e_{(g)}$ set.

A further requirement for the existence of orbital angular momentum owing to orbital rotation is that there must not be an electron in the second orbital with the same spin quantum number as that in the commencing orbital. With these rules, it is possible to deduce in which configurations of an ion orbital angular momentum is totally or only partly quenched in the presence of a cubic ligand field. The only $t_{2(g)}$ configurations (apart from $t_{2(g)}^0$) for which it is not possible to make the required transformations of the d orbitals are $t_{2(g)}^3$ (because there is no vacant, equivalent, degenerate position of the same spin as any of the three electrons) and $t_{2(g)}^6$ (same reasoning).

Now it is seen that the orbital degeneracies of the terms to which the configurations $t_{2(g)}^0 \cdot e_{(g)}^m, t_{2(g)}^3 \cdot e_{(g)}^m$, and $t_{2(g)}^6 \cdot e_{(g)}^m$ $(m = 0-4)$ correspond, or most nearly correspond, are unity or two. Thus *orbital angular momentum is quenched for A and E terms.* Conversely, for $t_{2(g)}^1 . e_{(g)}^m, t_{2(g)}^2 . e_{(g)}^m, t_{2(g)}^4 . e_{(g)}^m$, and $t_{2(g)}^5 \cdot e_{(g)}^m$ which lead to T terms, the orbital angular momentum is *not* completely quenched. Thus *orbital angular momentum remains for T terms.*

The position is summarized in Table 9.3. The exact amount to which the orbital contribution to the magnetic moment occurs in T terms is a matter for calculation in each individual instance and is investigated below.

The preceding deductions may also be put on a more formal and rigorous basis. Comparing Sections 6.3 and 5.4, we see that the orbital wavefunctions for the $A_{1(g)}$ (Eq. 6.32) and $A_{2(g)}$ (Eq. 6.35) terms are, respectively, $\{3,0\}$ and $2^{-1/2}$ $(\{3,2\} + \{3,-2\})$. In each case, the matrix element for the orbital angular momentum in the Z direction is zero:

$$\langle \psi_{A_{(1,2)(g)}} | \mathbf{L}_z | \psi_{A_{(1,2)(g)}} \rangle = 0 \tag{9.46}$$

$$\langle \{L,0\} | \mathbf{L}_z | \{L,0\} \rangle = \langle 2^{-\frac{1}{2}}(\{3,2\} + \{3,-2\}) | \mathbf{L}_z | 2^{-\frac{1}{2}}(\{3,2\} + \{3,-2\}) \rangle$$
$$= 0 \tag{9.47}$$

For the $E_{(g)}$ terms, the wavefunctions are (Section 6.2) $\{2,0\}$ and $2^{\frac{1}{2}}$ $(\{2,2\} + \{2,-2\})$.

The matrix elements of $\mathbf{L}_z$ between these wavefunctions are also zero. In all instances, the orbital angular momentum in the Z direction and, because the symmetry is cubic, in any other direction, is also zero. We confirm that the orbital angular momentum is absent for $A_{(g)}$ and $E_{(g)}$ terms.

T terms cannot be discussed easily in general, for reasons that become obvious later. As an example, however, the wavefunctions of one of the weak field T terms, as given in Section 6.3 may be used. One of the wavefunctions of the $T_{1(g)}$ terms in a weak ligand field is (Eq. 6.33):

$$24^{-\frac{1}{2}} (15^{\frac{1}{2}}\{3,3\} - 3\{3,-1\}) \tag{9.48}$$

TABLE 9.3. Ground terms for Which Orbital Angular Momentum Is or Is Not Quenched by a Ligand Field of Cubic Symmetry

		Stereochemistry					
		Octahedral			tetrahedral		
Number of d electrons	Free Ion Ground Term	Nearest $t_{2g}^n e_g^m$ Configuration	Ground Term in Complex	Orbital Contribution Expected	Nearest $t_2^n e^m$ Configuration	Ground Term in Complex	Orbital Contribution Expected
1	^2D	t_{2g}^1	$^2T_{2g}$	Yes	e^1	^2E	No
2	^3F	t_{2g}^2	$^3T_{1g}$	Yes	e^2	$^3A_2$	No
3	^4F	t_{2g}^3	$^4A_{2g}$	No	$e^2t_2^1$	$^4T_1$	Yes
4	^5D	$t_{2g}^3 e_g^1$	^5E_g	No	$e^2t_2^2$	$^5T_2$	Yes
		t_{2g}^4	$^3T_{1g}$	Yes			
5	^6S	$t_{2g}^3 e_g^2$	$^6A_{1g}$	No	$e^2t_2^3$	$^6A_1$	No
		t_{2g}^5	$^2T_{2g}$	Yes			
6	^5D	$t_{2g}^4 e_g^2$	$^5T_{2g}$	Yes	$e^3t_2^1$	^5E	No
		t_{2g}^6	$^1A_{1g}$	No			
7	^4F	$t_{2g}^5 e_g^2$	$^4T_{1g}$	Yes	$e^4t_2^3$	$^4A_2$	No
		$t_{2g}^6 e_g^1$	^2E_g	No			
8	^3F	$t_{2g}^6 e_g^2$	$^3A_{2g}$	No	$e^4t_2^4$	$^3T_1$	Yes
9	^2D	$t_{2g}^6 e_g^3$	^2E_g	No	$e^4t_2^5$	$^2T_2$	Yes

The matrix element of $\mathbf{L}_z$ for this wavefunction is nonzero:

$$24^{-\frac{1}{2}}(15^{\frac{1}{2}}\{3,3\} - 3\{3,-1\})\mathbf{L}_z 24^{-\frac{1}{2}}(15^{\frac{1}{2}}\{3,3\} - 3\{3,-1\})$$
$$= 24^{-1}(15 \cdot 3 - 6 \cdot 15^{\frac{1}{2}} \cdot 0 + 9 \cdot -1)h/2\pi \qquad (9.49)$$
$$= 1\tfrac{1}{2}\,h/2\pi$$

For the term chosen, orbital angular momentum is *not* completely destroyed by the ligand field. This, in fact, is a general result and applies also to strong field $T_{1(g)}$ terms and to $T_{2(g)}$ terms.

The matrix elements of $\mathbf{L}_z$ were evaluated above because the results are required later. Qualitatively, the problem may be studied by means of group theory with much less labor. The operator $\mathbf{L}\ (=\mathbf{L}_x + \mathbf{L}_y + \mathbf{L}_z)$ possesses the symmetry properties of rotations about the three Cartesian axes. It belongs to the irreducible representation T_{1g} of O_h. None of the direct products in O_h— $A_{1g} \times T_{1g} \times A_{1g}$, $A_{2g} \times T_{1g} \times A_{2g}$ and $E_g \times T_{1g} \times E_g$—contain A_{1g}. Therefore, the matrix elements $\langle \psi_{A_{1g}}|\mathbf{L}|\psi_{A_{1g}}\rangle$, $\langle \psi_{A_{2g}}|\mathbf{L}|\psi_{A_{2g}}\rangle$, and $\langle \psi_{E_g}|\mathbf{L}|\psi_{E_g}\rangle$ are all zero. On the other hand, the direct products $T_{1g} \times T_{1g} \times T_{1g}$ and $T_{2g} \times T_{1g} \times T_{2g}$ do contain A_{1g}. Therefore, the matrix elements $\langle \psi_{T_{1g}}|\mathbf{L}|\psi_{T_{1g}}\rangle$ and $\langle \psi_{T_{2g}}|\mathbf{L}|\psi_{T_{2g}}\rangle$, may be nonzero. The conclusion is the same as previously: Orbital angular momentum is absent for A_{1g}, A_{2g}, and E_g terms but can be present for T_{1g} and T_{2g} terms. The argument has been developed for octahedral symmetry. In the group T_d, $\mathbf{L}$ belongs to the irreducible representation T_2. On evaluating the relevant direct products, it is found that the corresponding conclusion is identical: Orbital angular momentum is absent for A_1, A_2, and E terms, and it may exist for T_1 and T_2 terms.

It should also be noted that octahedral complexes with E ground terms can undergo sizable geometric distortions owing to the Jahn–Teller effect (Section 7.1.2). The main effect is to introduce large low-symmetry components to the ligand field. This has little influence on the average value of the magnetic moment, but causes it to become anisotropic. Octahedral complexes with T ground terms are also formally subject to Jahn–Teller distortions, although the structural effects are normally quite small (Section 7.1.2). The low-symmetry components of the ligand field caused by the distortions do mean, however, that some quenching of the orbital angular momentum is to be expected for this class of complex; and in fact, as is noted in Section 9.5.3, the effect may be substantial.

9.4 THE MAGNETIC PROPERTIES OF A AND E TERMS

Despite the arguments presented in the previous section, the magnetic moments of complexes with $A_{(1,2)(g)}$ and $E_{(g)}$ terms usually differ appreciably from the spin-only values to which they are expected to conform in the absence of orbital angular momentum in the ground term. The reason for the departure from the spin-only value lies partly in the existence of the second-order Zeeman effect between the ground and the higher ligand field terms. It lies mainly, however, in the fact that, in the presence of spin–orbit coupling, the quenching effect of the ligand field cannot be complete. As mentioned in connection with spectra (Section 8.1.1.2), in the presence of spin–orbit coupling, it is not possible to factor accurately the total wavefunction into spin and orbital wavefunctions. For the present purposes, spin–orbit coupling may be supposed to mix in terms of different orbital quantum degeneracy so that their separation on the basis of their orbital quantum numbers in the ligand field is not entirely valid.

Consider, for instance, an F term that is split by a ligand field so that an $A_{2(g)}$ term lies lowest; the term wavefunction is $2^{-\frac{1}{2}}(\{3,2\}-\{3,-2\})$. The operator for spin–orbit coupling is:

$$\lambda \mathbf{L} \cdot \mathbf{S} = \lambda(\mathbf{L}_x\mathbf{S}_x + \mathbf{L}_y\mathbf{S}_y + \mathbf{L}_z\mathbf{S}_z) \qquad (9.50)$$

The spin for the two terms is the same, and that part of the operator involving $\mathbf{S}$ is treated separately, because the spin wavefunctions integrate out to a small number in the matrix elements. Considering only the orbital part of the matrix elements, we evaluate the way in which the two terms are connected by the operator $\mathbf{L}$. One of the matrix elements includes:

$$\langle 2^{-\frac{1}{2}}(\{3,2\} - \{3,-2\})|\mathbf{L}_z|2^{-\frac{1}{2}}(\{3,2\} + \{3,-2\})\rangle$$
$$= \frac{1}{2}(2 - 2 \cdot 2 \cdot 0 + 2)h/2\pi \qquad (9.51)$$
$$= 2(h/2\pi)$$

Indeed, this is the only matrix element that is nonzero as far as the Z direction is concerned. By first-order perturbation theory, then, the orbital wavefunction for the $A_{2(g)}$ term is mixed with a small amount of the wavefunction $2^{-\frac{1}{2}}(\{3,2\}-\{3,2\})$:

$$\psi_{A_{2(g)}} = (1 + c^2)^{-\frac{1}{2}}[2^{-\frac{1}{2}}(\{3,2\} + \{3,-2\}) - c2^{-\frac{1}{2}}(\{3,2\} - \{3,-2\})] \quad (9.52)$$

with

$$c = 2\lambda\langle[M_S]|\mathbf{S}_z|[M_S]\rangle/[E_{T_{2(g)}} - E_{A_{2(g)}}]$$
$$= 2M_S\lambda/\Delta \qquad (9.53)$$

If c^2 is neglected in comparison with unity, a neglect that is often justified because λ is usually much smaller than Δ, then:

$$\langle \psi_{A_{2(g)}}[M_S]|\mathbf{L}_z|\psi_{A_{2(g)}}[M_S]\rangle = -(2 \cdot 2 \cdot 2)M_S\lambda/\Delta \qquad (h/2\pi \text{ units}) \quad (9.54)$$

Thus, when the magnetic moment operator for the Z direction is applied to $\psi_{A_{2(g)}}$, the splitting factor g, instead of being 2.00 as for a term devoid of orbital angular momentum, is:

$$g = 2.00(1 - 4\lambda/\Delta) \qquad (9.55)$$

This may be seen by reference to, say, a $^3A_{2g}$ term, in which the spin wavefunctions are $[1, \pm 1]$ and $[1,0]$.:

$$\langle \psi_{A_{2g}}[1, \pm 1]|\mathbf{L}_z + 2\mathbf{S}_z|\psi_{A_{2g}}[1, \pm 1]\rangle = \pm 2\langle \psi_{A_{2g}}[1, \pm 1]|\mathbf{L}_z|\psi_{A_{2g}}[1, \pm 1]\rangle$$
$$= \pm 2 \mp 8\lambda/\Delta \qquad (9.56)$$
$$\langle \Psi_{A_{2g}}[1, 0]|\mathbf{L}_z + 2\mathbf{S}_z|\psi_{A_{2g}}[1, 0]\rangle = 0 \cdot \langle \psi_{A_{2g}}[1, 0]|\mathbf{L}_z|\psi_{A_{2g}}[1, 0]\rangle$$
$$= 0 \qquad (9.57)$$

Hence the magnetic moment from the first-order Zeeman effect terms is (Eq. 9.36 with $J = S$):

$$\mu_{\text{eff}} = (1 - 4\lambda/\Delta)[S(S + 1)]^{\frac{1}{2}}$$
$$= (1 - 4\lambda/\Delta)\mu_{\text{eff}}^{\text{spin only}} \qquad (9.58)$$

or

$$\chi_A = (1 - 8\lambda/\Delta)\chi_A^{\text{spin only}} \qquad (9.59)$$

The expressions indicate how orbital contribution is brought back into the spin-only moment through the effects of spin–orbit coupling.

Similar operations performed with $A_{1(g)}$ and $E_{(g)}$ terms give the following results. For $A_{1(g)}$ terms, there is no mixing, because there are no excited terms of the same multiplicity. For $E_{(g)}$ terms:

$$g = 2.00(1 - 2\lambda/\Delta) \qquad (9.60)$$
$$\mu_{\text{eff}} = (1 - 2\lambda/\Delta)\mu_{\text{eff}}^{\text{so}} \qquad (9.61)$$
$$\chi_A = (1 - 4\lambda/\Delta)\chi_A^{\text{so}} \qquad (9.62)$$

For use in these expressions, λ is given as a function of ζ in Table 9.4.

As well as the mixing of excited term wavefunctions into the ground term by spin–orbit coupling, there is also the second-order Zeeman effect with these higher terms to be considered. Once again, consider the orbital wavefunction of an $A_{2(g)}$ term and the $24^{-\frac{1}{2}}(\{3,2\} + \{3,-2\})$ wavefunction of the $T_{2(g)}$ term above it. The matrix element of $\mathbf{L}_z$ between the two wavefunctions is, as we have

TABLE 9.4. The Relationship Between the Spin–Orbit Coupling Parameters λ and ζ for Electron Configurations in Cubic Symmetry[a]

Number of d electrons	Symmetry	Ground Term	λ
1	Octahedral	$^2T_{2g}$	ζ
	Tetrahedral	^2E	ζ
2	Octahedral	$^3T_{1g}$	$\zeta/2$
	Tetrahedral	$^3A_2$	$\zeta/2$
3	Octahedral	$^4A_{2g}$	$\zeta/3$
	Tetrahedral	$^4T_1$	$\zeta/3$
4	Octahedral	^5E_g	$\zeta/4$
	Tetrahedral	$^5T_2$	$\zeta/4$
	Octahedral	$^3T_{1g}$	$\zeta/2$
5	Octahedral	$^6A_{1g}$	—
	Tetrahedral	$^6A_1$	—
	Octahedral	$^2T_{2g}$	$-\zeta$
6	Octahedral	$^5T_{2g}$	$-\zeta/4$
	Tetrahedral	^5E	$-\zeta/4$
	Octahedral	$^1A_{1g}$	—
7	Octahedral	$^4T_{1g}$	$-\zeta/3$
	Tetrahedral	$^4T_1$	$-\zeta/3$
	Octahedral	^2E_g	$-\zeta$
8	Octahedral	$^3A_{2g}$	$-\zeta/2$
	Tetrahedral	$^3T_1$	$-\zeta/2$
9	Octahedral	$^2E_{2g}$	$-\zeta$
	Tetrahedral	$^2T_2$	$-\zeta$

[a]The values for the A and E terms are for use in Eqs. 9.55, 9.58, 9.60, 9.61, 9.87, 9.89, 9.90, and 9.92–96.

seen, $2h/2\pi$. When the magnetic moment operator for the Z direction is applied between the two, the result is:

$$24^{-\frac{1}{2}}(\{3,2\} + \{3,-2\})|\mathbf{L}_z + 2\mathbf{S}_z|24^{-\frac{1}{2}}(\{3,2\} - \{3,-2\})\rangle = 2\mu_B \quad (9.63)$$

The part involving $\mathbf{S}_z$ contributes nothing, whatever the spin wavefunctions are, because the two orbital wavefunctions are orthogonal (Section 1.3). Thus the second-order Zeeman effect coefficient for the $A_{2(g)}$ term owing to this interaction is, summing over the $(2S+1)$ spin components:

$$\sum_{-S}^{S} \Sigma W_{A_{2(g)}}^{II} = (2\mu_B)^2(2S+1)/(E_{A_{2(g)}} - E_{T_{2(g)}})$$

$$= 4(2S+1)\,\mu_B^2/\Delta \quad (9.64)$$

Using this result in connection with Eq. 9.16 (or more directly with Eqs. 9.24 and 9.25) the susceptibility owing to the second-order Zeeman effect is:

$$\chi_A^{TIP} = 8 N_A \mu_B^2 / |\Delta| \tag{9.65}$$

The same procedure is carried out for the other ground terms. The results are as follows. For the $^6A_{1(g)}$ term of d^5 (and the $^1A_{1(g)}$ terms of $d^{0,10}$): zero. For the $^1A_{1g}$ term of d^6 (t_{2g}^6)

$$8 N_A \mu_B^2 / |\Delta| \tag{9.66}$$

For the $A_{2(g)}$ terms:

$$8 N_A \mu_B^2 / |\Delta| \tag{9.67}$$

For the $E_{(g)}$ terms:

$$4 N_A \mu_B^2 / |\Delta| \tag{9.68}$$

In this connection we note that $N_A \mu_B^2 = 0.261$ cm^{-1}, so that those quantities are on the order of $1 \cdot 10^{-9}$ m^3 mol^{-1}.

The magnetic behavior of complexes possessing $A_{(g)}$ and $E_{(g)}$ ground terms should be in the form predicted by Eq. 9.26: the Curie law plus a small constant term (the TIP). Also, on the basis of the preceding paragraphs, the Curie constant is seen to differ somewhat from the spin-only value.

9.5   THE MAGNETIC PROPERTIES OF T TERMS

9.5.1   Splitting by Spin–Orbit Coupling

The magnetic moments for complexes with T ground terms are obtained by performing the sums for the first- and second-order Zeeman effects among the states that arise from the splitting of the terms by spin–orbit coupling. The effect of spin–orbit coupling in T terms is found by operating with $\lambda \mathbf{L} \cdot \mathbf{S}$ on the appropriate wavefunctions. This rather laborious procedure, however, which cannot be elaborated on here, can be avoided by the observation that there is a correspondence between the wavefunctions of a T terms and those of a free-ion P term (4). In each case, there is threefold orbital degeneracy. The results for P terms are readily available (Sections 5.4 and 9.2). Attention is restricted mainly to strong-field configurations in which the wavefunctions of the T terms may be written in the form of $t_{2g}^n \cdot e_g^m$ configurations. The three members of the $t_{2(g)}$ orbital set correspond to the three free ion p orbitals (4). The relationships:

$$\langle \{1, \pm 1\} | \mathbf{L}_z | \{1, \pm 1\} \rangle = \langle \{2, \pm 1\} | \mathbf{L}_z | \{2, \pm 1\} \rangle = \pm 1 \cdot h/2\pi \tag{9.69}$$

and:

$$\langle\{1,0\}|\mathbf{L}_z|\{1,0\}\rangle = 2^{-1/2}(\{2,2\} - (2,-2)|\mathbf{L}_z|2^{-1/2}(2,2) - \{2,-2\})\rangle = 0$$
$$(9.70)$$

provide an obvious correspondence. The matrix elements for orbital angular momentum for the p orbitals are on the left side, and those for the t_{2g} orbitals are in the center. It is necessary, however, to consider the effects of $\mathbf{L}_x$ and $\mathbf{L}_y$ in addition to those of $\mathbf{L}_z$ (see Appendix 8). We find that the matrix elements between different wavefunctions are then important:

$$\langle\{1,\pm1\}|\mathbf{L}_{x,y}|\{1,\pm0\}\rangle = -\langle\{2,\pm1\}|\mathbf{L}_{x,y}|2^{-1/2}(\{2,2\} - \{2,-2\})\rangle$$
$$= 2^{1/2}h/2\pi \qquad (9.71)$$

The difference in sign for this matrix element for the p orbitals on the one hand and for the $t_{2(g)}$ orbitals on the other hand reverses the sign of the splitting of T terms of $t_{2(g)}^m$ configurations compared to p^n configurations. In other words, the wavefunction $2^{-\frac{1}{2}}(\{2,2\}-\{2,-2\})$ of the $t_{2(g)}$ orbitals is not entirely equivalent to the $\{1,0\}$ wavefunction of the p orbitals as far as orbital angular momentum is concerned. For the $^2T_{2(g)}$ terms and for the $^3T_{2g}$ terms at the strong field limit (where the wavefunctions are defined by t_{2g}^n configurations), the splittings by spin–orbit coupling are obtained by inverting those for the P terms arising from the same number of p electrons as of d electrons. Remembering that there is an inversion of the spin–orbit splitting pattern for a p shell more than half full (Section 5.5: $p^{6-n}\equiv p^n$, $n<3$) there results:

$$^2T_{2g}(d^1, t_{2g}^1) \equiv {}^2P(p^5) \qquad (9.72)$$

$$^3T_{1g}(d^2, t_{2g}^2) \equiv {}^3P(p^4) \qquad (9.73)$$

$$^3T_{2g}(d^4, t_{2g}^4) \equiv {}^3P(p^2) \qquad (9.74)$$

$$^2T_{2g}(d^5, t_{2g}^5) \equiv {}^2P(p^1) \qquad (9.75)$$

$$^3T_1(d^8, t_2^2) \equiv {}^2P(p^4) \qquad (9.76)$$

$$^2T_2(d^9, t_2^1) \equiv {}^2P(p^5) \qquad (9.77)$$

For the d^8 and d^9 examples, the e^4 filled orbitals may be ignored; the problem is attacked by considering the t_2 holes in the filled d^{10} shell.

The results for the $^4T_{1(g)}$ and $^5T_{2(g)}$ ground terms, which have no equivalent for p configurations, are obtained by the stratagem of *supposing* that there is a term with P orbital wavefunctions derived from the p orbitals, but with a spin quantum number S having the value of $1\frac{1}{2}$ or 2, as required. The splitting of such a hypothetical term by spin–orbit coupling is worked out by the rules given in Section 5.5 just as readily as for an ordinary P term of a p^{1-5} configuration. The results of these arguments are collected in Table 9.4 and Figure 9.4.

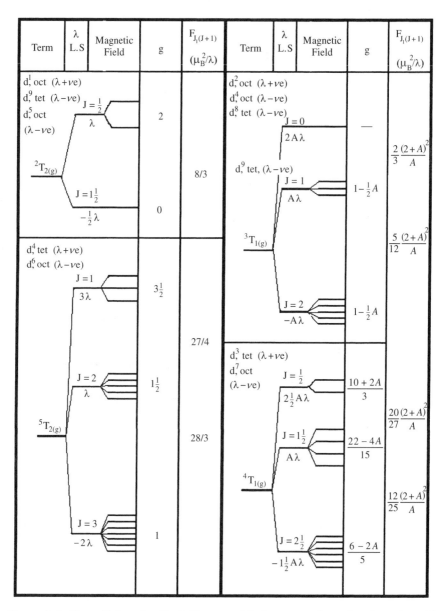

Figure 9.4. The splitting of cubic field terms of d electron configurations by spin–orbit coupling. The first- and second-order Zeeman effect coefficients are given, but the displacements owing to the second-order Zeeman effect are not included.

The problem of the $T_{1(g)}$ terms that are not at the strong field limit remains. The wavefunctions of such terms vary from those given in Section 5.4 (cf. Section 2.3) to those of the strong-field limit just treated, the $t_{2g}^n \cdot e_g^m$ configurations. The variation is connected with the mixing of the $T_{1(g)}$ terms from the F and P free ion terms of d configurations, as discussed in Section 6.8. The mixing is brought about by the presence of the ligand field. In Section 6.8, the secular determinant for the energies of the two $^3T_{1g}$ terms of d^2 is given (Eq. 6.49). With the substitution of the lower root of Eq. 6.52 into this determinant, a linear combination of the wavefunctions of the two weak-field $^3T_{1g}$ terms of d^2 is obtained. It corresponds to the $^3T_{1g}(F)$ ground term in other strengths of the ligand field:

$$\psi_{3T_{1g}}(F) = (1 + c^2)^{-1/2}[\psi_{3T_{1g}}(F)^0 + c\psi_{3T_{1g}}(P)^0] \qquad (9.78)$$

The wavefunctions for the free ion ^3F and ^3P terms—i.e., $\psi_{3T_{1g}}(F)^0$ and $\psi_{3T_{1g}}(P)^0$—are available in Section 5.4. In Eq. 9.78:

$$c = (6Dq + E)/4Dq \qquad (9.79)$$

Here the sign convention for Dq is that given along with Eq. 6.52, with E again being its lower root. Applying the operator L_z to $\psi_{3T_{1g}}(F)$:

$$\langle\psi_{3T_{1g}}(F)|L_z|\psi_{3T_{1g}}(F)\rangle = (1.5 - c^2)/(1 + c^2) \qquad (h/2\pi \text{ units}) = A \quad (9.80)$$

Because c lies between 0 and $-\frac{1}{2}$, A lies between 1.5 (weak field limit) and 1.0 (strong field limit). In other words, the matrix elements for orbital angular momentum are larger for the $^3T_{1g}(F)$ term in the weak- and intermediate-field regions than for the strong-field limit by the factor A. A may be evaluated from a knowledge of the parameters of interelectronic repulsion (B) and the ligand field magnitude (Δ). The splitting by spin–orbit coupling for the other regions is obtained by multiplying that for the strong-field limit by the factor A. This result is not confined to the $^3T_{1g}(F)$ term of d^2; it holds generally for all $T_{1(g)}(F)$ ground terms of d configurations. In Figure 9.4, the splitting of the various terms of d electron configurations by spin–orbit coupling is given. The compilation of this figure is aided by remembering the relationship between the spin–orbit coupling constants λ and ζ (Eq. 5.37), which holds among the cubic field terms as well as the terms of the free ion. The minus sign in Eq. 5.37 holds for a t_{2g} shell more than half filled. The relationship between λ and ζ for various configurations is summarized in Table 9.4.

9.5.2   The calculation of μ_{eff}

The calculation of the first- and second-order Zeeman effects for states of the cubic field terms from first principles is not a simple operation and is not elaborated on here. The magnitudes of these two effects may be obtained, however, from the results of the free ion p^n P terms, provided that the orbital angular momentum operator employed is $-AL_z$ rather than L_z. The minus sign allows for the inversion of the spin–orbit coupling splitting between the p^n and t_{2g}^n configurations discussed in Section 9.5.1. For T_{2g} ground terms, in which the orbital wavefunctions are given directly by the strong-field configurations $t_{2(g)}^n \cdot e_{(g)}^4$ (cf. Section 5.4 and Table 6.1), A may be taken to be unity for all values of the magnitude of the ligand field parameter. The equivalent to Eq. 9.37 for g becomes:

$$g = 1 - (\tfrac{1}{2})A + (2 + A)[S(S+1) - 2]/2\,J(J+1) \qquad (9.81)$$

and the equivalent of Eq. 9.40 is, again using $L = 1$:

$$F_{J,(J+1)} = [(2+A)^2/A]F_{J,(J+1)}^{\text{free ion P term}} \qquad (9.82)$$

The results of Eqs. 9.81 and 9.82 for the first- and second-order Zeeman effects applied to the states of T terms are included in Figure 9.4.

The evaluation of the magnetic susceptibility for the $^3T_{1g}$ ground term of d^2 is used as an example. For the ^3P term of p^2 the states $J = 0$, 1, and 2 lie at relative energies 0, 1, and 3λ. Consequently, for the $^3T_{1g}$ term of d^2, they lie at relative energies $3A\lambda$, $2A\lambda$, and 0, respectively. Applying Eqs. 9.81 and 9.82, we obtain the results set out in Figure 9.4, with the g values for the three levels:—, $1 - \tfrac{1}{2}A$, and $1 - \tfrac{1}{2}A$ respectively; the second-order Zeeman effects are between adjacent states, in order, $2(2+A)^2\mu_B^2/3A\lambda$ and $5(2+A)^2\mu_B^2/2A\lambda$. The application of Eq. 9.41 then gives, with $x = kT/\lambda$:

$$
\begin{aligned}
\chi_A = {} & (N_A\mu_B^2/3kT) \cdot 3\{5(2-A)^2/2 + 5(2+A)^2/6Ax + [(2-A)^2 \\
& + (2+A)^2/2Ax]\exp(-2Ax) - [4(2+A)^2/3Ax)] \\
& \times \exp(-3Ax)\}/\{5 + 3\exp(-2Ax) + \exp(-3Ax)\} \qquad (9.83)
\end{aligned}
$$

This expression applies to all $^3T_{1(g)}$ ground terms. The inversion of the splitting that takes place with, for example, the $^3T_1$ term of d^8 because of the reversal of the sign of λ (Table 9.4) is covered by the concomitant reversal in the sign of x for use in the expression. The expressions for the $^4T_{1(g)}$, $^2T_{2(g)}$, and $^5T_{2(g)}$ ground terms are given in Appendix 9.

The results of calculations based on these expressions are put conveniently in the form of graphs of μ_{eff} against temperature, as measured by the parameter kT/λ. Two graphs are necessary for each term, one for λ positive, and one for λ negative. In these expressions μ_{eff} is the square root of the portion that follows $N_A\mu_B^2/3kT$. The relevant graphs are shown in Figure 9.5. In the cases of the $^3T_{1(g)}$ and $^4T_{1(g)}$ ground terms, an infinite series of plots would be required, because A varies from 1 to 1.5. Only the plots for the extremes of $A = 1.0$ and 1.5 are, in fact, given, corresponding to the strong- and weak-field limits, respectively.

The fact that the Co^{2+} ion usually does not correspond to $A = 1.0$ or $A = 1.5$ can be seen by evaluating A for a typical set of donor atoms with the ion. Taking

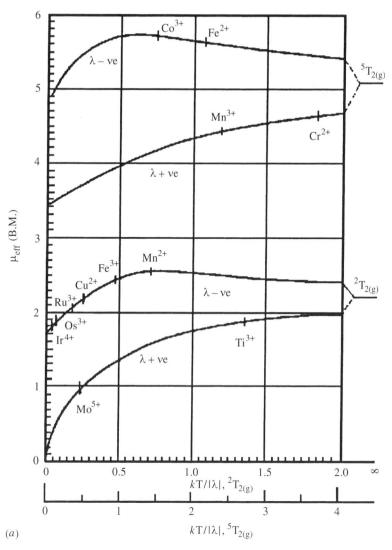

(a)

Figure 9.5. The magnetic moment as a function of temperature, as defined by the parameter $kT/|\lambda|$, for d electron T terms in cubic stereochemistries. Ions are marked at the free-ion values of spin–orbit coupling and at 300 K. (**a**) The $^2T_{2(g)}$ terms of d^1, d^5, and d^9 and the $^5T_{2(g)}$ terms of d^4 and d^6. (**b**) The $^3T_{1(g)}$ terms of d^2, d^4 (spin paired), and d^8 for first-row metal ions (**c**) second- and third-row metal ions. (**d**) The $^4T_{1(g)}$ terms of d^3 and d^7.

water as the ligand, Δ is known to be $9200 \, \text{cm}^{-1}$ and B to be $850 \, \text{cm}^{-1}$ (Section 8.4.1). Substituting these values into Eq. 6.52, the lower root is:

$$E = -6290 \, \text{cm}^{-1} \tag{9.84}$$

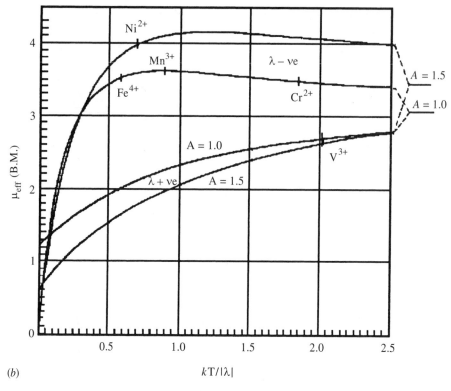

Figure 9.5. (*Continued.*)

Substitution of this value into Eq. 9.79 gives:

$$c = -0.208 \tag{9.85}$$

Then, by Eq. 9.80:

$$A = 1.40 \tag{9.86}$$

9.5.3   Departure from Cubic Symmetry

The Jahn–Teller theorem suggests that even when identical ligands are present, complexes with T ground terms are likely to adopt distorted structures (Section 7.1.2). This introduces ligand field components of symmetry lower than cubic, which lift the orbital degeneracy of the T terms. Such components markedly affect the magnetic properties of the T terms, because they alter the arrangements of the energy levels among which thermal distribution occurs. The problem of finding the magnetic moment of a term subject to the influence of both spin–orbit coupling and a low-symmetry ligand field component is one of some complexity. No attempt can be made here to indicate how the problem is

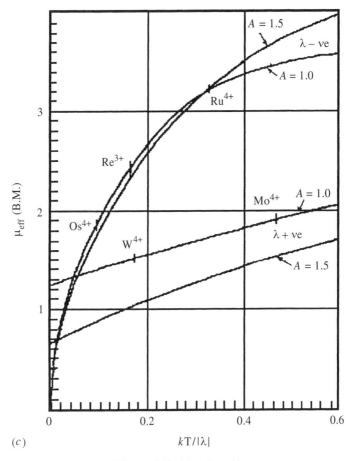

Figure 9.5. (*Continued.*)

dealt with or many details of the results. Some general comments on the effects of such components can, however, be made.

If the ligand field component creates a splitting of the T term that, in the absence of spin–orbit coupling, would be of energy a good deal less than the spin–orbit coupling constant λ, then the effect on the magnetic moment is fairly small. As the splitting by the low-symmetry ligand field component becomes larger than the spin–orbit coupling constant, the magnetic moment tends toward the spin-only value, and the temperature dependence tends to decrease. Finally, as the component becomes very large, the moment approaches the spin-only value and is independent of temperature. These effects occur because the low-symmetry ligand field component further quenches orbital angular momentum by destroying the degeneracy of the $t_{2(g)}$ orbital set. In Figure 9.6, the effect of a low-symmetry ligand field component—the magnitude of which is expressed as the ratio of the splitting of the T orbital energies to λ—is given for the $^2T_{2(g)}$ ground term.

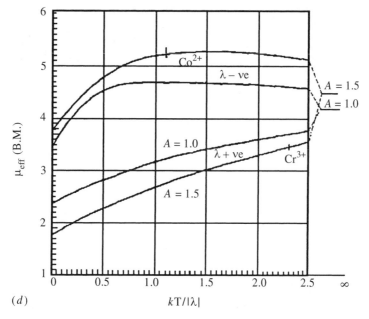

(d)

$kT/|\lambda|$

Figure 9.5. (*Continued.*)

9.6 $t_{2(g)}$ ELECTRON DELOCALIZATION

9.6.1 General

It is necessary to consider the effective reduction of the orbital angular momentum of a central metal ion, particularly when it is caused by delocalization of electrons from the $t_{2(g)}$ orbitals of the ion onto the donor atoms of the ligands. Such delocalization takes place, of course, whenever the wavefunctions of the metal ion mix with those of the ligands to form the molecular orbitals of the complex (Sections 3.1.1 and 8.4.2). It is customary to suppose that the matrix elements of orbital angular momentum are reduced by the factor k below those calculated by using the operator **L**. In other words, the operator for orbital angular momentum in the X direction, for example, is $k\mathbf{L}_x$. The value of k is unity for no delocalization but is generally somewhat less than unity; $2(1-k)$ may be taken to be the approximate weight with which the p_π orbital set enters into the molecular orbitals for the d electrons; for instance, the weight with which the p_π ligand t_{2g} orbital set enters into the t^b_{2g} orbitals shown in Fig. 3.2. At first sight it appears that the weight should be $(1-k)$; however, the unpaired electrons in the ligand p_π orbitals make a contribution to the orbital angular momentum of the system. About half of the total orbital angular momentum of the system comes from this contribution, which is about half of that which would be made if the unpaired electrons were on the metal ion. The origin of k may be more complex than the electron delocalization just discussed

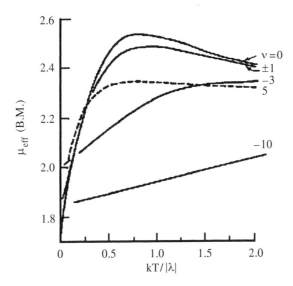

Figure 9.6. The effect of a low-symmetry ligand-field component on the magnetic moment of the $^2T_{2(g)}$ term, with λ −ve. The component is expressed as the ratio of the splitting of the T orbital wavefunctions to the spin–orbit coupling constant λ; $v = \delta/\lambda$. The curves are constructed for $k = 1$ (Section 9.6).

(5), and care must be taken to see that other contributions are not serious before attributing a reduction in k to this cause. Cases for T terms are known for which, owing to interaction with a nearby higher configuration, the effective value of k is greater than unity see also Section 10.3.1.

Another mechanism by which orbital angular momentum can be reduced for T terms can be particularly effective. A *dynamic* Jahn–Teller effect results when the associated structural distortion is so small that the resultant energy splitting is less than kT at the temperature involved. When there are certain relationships between the dynamic Jahn–Teller effect and vibrations, strong quenching of orbital angular momentum can result, which shows itself as large reductions in the apparent value of k. For example, the EPR g$_\parallel$ value for the V^{3+} ion in Al_2O_3, 1.11, appears to correspond to $k \approx 0.5$, owing to this phenomenon. The phenomenon is commonly known as the *Ham effect* (6,7). The relationships involved in the Ham effect are complex and cannot be elaborated here.

The effect of $t_{2(g)}$ electron delocalization, or other cause of reduction in k below unity, is usually to bring the magnetic moment closer to the spin-only value, because it corresponds to an additional quenching of the orbital angular momentum.

9.6.2   A and E Terms

For $A_{(g)}$ and $E_{(g)}$ terms, the effect of $t_{2(g)}$ electron delocalization is taken into account by using Eqs. 9.55, 9.58, 9.60, and 9.61, but with a value of λ that is

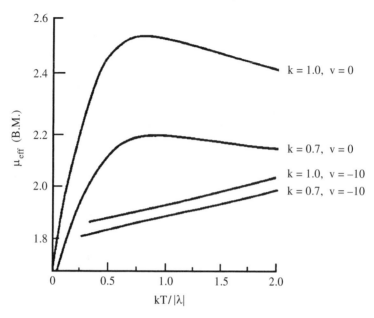

Figure 9.7. The effect of t_{2g} electron delocalization on the magnetic moment of the $^2T_{2g}$ term, λ $-$ve. The results are given for $k = 0.7$, for cubic symmetry ($v = 0$), and for a highly distorted arrangement ($v = -10$).

reduced below that of the free ion by the factor k. The operator $k\mathbf{L}_z$ is thus used to evaluate the orbital angular momentum. The effective reduction in λ comes about because the spin–orbit coupling operator becomes $\lambda_0 k\mathbf{L} \cdot \mathbf{S}^*$. Henceforth we refer to the free ion spin–orbit coupling as λ_0, employing λ to mean the value that is effective in the complexed metal ion. There results, on the modification of the equations for the first-order Zeeman effect, for $A_{2(g)}$ terms:

$$\chi_A = \chi_A^{so}(1 - 8k^2\lambda_0/\Delta) \tag{9.87}$$

$$g = 2(1 - 4k^2\lambda_0/\Delta) \tag{9.88}$$

$$\mu_{eff} = \mu_{eff}^{so}(1 - 4k^2\lambda_0/\Delta) \tag{9.89}$$

For $E_{(g)}$ terms:

$$\chi_A = \chi_A^{so}(1 - 4k^2\lambda_0/\Delta) \tag{9.90}$$

$$g = 2(1 - 2k^2\lambda_0/\Delta) \tag{9.91}$$

$$\mu_{eff} = \mu_{eff}^{so}(1 - 2k^2\lambda_0/\Delta) \tag{9.92}$$

---

* k for use in connection with $\lambda\mathbf{L} \cdot \mathbf{S}$ partly arises from e_g electron delocalization; however, we can include the effect under the t_{2g} delocalization with good enough approximation for the present purposes see also Section 10.3.1.

The expressions for the second-order Zeeman effect are influenced in much the same way: Eqs. 9.67 and 9.68 become for the $^1A_{1g}$ term of t^6_{2g} and $A_{2(g)}$ terms:

$$\chi_A = 8k^2 N_A \mu_B^2 / |\Delta|, \qquad (9.93)$$

for the $E_{(g)}$ terms:

$$\chi_A = 4k^2 N_A \mu_B^2 / |\Delta| \qquad (9.94)$$

In total, the susceptibilities for the $A_{(g)}$ terms are:

$$\chi_A = \chi_A^{so}(1 - 8k^2\lambda_0/\Delta) + 8kN_A\mu_B^2/|\Delta| \qquad (9.95)$$

and for the $E_{(g)}$ terms:

$$\chi_A = \chi_A^{so}(1 - 4k^2\lambda_0/\Delta) + 4k^2N_A\mu_B^2/|\Delta| \qquad 9.96$$

9.6.3 T Terms

Electron delocalization is introduced in the results for the magnetic moments of T terms by writing kA everywhere A appears in Eq. 9.83. Approximately, one may take $\lambda = k\lambda_0$. In the case of T terms the complicated nature of the interaction between spin and orbital angular momenta brought about by spin–orbit coupling makes generalizations on the effect of the introduction of k difficult. It may be noted that the most usual effect is that, as might be expected, the magnetic moment is brought nearer to the spin-only value on the reduction of k, but exceptions occur. In Figure 9.7 the effect of a reduction in k to below unity is given for the $^2T_{2g}$ term, both in cubic and in lower symmetry. No further detail of the electron delocalization problem is considered here.

9.7   THE MAGNETIC PROPERTIES OF COMPLEXES WITH A AND E GROUND TERMS

9.7.1 Octahedral Complexes

The discussion that follows for this and the succeeding sections is necessarily rather approximate in nature, because a number of secondary features of magnetic behavior are not considered. Also, most effects of magnetic exchange are ignored, although certain limited aspects are taken up in Section 9.11. It is seen by inspection of the equations in Sections 9.4 and 9.6.2 that, apart from the influence of the TIP term in the susceptibility, the magnetic moments of complexes possessing $A_{(g)}$ or $E_{(g)}$ ground terms should not vary with temperature. The magnetic properties of some typical complexes of transition-metal ions with A_{2g} and E_g ground terms are listed in Table 9.5. For reasons given in the footnotes to the table, we restrict comments to the first-row transition metals. The departure of the Curie law component from the spin-only value can be more accurately measured by the EPR g value, as is shown in Chapter 10, so the discussion here is limited to essentials.

In the first-row metals on the left side of the series, λ is small and Δ_{oct} is fairly large, so that the Curie law component is little different from the spin-only value and the TIP component is almost negligible. Consequently, for the cases of d^3 ($^4A_{2g}$), d^4 (^5E_g), and particularly d^5 ($^6A_{1g}$) the magnetic moment is close to the spin-only value and is independent of temperature.

For the metals on the right side of the transition series, λ is larger and Δ_{oct} is smaller, and the departures from the ideal behavior mentioned above are noticeable. The cases involved are d^7 (^2E_g), d^8 ($^3A_{2g}$), and d^9 (^2E_g). A small temperature dependence of the magnetic moment may be observable, and it may be possible to extract chemically relevant information with worthwhile confidence. As an example, we use the $Ni(H_2O)_6^{2+}$ ion, d^8 ($^3A_{2g}$) and refer to the data in Table 9.5.

The TIP here is not quite negligible compared to the Curie law susceptibility and introduces a small variation of the magnetic moment with temperature. The spin–orbit coupling constant in the Ni^{2+} ion is sufficiently large to make possible a useful comparison between the experimental data and the predicted results. With $\Delta_{oct} = 8900\,cm^{-1}$, $\lambda_0 = -315\,cm^{-1}$, and $\chi_A = 54.5 \cdot 10^{-9}\,m^3\,mol^{-1}$, for the $Ni(H_2O)_6^{2+}$ ion, one has from Eq. 9.95:

$$54.5 = 41.9(1 + 0.28k^2) + 3.0k^2 \qquad (9.97)$$

giving $k = 0.92$. The estimate of k is of limited accuracy, but it does provide evidence for some small degree of delocalization.

With similar treatments, the $Co(NO_2)_6^{4-}$ ion [d^7 (^2E_g), Eq. 9.96] and the $Cu(H_2O)_6^{2+}$ ion [d^9 (^2E_g), Eq. 9.96] yield k as 0.7 and 0.8, respectively. Again, the accuracy of these estimates is limited, but there is evidence of delocalization effects.

The case of d^6 ($^1A_{1g}$) is an exception. Here the Curie law susceptibility is absent, and TIP is the only paramagnetic term. Many of the relevant complexes in Table 9.5 do indeed show the expected small paramagnetism at ambient temperature.

9.7.2 Tetrahedral Complexes

The magnetic properties of some typical complexes of transition-metal ions with A and E ground terms in tetrahedral coordination are given in Table 9.6. In connection with Eqs. 9.59, 9.60, 9.61, 9.67, 9.68, 9.95, and 9.96, it is necessary to take Δ_{tet} without regard to its sign.

Values of Δ_{tet} are only about one-half those for Δ_{oct}, so the departures from the spin-only magnetic moment and variation with temperature are more marked

TABLE 9.5. The Magnetic Moments of Some Typical Octahedral Transition-Metal Complexes with A or E Ground Terms[a]

d^n	Ground State	$[4S(S+1)]^{1/2}$ (μ_B)	Ion	Compound	μ_{eff}^{90K} (μ_B)	μ_{eff}^{300K} (μ_B)	μ_{eff}^{calc} (μ_B)	$10^9 \chi_A^{300K}$ ($m^3\ mol^{-1}$)	TIP
3	$^3A_{2g}$	3.87	Cr^{3+}	$KCr(SO_4)_2 \cdot 12H_2O$	3.84	3.84	3.79	77.3	1.5
			Mo^{3+}	K_3MoCl_6	3.62	3.79	3.66	75.4	1.5
			Mn^{4+}	$BaMnF_6$	3.80	3.80	3.76	75.9	1.2
			Re^{4+}	Cs_2ReCl_6	2.87	3.35	3.58^b	58.7	1.0
4	^5E_g	4.90	Cr^{2+}	$CrSO_4 \cdot 6H_2O$	4.84	4.82	4.87	121.3	0.9
			Mn^{3+}	$Mn(acac)_3$	4.75	4.86	4.87	123.8	0.5
5	$^6A_{1g}$	5.92	Mn^{2+}	$K_2Mn(SO_4)_2 \cdot 6H_2O$	5.92	5.92	5.92	183.5	0.0
			Fe^{3+}	$KFe(SO_4)_2 \cdot 12H_2O$	5.89	5.89	5.92	182.2	0.0
6	$^1A_{1g}$	0.0	Fe^{2+}	$K_4Fe(CN)_6$	0.16	0.35	0.0	0.6	1.5
			Co^{3+}	$Co(NH_3)_6 \cdot Cl_3$	0.25	0.46	0.0	1.1	1.1
			Rh^{3+}	$Rh(NH_3)_6 \cdot Cl_3$	0.16	0.35	0.0	0.6	0.8
			Ir^{3+}	K_3IrCl_6	0.0	0.0	0.0	0.0^c	1.0
			Pt^{4+}	K_2PtCl_6	0.0	0.0	0.0	0.0^c	0.9
7	^2E_g	1.73	Co^{2+}	$K_2BaCo(NO_2)_6$	1.74	1.81	1.86	17.2	0.9
8	$^3A_{2g}$	2.83	Ni^{2+}	$K_2Ni(SO_4)_2 \cdot 6H_2O$	3.20	3.23	3.23	54.5	1.5
9	^2E_g	1.73	Cu^{2+}	$K_2Cu(SO_4)_2 \cdot 6H_2O$	1.91	1.91	1.97	19.1	1.1

[a] $\mu_{eff}^{calc} = \mu_{eff}^{so}(1-4\lambda/|\Delta_{oct}|)$ for A terms. The TIP of $8N_A\mu_B^2/|\Delta_{oct}|$ is ignored. $\mu_{eff}^{calc} = \mu_{eff}^{so}(1-2\lambda_0/|\Delta_{oct}|)$ for E terms. The TIP of $4N_A\mu_B^2/|\Delta_{oct}|$ is ignored. The value of Δ_{oct} used is that for the set of ligands in the compound listed, evaluated in Table 8.5. Values of λ_0 are from Table 5.5.

[b] For this compound, the effect of magnetic exchange and the secondary features of magnetic behavior (particularly $j \cdot j$ coupling; see Section 5.5) are important.

[c] Experimentally, χ_A is slightly negative, but such a result is meaningless in the present connection. It is probably owing to experimental inaccuracy and to uncertainty in the diamagnetic contribution to the susceptibility.

TABLE 9.6. The Magnetic Properties of Some Typical Tetrahedral Complexes with A and E Ground Terms[a]

d^n	Ion	Compound	Ground State	$[4S(S+1)]^{1/2}$ (μ_B)	μ_{eff}^{90K} (μ_B)	μ_{eff}^{300K} (μ_B)	μ_{eff}^{calc} (μ_B)	$10^9 \chi_A^{300K}$ (m^3 mol^{-1})	TIP
1	V^{4+}	VCl_4	^2E	1.73	1.71	1.69	1.62[b]	15.1	1.6
5	Mn^{2+}	$[(C_2H_5)_4N]_2\ MnCl_4$	$^6A_1$	5.92	—	5.94	5.92	184.7	0.0
	Fe^{3+}	$[(C_2H_5)_4N]\ FeCl_4$			—	5.88	5.92	182.2	0.0
6	Fe^{2+}	$[(C_2H_5)_4N]_2\ FeCl_4$	^5E	4.90	5.40	5.40	5.2	152.1	4.0
7	Co^{2+}	Cs_3CoCl_5	$^4A_2$	3.87	4.48	4.71	4.70	116.2	8.6

[a]Calculated from Eqs. 9.58, 9.60, and 9.61. The values of λ_0 are from Table 9.4; the values of Δ_{tet} from Table 8.3.
[b]$\Delta_{tet} = -8000$ cm^{-1}.

than for octahedral complexes, as may be seen by comparing Tables 9.5 and 9.6. For the d^1 (^2E) ground state case, λ is so small that departure from the spin-only moment is scarcely observable. Of course, there is also none for the d^5 ($^6A_1$) ground term.

On the right side of the first-row metals, λ is larger and Δ_{tet} is smaller, and appreciable effects can be seen in the d^6 (^5E) and d^7 ($^4A_2$) cases. As an example of the latter, the $CoCl_4^{2-}$ ion is treated.

The TIP for the cobaltous complex of Table 9.6 is important, even in the presence of the large Curie law susceptibility. It is largely responsible for the manner in which the magnetic moment varies with temperature. Proceeding as before, with $\Delta_{tet} = -3200\,cm^{-1}$, and $\lambda_0 = -172\,cm^{-1}$, we write:

$$116.2 = 78.5(1 + 0.43k^2) + 8.6k^2 \tag{9.98}$$

showing that for the $CoCl_4^{2-}$ ion, $k \simeq 0.9$. Here within the limited accuracy of the estimation, there is again evidence of some delocalization. In this instance, the TIP contribution is so large that it may be estimated from the temperature dependence of the susceptibility with a certain amount of accuracy. Experimentally, it is found to be about $6.3 \cdot 10^{-9}\,m^3\,mol^{-1}$, in fair agreement with the calculated value of $8.6 \cdot 10^{-9}\,m^3\,mol^{-1}$.

9.8   THE MAGNETIC PROPERTIES OF COMPLEXES WITH T GROUND TERMS

The ensuing discussion of the magnetic properties of complexes with T ground terms can be only rather general, because the details of the behavior are complicated. They depend critically on the presence of low-symmetry ligand field components and of t_{2g} electron delocalization or other cause of the reduction of the value of k. These are matters that we cannot treat here in any exact fashion. In general, the magnetic moments of complexes possessing a T ground term are expected to be a function of temperature. The magnetic behavior of some typical compounds that give rise to T terms are summarized in abbreviated form in Table 9.7.

9.8.1 d^1, $^2T_{2g}$

As anticipated for a compound possessing a T ground term, the magnetic moment of cesium titanium alum varies with temperature. It does not, however, vary as much as required by Figure 9.5a. Although the moment at ambient temperature is fairly close to the value given by Figure 9.5a, it is not possible to account accurately for the magnetic behavior of the compound with temperature unless electron delocalization and a low-symmetry ligand-field component are taken into account. In Figure 9.8, the behavior is compared to the theory that includes these effects, as derived from the type of information set out in Figures

Table 9.7. The Magnetic Properties of Some Typical Transition-Metal Complexes with T Ground Terms[a]

d^n	Ground State	$[4S(S+1)]^{1/2}$ (μ_B)	Ion	Compound	μ_{eff}^{90K} (μ_B)	μ_{eff}^{300K} (μ_B)	$\mu_{\text{eff}}^{\text{calc}\,a}$ (μ_B)
1	$^2T_{2g}$	1.73	Ti^{3+}	$CsTi(SO_4)_2 \cdot 12H_2O$	1.55	1.84	1.9
2	$^3T_{1g}$	2.83	V^{3+}	$(NH_4)V(SO_4)_2 \cdot 12H_2O$	2.78	2.80	2.9^b
4	$^3T_{1g}$	2.83	Mn^{3+}	$K_3Mn(CN)_6$	3.31	3.50	3.6
			Ru^{4+}	K_2RuCl_6	1.63	2.67	3.1
			Os^{4+}	K_2OsCl_6	0.78	1.50	1.4^c
5	$^2T_{2g}$	1.73	Mn^{2+}	$K_4Mn(CN)_6 \cdot 3H_2O$	2.03	2.18	2.5
			Fe^{3+}	$K_3Fe(CN)_6$	1.90	2.25	2.4
			Ru^{3+}	$Ru(NH_3)_6Cl_3$	1.85	2.13	2.1
			Os^{3+}	$Os(NH_3)_6Br_3$	1.62	1.62	1.9
6	$^5T_{2g}$	4.90	Fe^{2+}	$(NH_4)_2Fe(SO_4)_2 \cdot 6H_2O$	5.37	5.47	5.6
7	$^4T_{1g}$	3.87	Co^{2+}	$(NH_4)_2Co(SO_4)_2 \cdot 6H_2O$	4.60	5.10	5.1^d
8	$^3T_1$	2.83	Ni^{2+}	$[(C_2H_5)_4N]_2NiCl_4$	3.25	3.89	4.0^e

[a] Calculated employing the spin–orbit coupling constant for the free ion and ignoring electron delocalization and low-symmetry ligand field components.
[b] Calculated for $A = 1.27$. (Section 9.5) and $\Delta/B = 2.8$, estimated from Tables 8.4 and 5.3.
[c] Second-order effects of the large spin–orbit coupling must be included here.
[d] Calculated for $A = 1.4$ (Section 9.5)
[e] Calculated for $A = 1.5$ (Section 9.5) and Δ/B estimated from Tables 8.4 and 5.3.

9.6 and 9.7 for the $^2T_{2g}$ terms. k has been deduced to be 0.7, and the splitting of the $^2T_{2g}$ orbital wavefunctions to be about $350\,\text{cm}^{-1}$.

9.8.2 d^2, $^3T_{1g}$

Contrary to initial expectation, the near spin-only magnetic moment of ammonium vanadium alum does not fall appreciably as the temperature is lowered (Fig. 9.5b). This behavior is the result of the presence of a large low-symmetry component in the ligand field. The component is so large it overwhelms almost completely the effects of spin–orbit coupling (Section 9.5.3). It is estimated that the splitting of the orbital degeneracy of the $^3T_{1g}$ term by this low-symmetry component must be at least 10 times the magnitude of the spin–orbit coupling constant λ for the V^{3+} ion. In fact, the splitting has been estimated to be about $2000\,\text{cm}^{-1}$, with k = 0.8. At low temperatures, the magnetic behavior is more typical of a T term—the moment then falls rapidly as the temperature is lowered.

9.8.3 d^4, $^3T_{1g}$

To force spin pairing, the octahedral ligand field must be strong. Therefore, one may take A to be 1.0 for use with the magnetic results for the $^3T_{1g}$ term of d^4 (Section 9.5).

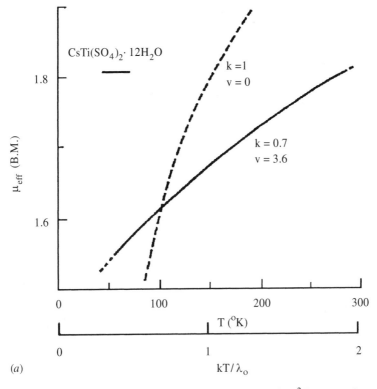

(a)

Figure 9.8. The magnetic moments of two complexes possessing $^2T_{2g}$ ground terms as a function of temperature, compared to theory. *Solid lines*, experimental results; *dashed lines*, calculated from the theory without allowance for delocalization or low-symmetry ligand field components. The experimental results may be closely reproduced by using the values of the parameters used in Figures 9.6 and 9.7.

The magnetic moment of the spin-paired trivalent manganese compound in Table 9.7 goes through a broad maximum between ambient temperature and 80 K and falls rapidly below this to zero at 0 K. As may be seen, that is the form of behavior expected on the basis of Figure 9.5b. The moment at room temperature is close to that required by Figure 9.5b, with $\lambda_0 = -177\,\mathrm{cm}^{-1}$ (Tables 5.5 and 9.4); however, the position of the maximum in the moment and the rate of change of moment with temperature are not closely in agreement with the Figure. Electron delocalization and a low-symmetry ligand field component must be taken into account in a more careful treatment.

The magnetic moment of the tetravalent ruthenium complex of Table 9.7 varies a good deal with temperature, as expected. In fact, although the moment required by Figure 9.5c is close to the spin-only value for a spin-paired d^4 compound, the susceptibility is almost entirely owing to the second-order Zeeman effect between the $J = 0$ and $J = 1$ states shown in Figure 9.4. The spin–orbit coupling is not so large that the susceptibility is independent of temperature

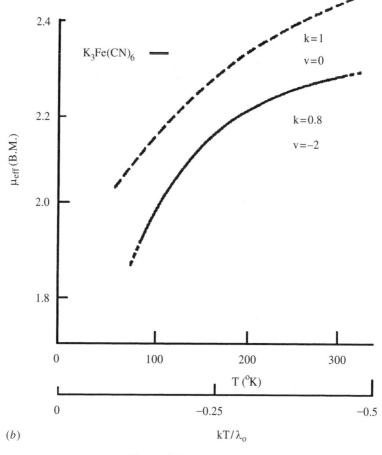

Figure 9.8. (*Continued.*)

(Section 9.1.3, application 2). It is found that the susceptibility can be accounted for on the basis of Eq. 9.31 ($n = 1$; $m = 3$), but it is probable that some electron delocalization and a low-symmetry ligand field component are present.

The magnetic moment of the tetravalent osmium compound of Table 9.7 varies as $T^{1/2}$, a behavior characteristic of TIP. Indeed, the spin–orbit coupling constant for the Os^{4+} ion is so large that the susceptibility, which is entirely owing to the second-order Zeeman effect, is independent of temperature. The susceptibility as derived from the moment given in Figure 9.5c, however, is a good deal larger than the experimental value. The difference lies in the large value of the spin–orbit coupling constant. It is so large that it cannot be considered to be a perturbation on the effects of the interelectronic repulsion terms. An intermediate coupling scheme must be invoked, as outlined in Section 5.5. The magnetic moment included in the μ_{eff}^{300K} column in Table 9.7 has been corrected for that effect.

9.8.4 d^5, $^2T_{2g}$

The magnetic moments of the spin-paired bivalent manganese and trivalent iron complexes listed in Table 9.7 fall quite appreciably with temperature. Such a result is expected in view of the values of their spin–orbit coupling constants in relationship to temperature. A comparison of the variation of moment with temperature, given for $K_3Fe(CN)_6$ in Figure 9.8, allows an estimate of the extent of electron delocalization and the magnitude of the low-symmetry ligand field component to be made. k is found to be about 0.8 in each compound. There is no appreciable low-symmetry ligand field component relative to the spin–orbit coupling constant in the iron complex. The orbital splitting of the $^2T_{2g}$ ground term is estimated to be about 400 cm^{-1} in the manganocyanide salt.

The spin–orbit coupling constants for the Ru^{3+} and Os^{3+} ions are so large that the compounds of trivalent ruthenium and osmium involve the use of the left side of Figure 9.5a. The magnetic moments are not very different from the spin-only values and their variations with temperature are small. Indeed, the ruthenium compound in Table 9.7 is compatible with the figure. The insensitivity of the magnetic moment to temperature changes and its proximity to the spin-only value make it difficult to obtain information about the magnitudes of electron delocalization and low-symmetry ligand field components. The magnetic moment of the osmium compound at ambient temperature is, as expected, quite close to the spin-only value and independent of temperature.

9.8.5 d^6, $^5T_{2g}$

The magnetic moment of the bivalent iron compound listed in Table 9.7 is, at ambient temperature, close to that predicted with the aid of Figure 9.5a. The variation of the moment with temperature, however, can be accounted for in detail only if electron delocalization and a low-symmetry ligand field component are included in the treatment. k has been deduced to be about 0.7, and the orbital splitting of the $^5T_{2g}$ term to be about 200 cm^{-1}.

9.8.6 d^7, $^4T_{2g}$

The magnetic moment of the bivalent cobaltous complex listed in Table 9.7 is close to that predicted on the basis of Figure 9.5d, when the calculation is performed with $A = 1.4$ (Section 9.5.2). To account accurately for the experimental results, it is necessary to consider electron delocalization and a low-symmetry ligand field component. The former is small—k ≈ 1.0—but the orbital splitting is about 1000 cm^{-1}.

9.8.7 d^8, $^3T_1$

The magnetic moment of the tetrahedral bivalent nickel complex listed in Table 9.7 is close to that predicted on the basis of Figure 9.5b. The cubic ligand field parameter in this tetrahedral complex is so small ($\Delta_{tet} \approx -3500\,\mathrm{cm}^{-1}$) that it may be considered to be weak, and $A = 1.50$. It is with the weak-field limit that comparison is made in Figure 9.5b. The variation of the moment with temperature cannot be reproduced well without the inclusion of the effects of electron delocalization and a low-symmetry ligand field component. It is found that there is comparatively little delocalization—$k \approx 0.95$—but that the orbital splitting of the $^3T_1$ term is fairly large, $\sim 700\,\mathrm{cm}^{-1}$.

9.9 SUMMARY

To summarize the discussion of the magnetic results of the last two sections the following statements can be made:

1. As predicted in Section 9.4, the magnetic moments of complex ions possessing $A_{2(g)}$ and $E_{(g)}$ ground terms depart from the spin-only value by a small amount. The amount depends on the relationship between the spin–orbit coupling constant and the magnitude of the cubic ligand field parameter Δ. The moments are substantially independent of temperature, a result consistent with the theory. There is evidence of some degree of electron delocalization in most instances, although the amounts are not well defined.

2. The magnetic moments of complexes possessing T ground terms usually depart appreciably from the spin-only value and vary with temperature. Such behavior is anticipated on the basis of Section 9.8, but the agreement in detail between the experimental results and the theory outlined there is not good. Although the effects of $t_{2(g)}$ electron delocalization and low-symmetry ligand field components cannot be considered here in more than a passing fashion, it has been made obvious that their presence is responsible for much of the lack of agreement. In many instances, the results of more sophisticated treatments of the T term data have been quoted. They give an idea of the extent of $t_{2(g)}$ delocalization that is found and the magnitudes of the low-symmetry ligand field components that are present.

3. The magnetic moments of the complexes possessing $^6A_{1(g)}$ ground terms are close to their spin-only values and are independent of temperature. Some complexes with $^1A_{1(g)}$ ground terms show small values of TIP.

9.10 SPIN-FREE–SPIN-PAIRED EQUILIBRIA

On occasion, the ligand field acting on relevant ions may make the ground terms of the spin-free (high-spin) and the spin-paired (low-spin) configurations almost equal in energy. In other words, the ligand field magnitude may cause the energy levels to lie close to the cross-over points in Figures. 6.5 to 6.8, or the points where there is a change of ground term in Figures 6.11 to 6.14. In each case, the low-spin state has fewer electrons in the antibonding e_g^* orbital set than has the high-spin state. For this reason, the change in ground term is almost always accompanied by a substantial change in the metal–ligand bond length; this aspect was discussed in detail in Section 7.1.3. Because the ligand-field splitting Δ depends on the metal–ligand bond distance, it decreases when a complex goes from low spin to high spin. Spin equilibria must, therefore, always be interpreted using the energy levels of two different Δ values, one on each side of the cross-over point of the appropriate Tanabe–Sugano diagram. If the energies of the two spin states do not differ by more than about kT, then their relative populations are of comparable magnitudes and so vary with temperature.

If the two terms of different multiplicities have zero matrix element of spin–orbit coupling between them then, as a first approximation, a fairly simple expression holds for the susceptibility of the system. It can be written as the population-weighted average of the susceptibilities for the two terms. The energy separation between the two terms related to kT is defined as:

$$y = (E_{S2} - E_{S1})/kT \tag{9.99}$$

Then:

$$\chi_A = [(2S_1+1)\chi_{A1} + (2S_2+1)\chi_{A2}\exp(-y)]/[(2S_1+1) + (2S_2+1)\exp(-y)] \tag{9.100}$$

where E_{S1} and E_{S2} are the energies of the terms with degeneracies of $(2S_1 + 1)$ and $(2S_2 + 1)$, respectively (Section 9.2); χ_{A_1} and χ_{A_2} are the respective atomic susceptibilities for these terms and may themselves also may be functions of temperature.

More properly, the study of the equilibrium must take into account the effects of the entropy term that arises because of the different numbers of components in the S_1 and S_2 terms and because of the thermal vibrations, which affect the terms differently (8). Such considerations are beyond the scope of this work.

Spin-free–spin-paired equilibria are possible in octahedral complexes with the following configurations and ions:

$$d^4, Cr^{2+}, Mn^{3+} \quad {}^5E_g \leftrightarrow {}^3T_{1g}$$
$$d^5, Mn^{2+}, Fe^{3+} \quad {}^6A_{1g} \leftrightarrow {}^2T_{2g}$$
$$d^6, Fe^{2+}, Co^{3+} \quad {}^5T_{2g} \leftrightarrow {}^1A_{1g}$$
$$d^7, Co^{2+} \quad {}^4T_{1g} \leftrightarrow {}^2E_g$$

Of these, only the d^5 and d^6 cases fall into the category that has zero matrix element of $\mathbf{L} \cdot \mathbf{S}$ between the two terms. They are accounted for, in the first case for example, by:

$$\chi_{{}^6A_{1g}} = [6\chi_{{}^6A_{1g}} + 6\chi_{{}^2T_{2g}}\exp(-y)]/[6 + 6\exp(-y)]$$
$$= [\chi_{{}^6A_{1g}} + \chi_{{}^2T_{2g}}\exp(-y)]/[1 + \exp(-y)] \tag{9.101}$$

with:

$$y = (E_{^2T_{2g}} - E_{^6A_{1g}})/kT \tag{9.102}$$

In this expression, $\chi_{^2T_{2g}}$ itself is a function of temperature and may be introduced as a calculated quantity (Section 9.5) or as an experimentally determined result taken from a $^2T_{2g}$ complex with similar ligands around the same metal atom. In the latter method, some allowance for the effects of electron delocalization and a low-symmetry ligand field component might be included.

It has been established that high-spin–low-spin equilibria are in fact present in a number of systems. The tris(dithiocarbamato)Fe(III) (d^5) complexes form a well-defined example (8), as do those of some bis(bipyridyl)Fe(II) (d^6) and bis(terpyridyl)Co(II) (d^7) salts (9,10). In figure 9.9 the variation with temperature of the magnetic moment of four complexes of the type $(R_2N \cdot CS_2)_3Fe$ is shown. It is seen that for $R = py$ the system is always high-spin $S = 2\frac{1}{2}$, but for $R = nBu$, Me, and iBu, there is an increasing presence of the low-spin, $S = \frac{1}{2}$, form, which increases as the temperature becomes lower. The data do not fit Eq. 9.100 at all well; the effective sextet–doublet energy separation varies with temperature. Entropy, vibrations, and possibly other effects confuse the issue.

For the remaining two configurations, d^4 and d^7, the position is complicated by the fact that the matrix elements of spin–orbit coupling between the two terms may be appreciable. In the region where the two terms are separated by an

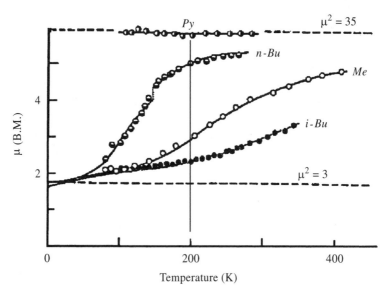

Figure 9.9. Variation of the magnetic moments of $[Fe(S_2CNR_2)_3]$ compounds with temperature (8). *Solid lines*, calculated from Eq. 9.100, with allowance for entropy and vibrational considerations. [From Ref. 8. Used with permission.]

amount that is not large compared to the spin–orbit coupling splitting, the arrangement of the energy levels, and consequently the expressions for the susceptibility, can be complicated. If spin–orbit coupling connects the two terms strongly, the resultant ground state is of "mixed spin" type, belonging to neither of the spin quantum numbers of the individual terms. The distinction between this and fractional populations of terms with different spins can be seen in the experimental data in some instances, but the effects are too subtle to be treated here.

In fact, the position for all the configurations to be considered is often even more complicated, because *cooperative* effects are not uncommon, as seen for example in $Fe(phen)_2(NCS)_2$ (11). These occur because a change in metal–ligand bond length accompanies the spin change owing to the different occupation of the antibonding e_g^* orbitals in the two spin cases (Section 7.1.3). In solids, in which the complexes are in close contact, when one metal ion changes spin it will change the interactions with neighboring complexes and hence the energy separation between their spin states. Then the presence of enough of one spin form can cause the whole crystal to adopt that form. The result is a large change of magnetic properties over a small temperature range, a form of magnetic phase separation. The change may be sharp, being spread out over only a few degrees. Also, hysteresis effects may occur, so that the susceptibility versus temperature curve with cooling is offset when the sample is warmed. These phenomena are beyond the scope of this text.

9.11 MAGNETIC EXCHANGE

In the introduction to this chapter, it was stated that the interactions between atomic magnetic dipoles in *magnetically concentrated* substances leads to complicated magnetic behavior, beyond the scope of this work. There is, however, a limited class of substances for which magnetic exchange is present but does not develop the cooperative effects that bring complicated field dependent behavior patterns. In this class, only a few atomic dipoles are coupled together, and this group does not interact magnetically with any other. Such a group is referred to as a *cluster*. We now look at some features of clusters of interacting atomic magnetic centers.

9.11.1 The Heisenberg Hamiltonian

Magnetic exchange between two dipoles on atoms 1 and 2 specified by spin quantum numbers S_1 and S_2 is usually parametrized in terms of the Heisenberg Hamiltonian (1):

$$\mathbf{H}_{ex} = -2J_{12}S_1 \cdot S_2 \tag{9.103}$$

where J_{12} is the *exchange integral*. The result of this Hamiltonian is to create a number of spin states for the combined system, with spin quantum numbers S'. If

J_{12} is positive, the ground state has total spin $S_1 + S_2$. The highest state then has $S' = |S_1 - S_2|$ and the interaction is said to be *ferromagnetic* in nature, because it corresponds to the alignment of the spin vectors S_1 and S_2 parallel (although the actual ferromagnetic properties developed are modest because the effects are limited to isolated sections of the lattice). If J_{12} is negative, the lowest state is that of lowest S', often 0 or $\frac{1}{2}$ and the interaction is said to be *antiferromagnetic*, with S_1 and S_2 antiparallel.

For a cluster, one must sum over all the pairs of atoms involved:

$$\mathbf{H}_{ex} = -2 \sum_{i,j} J_{ij} S_i \cdot S_j \qquad (9.104)$$

and the states specified by S' usually ranging from 0 or $\frac{1}{2}$ up to $\sum S_i$ occur, each possibly more than once. The simplest and commonest case is for two spins each of $S = \frac{1}{2}$. Then only the single exchange interaction J_{12} is to be considered. The best-known example is the dimeric cupric acetate hydrate $[Cu(CH_3 \cdot COO)_2 \cdot H_2O]_2$, with tetra ($\mu$-acatato) bridging, as shown in Figure 9.10. Each d^9 cupric ion has an unpaired electron in its $d_{x^2-y^2}$ orbital, and they are magnetically coupled with $J_{12} = -280 \, cm^{-1}$. For that case, the Heisenberg Hamiltonian leads to states with $S' = \frac{1}{2} + \frac{1}{2} = 1$ and $\frac{1}{2} - \frac{1}{2} = 0$, with separation $2J_{12}$. The interaction between the cupric ions is antiferromagnetic, and J_{12} is negative. The state with $S' = 0$ is lowest in energy, and that with $S' = 1$ lies $-2J_{12}$ higher. It is relatively easy to write down the magnetic susceptibility expected for such as system; it is

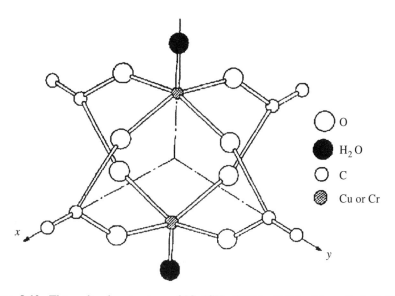

O   O

● H_2O

O   C

⊗   Cu or Cr

Figure 9.10. The molecular structure of $[Cu(CH_3 \cdot COO)_2 \cdot H_2O]_2$. [From Ref. 13. Used with Permission.]

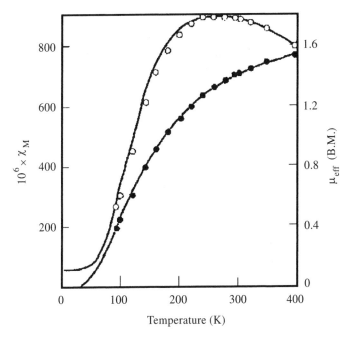

Figure 9.11. Calculated (*curves*) and experimental magnetic susceptibilities χ_{Cu} (○) and moments μ_{eff} (●) for $[Cu_2(CH_3 \cdot COO)_4 \cdot (OH_2)_2]$. [From Ref. 13. Used with permission.]

the Boltzmann–weighted sum of those for individual terms:

$$\chi_{Cu} = 12(N_A\mu_B^2/3k)(g^2/4)\exp(-2J_{12}/kT)/[1+3\exp(-2J_{12}/kT)]$$
$$+ 4N_A\mu_B^2/\Delta \tag{9.105}$$

The magnetic behavior of the complex conforms to that expression very closely, as is shown in Figure 9.11. Dimeric cupric complexes with a wide variety of bridging ligands other than acetate are readily obtained. The host of results for such compounds all follow Eq. 9.105, with J_{12} varying from negative values of a few hundred cm^{-1} to small positive values (Section 9.11.3).

9.11.2   Mechanism of Magnetic Exchange: Superexchange

Although the Heisenberg exchange Hamiltonian successfully parametrizes exchange interactions by a formal dipole coupling, it does not concern itself with the *mechanism* of the exchange coupling. Contrary perhaps to initial expectation, the most important source of magnetic coupling is *not* direct through-space overlap of orbitals on the magnetic centers. Rather it is through the bridging ligand atoms, in a phenomenon known as *superexchange* (12,13). It can readily be shown that through-space magnetic exchange falls off rapidly with distance between the magnetic centers, so that in the cupric acetate case mentioned above, the Cu-Cu separation (a little over 260 pm) is such that direct exchange must be negligible. Superexchange, on the other hand, can be carried through a number of intervening chemically bonded atoms in a ligand without drastic attenuation. Because of superexchange, then, magnetic exchange between metal atoms 1 nm or more apart can be appreciable, even when the pathway involves weak hydrogen bonds.

In principle, the energies of the states that arise in a "molecule" with several centers each with unpaired electrons (e.g., transition-metal atoms) can be calculated by molecular orbital theory. In practice, the state separations are so small compared to other bonding effects that this fails in all but the simplest cases. For the cupric acetate system mentioned above such a calculation is useful, but not really quantitative. For larger systems, not even qualitatively correct results are available by this approach.

A more generally successful, although at best semiquantitative, description of the superexchange process comes from a perturbation approach that draws heavily on the symmetry properties of atomic orbitals involved in the pathway.

First, however, it is illuminating to examine an expression that was given in the earliest approach to the magnetic exchange problem. For two electrons 1 and 2 assigned to orbitals ϕ_a and ϕ_b attached to, but not necessarily confined to, different atoms, the exchange integral is readily written rather generally as:

$$J_{12} = (e^2/r_{ab})S^2 - 2e^2S\langle\phi_a^1|\phi_b^1\rangle/r_{a,b} + \langle\phi_a^1\phi_b^2|e^2/r_{12}|\phi_a^2\phi_b^1\rangle \qquad (9.106)$$

Although the quantitative evaluation of this expression is rarely possible, it does lead to some valuable generalizations:

1. The last, "exchange," term of Eq. 9.106 is smaller than the first and second ones; it is always positive and is likely to be finite even for reasonable separations between the origins of orbitals ϕ_a and ϕ_b. This small ferromagnetic contribution to J_{12} is omnipresent.

2. The first and second terms in Eq. 9.106 are of opposite sign, but both are modulated by the $\phi_a - \phi_b$ overlap integral S and so vanish if those orbitals are orthogonal. In that case, the residual ferromagnetic third term dominates.

3. For nonorthogonal orbitals ϕ_a and ϕ_b, the second term in Eq 9.106 is usually larger than the first, and J_{12} is negative; i.e., it is antiferromagnetic in nature. The relationship between the values of the first and second terms is critical, and it is not possible to obtain the magnitude of J_{12} from simple considerations.

As indicated, ϕ_a and ϕ_b may be each be delocalized over two or more atoms. Thus the magnetic coupling paramatrized by J_{12} may be transmitted by the overlap of ϕ_a and ϕ_b on atoms forming a bridge between their formal centers, which are generally transition metal ions. Even the weak hydrogen bond may participate successfully in such a bridge.

With the above features in mind, it has been possible to formulate some relatively simple rules with predictive power about the sign and, to a limited

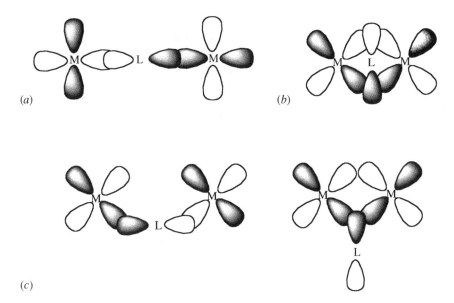

Figure 9.12. Superexchange pathways at 180°, σ (**a**); 180°, π (**b**); and 90° (**c**).

extent, the relative magnitude of J_{12} in different situations. Consider the following example of two transition metal ions with an intervening F^- or O^{2-} ion at the center of the line joining them. This is the 180° superexchange pathway shown in Figure 9.12.

9.11.2.1 *180° Superexchange.* First, suppose that the relevant metal orbitals are both half filled, whereas that of the bridging ion is filled, thus giving a total of four electrons. They are distributed over three orbitals, and it is possible to write four essentially different configurations, in two of which an electron is transferred from the bridge orbital to a metal orbital. They are:

Configuration	Atom a	Bridge	Atom b
1	↑	↑↓	↓
2	↑	↑↓	↑
3	↑↓ ↑	↑ ↓	↓ ↑↓
4	↑↓ ↑	↓ ↑	↓ ↑↓

Of these configurations, 1 and 2 represent, respectively, antiferromagnetic (J_{12} negative) and ferromagnetic (J_{12} positive) magnetic coupling between the metal atoms and are expected to differ only slightly in energy. From elementary considerations of chemical bonding, it is expected that the singlet excited state configurations 3 should lie a good deal lower than the triplet configurations 4.

Thus the perturbation mixing of configuration 3 into 1 lowers the energy of 1 more that does the mixing of configuration 4 into 2 lower the energy of 2.

Thus the prediction is that configuration 1 is the ground state and the 180 superexchange interaction is antiferromagnetic for metal half-filled nonortho-gonal orbitals. For the case of σ bonding, it also can be shown that this is the largest likely interaction. The π bonding pathway shown in Figure 9.12b gives the same result, but the magnitude is less owing to the poorer overlap of orbitals.

Second, consider the case of a half-filled orbital on one metal atom and an empty orbital on the other. Distributing the three electrons over the three orbitals by an electron transfer from the bridge gives three configurations:

Configuration	Atom a	Bridge	Atom b
1	↑	↑↓	
2	↑	↓	↑
3	↑	↑	↓

Configuration 1 is by definition the ground state. Of the other two, from elementary considerations of chemical bonding, configuration 2 is expected to lie a good deal lower in energy than 3. Thus by perturbation theory, configuration 2 is mixed into the ground state to a larger extent than is 3, and so gives it a net ferromagnetic component. Although this component reinforces the third, exchange, term in Eq. 9.106, the net result is still expected to be distinctly smaller than the half-filled–half-filled orbital interaction. A similar, but somewhat more lengthy argument produces the same result for the equivalent case of a half-filled orbital on one metal atom and a *filled* orbital on the other.

The conclusion is that 180° superexchange between a half-filled orbital and an empty or a filled orbital should be ferromagnetic, although relatively small in magnitude.

It seems that there is no requirement for strict linearity: The conclusions drawn for the 180° pathway appear to hold, at least in regard to sign, when the molecule is bent at the bridge atom, even down to 120°, which occurs in an important classes of clusters mentioned below. Of course, for the σ pathway, there is a reduction in magnitude owing to poorer overlap, but this is not obviously the case for the π route.

9.11.2.2 90° Superexchange. When the metal–bridge–metal angle is 90°, the position becomes more complex because several possible pathways can be written. The two pathways that are considered to be the most important are given in Figure 9.12c where, as an example, two metal d_{xy} orbitals overlap with a bridge p orbital. There is an important difference between the two cases given. For the bridge p_x orbital pathway, the two metal d_{xy} orbitals have the *same* phase. For the bridge p_y pathway, they have *opposite* phases.

If either of the two pathways shown in Figure 9.12c is dominant and both metal orbitals are half-filled, then the magnetic exchange is expected to be relatively large and to be antiferromagnetic in nature. If, however, the two pathways are equally important, then they require equal contributions of the metal d_{xy} with the same and with opposite phases. That situation leads to orthogonality. In such an event, the magnetic exchange would be relative small and ferromagnetic, arising from the exchange term of Eq. 9.106.

The conclusion is, then, that generalization about magnetic exchange by the 90° pathway cannot be made as readily as it can about the 180° pathway. Any result from relatively large antiferromagnetism through to relatively small ferromagnetism is possible, and the outcome is likely to depend on a quite fine balance between competing possibilities.

9.11.3   Some Examples of Superexchange in Clusters

We now discuss some specific cases of magnetic exchange within clusters in terms of the occupation of d orbitals on metal centers, which occupations, of course, are determined by the considerations of ligand field theory.

1. $[(NH_3)_5Cr-O-Cr(NH_3)_5]^{4+}$, d^3-d^3. The salts of the classic linear rhodo-pentaamminechromium(III) ion have low magnetic moments, or are diamagnetic. This reflects antiferromagnetic coupling between the chromium atoms, and J_{12} is a good deal larger than kT at ambient temperature, which is $\sim 200\, cm^{-1}$. This exchange can be understood as 180° π pathway superexchange involving half-filled t_{2g} orbitals on each Cr atom overlapping with a p_π orbital on the bridging oxygen atom. Even though the π pathway overlap with the central oxygen atoms is not great, the efficiency of the single atom 180° mechanism leads to a large magnitude for J_{12}.

2. $[L_2Cu(OH)_2CuL_2]^{2+}$, d^9-d^9. The range of values for $J_{12} \approx -550$ to $180\, cm^{-1}$ according to the nature of L, as mentioned above can be rationalized as the result of the balance between the 90° pathways involving a half-filled $d_{x^2-y^2}$ orbital on each Cu atom (Fig. 9.12c). The balance is critically dependent on the Cu–O–Cu angle (Fig 9.13), and this angle is determined by the group L.

3. $[Cu(CH_3.COO)_2 \cdot H_2O]_2$, d^9-d^9. This classic case of cluster magnetic exchange was first believed to involve direct sideways overlap of two half-filled $d_{x^2-y^2}$ orbitals, one on each Cu center— a δ bond. More recent analysis shows, however, that such overlap is negligible and the interaction occurs via superexchange between the same orbitals but involving the O=C–O ligand pathway. The moderate magnitude of J_{12} ($-280\, cm^{-1}$) probably reflects some attenuation by the longer bridge relative to the chromium complex (case 1),combined with reinforcement because four parallel pathways are available that involve σ rather than π overlap.

4. $Ni_3(acetylacetone-H)_6$, d^8-d^8. This complex was the first in which a ferromagnetic nearest-neighbour metal–metal interaction was clearly established and the role of the two superexchange pathways defined. The structure of

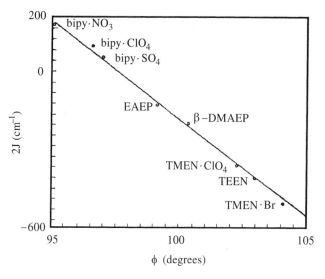

Figure 9.13. Dependence of $2J_{12}$ on the bridging angle for bis(hydroxo)-bridged copper(II) dimers. The curve was calculated using $2J_{12} = -74.53\phi + 7270 \text{ cm}^{-1}$. [From Ref. 14. Used with permission.]

the complex is shown in Figure 9.14. There are two exchange integrals to be considered: that between the central Ni^{2+} ion of the trimer and the terminal members J_{ct} and that between the terminal members themselves J_{tt}. Both are appreciable, with J_{ct} at 26 cm^{-1} and J_{tt} at -7 cm^{-1}. The positive nature of J_{ct} can be understood as a case of $90°$ superexchange through the shared O atom of the acetylacetonato group similar to and of about the same magnitude as some instances in case 2. The negative and smaller value for J_{tt} is seen in terms of an extended pathway through an acetylacetonato group. Of course, the competing effects of the two interactions with opposing signs lead to complex magnetic behavior. As in all such cases, the antiferromagnetic component must dominate at the lowest temperatures, and the magnetic moment maximizes at $4.2 \mu_B$ and then falls away rapidly below ~ 5 K to zero.

5. $(L_4Cu \cdot Fe(TTP) \cdot CuL_4)^+$ $d^9\text{-}d^5\text{-}d^9$. Here L_4 is a tetradentate Schiff's base ligand and TTP is *meso*-tetraphenylporphine. The Fe(III) site is low-spin, with the unpaired electron in a degenerate $d_{xz,yz}$ orbital pair. The unpaired electrons of the Cu atoms are in an orbital directed at the local N atoms and that is perpendicular to the z axis, which is, of course, normal to the porphyrin plane. The result is that the Fe–Cu magnetic orbital overlap is zero because of orthogonality, and the exchange is of the weak ferromagnetic type: $J_{Fe-Cu} = 44 \text{ cm}^{-1}$.

6. $[(L_4Ni)_3Cr]^{3+}$, $d^8\text{-}d^3\text{-}d^8$, and $[(L'_4Cu)_3Mn]^{2+}$, $d^9\text{-}d^5\text{-}d^9$. Here L and L' are two rather similar Schiff's base ligands, and the structures of the complexes

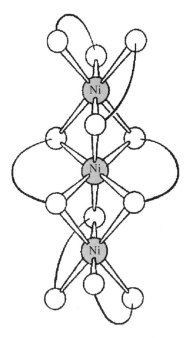

Figure 9.14. Structure of $Ni_3(acetylacetone-H)_6$.

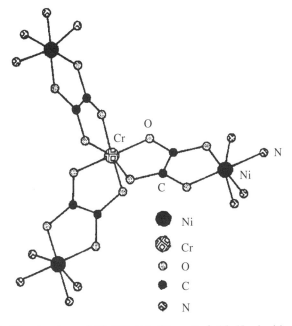

O

Cr

N

Ni

C

● Ni

Cr

O

● C

N

Figure 9.15. The structure of $[(L_4Ni)_3Cr]$. [From Ref. 15. Used with permission.]

have the framework shown in Figure 9.15. The two cases make an interesting comparison. Each Cu has one unpaired electron in an orbital directed at its local N atoms of the Schiff's base, and this is orthogonal to the π t_{2g} orbitals of the Mn ion, which are half-filled. As a result, the exchange is of the weak ferromagnetic type, $J_{Cu-Mn} = 5.3\ cm^{-1}$. On the other hand, Ni has two unpaired electrons, and orthogonality with π (t_{2g}) half-filled orbitals on the Cr ion is not possible. There the exchange is antiferromagnetic and larger, $J_{Ni-Cr} = -26\ cm^{-1}$.

7. $(L_5M)_3O$. A series of systems occurs in which three metal atoms, often with other ligands bridging between them, share a central oxygen atom, so that each M–O–M angle is 120°. Classical members are the basic acetate salts of Cr(III) and Fe(III); e.g., $[Cr_3(OOC.CH_3)_6(OH_2)_3O]Cl$. Exchange in these compounds is small and antiferromagnetic, $J_{Cr-Cr} = -10.4\ cm^{-1}$ for the Cr case. The result can be rationalized by looking at it as a distortion of the 180° pathway, with reduction of σ type overlap, if applicable, owing to the bending, leading to reduced magnitudes for J_{M-M}.

These simpler cases disguise the complexities that arise in many more common situations in which, like case 4, both ferromagnetic and antiferromagnetic pathways are present. These cases also misrepresent the balance between the occurrence of ferromagnetic and antiferromagnetic examples. For illustrative purposes about equal numbers of examples for each were presented. In practice, antiferromagnetic cases greatly outnumber ferromagnetic ones.

REFERENCES

1. Van Vleck, J. H., *Electric and Magnetic Susceptibilities*, Oxford University Press, Oxford, U.K. 1932.
2. Gerloch, M., *Magnetism and Ligand Field Analysis*, Cambridge University Press, Cambridge, U.K. 1983.
3. Koenig, E., Landolt-Bornstein, Hellwege, K. H., and Hellwege, A. M. eds., in *Tables of Magnetic Susceptibilities*, Springer-Verlag, Berlin, 1965.
4. Ballhausen, C. J., *Introduction to Ligand Field Theory*, McGraw-Hill, New York, 1962, chap. 4.
5. Gerloch, M., and Miller, J. R., *Prog. Inorg. Chem.*, 1968, **10**, 1.
6. Ham, F. S., *Phys. Rev.*, 1968, **166**, 307.
7. Ham, F. S., in Geschwind, S. ed., *Electron Paramagnetic Resonance*, Plenum Press, New York, 1971.
8. Ewald, E. H., Martin, R. L., Sinn, E., and White, A. H., *Inorg. Chem.*, 1969, **8**, 1817.
9. Judge, J. S., and Baker, W. A,, *Inorg. Chim. Acta*, 1967, **1**, 245.
10. Kremer, S., Henke, W., and Reinen, D., *Inorg. Chem.*, 1982, **21**, 3013.
11. Koenig, E. and Madeja, K., *Inorg. Chem.*, 1967, **6**, 48.

12. Willett, R. D., Gatteschi, D., and Kahn, O., eds., *Magneto-Structural Correlations in Exchange-Coupled Systems*, Academic Press, New York, 1985.

13. Martin, R. L., in Ebsworth, A. V. and Sharpe, A. G., eds., *New Pathways in Inorganic Chemistry*, Cambridge University Press, Cambridge UK, 1968.

14. Crawford, V.-H, Richardson, W. H., Wasson, J. R., Hodgeson, D. J., and Hatfield, W. E. *Inorg. Chem.*, 1976, **15**, 2107.

15. Pei, Y., Journaux, Y., and Kahn, O., *Inorg. Chem.*, 1989, **28**, 100.

CHAPTER 10

ELECTRON PARAMAGNETIC RESONANCE SPECTRA OF COMPLEXES

10.1 NATURE OF THE EPR EXPERIMENT

10.1.1 Introduction

In Chapter 9, the effect of a magnetic field on the energy levels of a transition-metal ion was discussed. The magnetic field lifts the degeneracy of a state, either just by the simple splitting pattern of the first-order Zeeman effect as in Figure 9.2 or in a more complicated fashion involving also the second-order Zeeman effect, as in Figure 9.3. How that leads to the magnetic susceptibility of a substance and the magnetic moment of a metal ion was developed there. As an alternative to measuring the magnetic properties, in some circumstances it is possible to observe directly absorption spectra owing to transitions between the energy levels represented in these figures. This is the basis of the electron paramagnetic resonance (EPR) experiment (also often called electron spin resonance; ESR). The transitions observed are those allowed by the selection rule $\Delta M_S = \pm 1$, as shown for a single unpaired electron in Figure 10.1. For readily obtainable laboratory magnetic fields, the first-order Zeeman effect separations are on the order of $1 \, \text{cm}^{-1}$, which corresponds to radiation in the microwave region of the electromagnetic spectrum. Corresponding frequencies are typically in the range of 3 to 100 GHz. The EPR experiment contains essentially all the information about the symmetry of the site of a paramagnetic metal atom center that determines its magnetic behavior and so is dominated by ligand field effects. That is what makes it so valuable for the present purposes. It also contains other information very relevant to ligand field considerations, and we shall expand on that below.

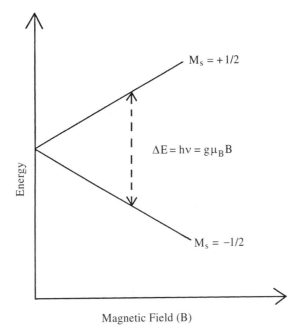

Figure 10.1. EPR transitions between the energy levels owing to the first-order Zeeman splitting of an $S = 1/2$ state.

The EPR experiment is more powerful, but is more limited, than is the measurement of magnetic properties. The two methods overlap, as we shall see, but also complement each other. EPR has a historical place in coordination chemistry, beacause it was there that, through superhyperfine coupling with ligand nuclei, direct evidence of covalence in metal–ligand bonding was first seen. Before developing the subject of EPR in more detail, it is instructive to point out some generalities of the previous statements. We now compare and contrast the two techniques.

1. The magnetic susceptibility of *any* compound can be measured, including diamagnetics.

The EPR spectrum can be measured only for some systems, mainly when there is an odd number of unpaired electrons (S half-integral), in particular with one unpaired electron ($S = 1/2$). Most information is obtained when the metal ion is in high magnetic dilution, especially in liquid or solid solutions. The first-row ions that have been studied most widely by EPR are Ti^{3+} (distorted octahedral complexes), V^{4+} ($V = O^{2+}$ compounds), Cr^{3+}, Mn^{2+}, Fe^{3+}, Co^{2+}, Ni^{2+} (octahedral complexes), and Cu^{2+}.

2. When, as is usual, there is more than one type of paramagnetic site in a crystal unit cell (different metal ions or, more often, the same ion in a different orientation), the magnetic susceptibility is the *average* of those of the individual

sites. Measurements are generally made using powdered samples, and information on the anisotropy of the magnetic moment is not obtained.

Except when electron exchange between the sites is faster than the frequency difference between their EPR signals, the EPR spectrum is the *superposition* of the spectra of the individual sites, and under suitable conditions it is possible to study these separately. Measurements are commonly made using solutions, powders, and crystals, and information on the anisotropy of the parameters of interest is often revealed.

3. The measurement of magnetic susceptibility generally involves relatively insensitive techniques, requiring at least several milligrams of substance and yielding an accuracy rarely better than a few percent.

The EPR experiment can be very sensitive, and samples of only micrograms or less of the metal ions are sufficient. The accuracy with which parameters are determined can be high.

4. The magnetic susceptibility is unaffected by the spin of the nuclei of the component atoms of a molecule or ion.

The EPR experiment can be strongly influenced by the nuclear spin of the metal ion housing the unpaired electron(s) and even by those of attached ligand atoms. This phenomenon adds considerably to the power of the technique.

5. When measured by usual techniques, the magnetic susceptibility is not affected by physical processes occurring in the crystal or a solution, ferromagnetism excepted.

The EPR experiment can be strongly affected by, for example, relaxation processes that involve molecular reorientation, and by molecular distortions.

10.1.2   Features of EPR Spectra

The EPR experiment is a complex one, rich in detail and reflecting a range of physical phenomena. This chapter cannot even start to consider many of these effects; only principal features can be developed, and then but briefly. The important fields of lanthanide and actinide element compounds are not considered, except to a minor extent in Chapter 11.

Before looking at the interpretation of EPR spectra, it is instructive to make some remarks about their form and accessibility.

10.1.2.1 Continuous Wave Versus Pulsed techniques. EPR transitions are usually characterized by measuring the absorption of a sample placed in a continuous beam of radiation. This is sometimes referred to as the continuous wave form of the technique (CW-EPR). Spectra may also be measured by subjecting the sample to an intense pulse of radiation and monitoring the emission as the complex relaxes to the ground state. This has the advantage of allowing the lifetime in the excited state to be monitored, because the relaxation rate depends on it. Pulsed EPR spectrometers have only recently become commercially available, and it is still too early to say whether this way of measuring spectra will replace the traditional method, as has occurred in the related technique of nuclear magnetic resonance (NMR) spectroscopy. Because the vast majority of experiments still employ continuous wave EPR, the present discussion will be limited to this form of the experiment, except for a brief introduction to the technique known as ENDOR.

10.1.2.2   Line Shapes. Although EPR is an absorption experiment, for technical reasons spectra are usually presented as the first derivatives of the absorption bands. This form of presentation has the advantage that small features which may amount to only shoulders or inflections of absorption bands, are made more obvious. This may be seen in Figure 10.2, where part of the EPR spectrum of $[Cu(CH_3COO)_2 \cdot H_2O]_2$ is shown in both forms. For similar reasons, it is sometimes advantageous to record second-derivative spectra.

10.1.2.3   Magnetic Field Sweep. In the microwave region, techniques for the generation of electromagnetic radiation lend themselves to providing fixed frequencies but not to varying them continuously over an appreciable range.

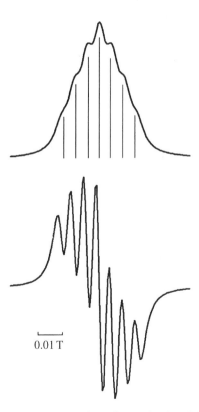

0.01 T

Figure 10.2. Part of the EPR spectrum of the dimer $[Cu(CH_3CO_2)_2 \cdot H_2O]_2$ measured in absorption (*top*) and as the first derivative (*bottom*)

Consequently, the EPR experiment is conducted by varying the magnetic field so that, in Figure 10.1 for example, the resonance condition:

$$g\mu_B \mathbf{B} = h\nu \qquad (10.1)$$

is met by an appropriate value of **B**. EPR spectra are, then, mostly presented as the derivatives of absorption bands versus magnetic field strength.

10.1.2.4   X- and Q-Band Frequencies. In the microwave region, for technical reasons, certain frequencies are much more readily available than are others. At present, the most convenient of these frequencies is the X band at $\sim 9.2\,\text{GHz}$, and most EPR measurements are carried out there. The next most readily available suitable frequency is the Q band at $\sim 35\,\text{GHz}$.

10.1.2.5 $g_e = 2.0023$. The reference point in EPR spectroscopy is the splitting of the energy levels of a free electron by a magnetic field. The free electron is a pure spin system $S = 1/2$ with no orbital angular momentum. The energy splitting is:

$$\Delta E = g_e \mu_B \mathbf{B} \qquad (10.2)$$

where g_e is the gyromagnetic ratio of the free electron and has the value 2.0023, to sufficient accuracy for our purposes (it is, in fact, one of the most accurately known of the fundamental physical constants). With $\mu_B = 9.2740 \times 10^{-24}\,\text{J}\,\text{T}^{-1}$ and $h = 6.6261 \times 10^{-34}\,\text{Js}$, resonance for g_e is obtained for a magnetic field of $0.3283\,\text{T}$ in the X band and for $1.249\,\text{T}$ in the Q-band ($1\,\text{T} = 10^4\,\text{G}$).

10.1.2.6 $g \neq 2.0023$. For transition metal ions, orbital angular momentum is never completely absent, although it may be largely quenched, as is discussed below and at some length in Chapter 9. Its effect is to alter g from the free electron value, because g is determined by the magnetic moment operator, $\boldsymbol{\mu}_i = (\mathbf{L}_i + g_e \mathbf{S}_i)\mu_B \cong (\mathbf{L}_i + 2\mathbf{S}_i)\mu_B, i = x, y, z$. Much of the interest in EPR, for the present purposes, comes from the deviation of g from g_e; i.e., from the effects of orbital angular momentum mixed into the ground state from higher lying levels, as discussed in Chapter 9. In this area, the study of magnetic properties of complexes overlaps with that of their EPR spectra.

It must also be recognized that sometimes the effects of interactions other than simply angular momentum are incorporated into the g value, making this an 'effective' parameter used simply to denote the position of a resonance signal. This aspect is discussed further in connection with the spin Hamiltonian (Section 10.2).

10.1.2.7   Crystals Versus Powders and Solutions. In a crystal, a paramagnetic metal ion is surrounded by a ligand arrangement which is almost always of symmetry lower than cubic. This means that the orbital angular momentum mixing, and hence the g value, is anisotropic. For the study of magnetic properties, in which measurements are normally made on powders (Section 10.1.1, statement 2), it is reasonable to ignore such anisotropy if it is small, and the effect receives little attention in Chapter 9. For the EPR experiment, however, with its greater accuracy, the effects are much more obvious. Very often, the complex has close to tetragonal symmetry, and the g value parallel to the symmetry axis $g_{\parallel}$ is quite different from those perpendicular to it $g_{\perp}$, which are equal. These anisotropic effects may be studied in great detail by taking a single crystal and carrying out the EPR experiment with different orientations of the crystal axes relative to the magnetic field. This is a rewarding, although time-consuming, process.

It is also possible to carry out the EPR experiment on powdered specimens, and in frozen solutions, and still retain much useful information. This is because the spectrum obtained is the *superposition* of the spectra of the individual molecules, not the average. Because the probability that a molecule is oriented such that a $g_{\perp}$ direction along the magnetic field is much greater than it being oriented in the $g_{\parallel}$ direction, the EPR spectrum is dominated by the $g_{\perp}$ feature, whereas $g_{\parallel}$ is minor and may not be easy to identify.

The situation for liquid solutions is more complicated, because the rate of tumbling of molecules may be slower than, comparable with, or faster than the microwave frequency. In the first case, the results are much as for powders, but for fast tumbling, averaging takes place, and anisotropy information is lost. The tumbling rate needs to exceed the frequency difference between the relevant lines for the averaging to take place. Rates greater than about $10^9\,s^{-1}$ in the X band and about four times higher in the Q band experiments are required for the averaging to be effective. This ratio is sufficiently large that spectra may be distinctly different at the two frequencies. Some of the features that are observed in powder and solution EPR spectra are shown in Figures 10.3 and 10.4. In the frozen solution spectrum (Fig. 10.4a), high resolution is achieved because the paramagnetic ions are well separated, and hyperfine splittings owing to coupling with the metal nuclear spin are resolved. In the liquid solution spectrum (Fig. 10.4b), some lines are broadened because the tumbling rate of the complex is comparable with the EPR frequency.

10.1.2.8 *Zero-Field Splittings and Their Effect on EPR*.
For systems with $S > 1/2$ complications develop. The most important of these is that the degeneracy of the ground state is lifted even without the presence of a magnetic field. The effect is termed the zero-field splitting (ZFS). It arises from the effects of spin–orbit coupling between the ground and excited states, taken in second order, and critically depends on the symmetry of the ligand environment of the metal ion concerned. This ZFS can vary in magnitude from almost zero for complexes with essentially cubic symmetry to many cm^{-1} for a low-symmetry environment and a large spin–orbit coupling constant. The allowed EPR transitions for a complex with an $S = 1$ ground term are shown in Figure 10.5a

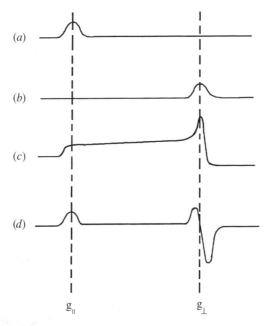

Figure 10.3. Absorption EPR spectra of an axially symmetric complex with the magnetic field parallel (*a*) and perpendicular (*b*) to the symmetry axis. The spectra of a powder of the complex in absorption mode (*c*) and as the first derivative (*d*).

and those for a complex with an S = 3/2 ground term are shown in Figure 10.5b. In both cases the ZFS corresponds to a low symmetry ligand field component of tetragonal symmetry.

When the ZFS is large and there is an even number of unpaired electrons (S integral), it may well be that no available magnetic field can bring an energy level separation to match the microwave radiation. Although the problem has to a certain extent been alleviated by the increasing availability of EPR spectrometers operating at very high frequency, this situation is not at all uncommon and presents a major limitation on the applicability of the EPR experiment. A system with unpaired electrons but for which EPR cannot be observed is often said to be "EPR silent". For an odd number of electrons (S half-integral), it can be shown that in the absence of a magnetic field there must be at least two-fold degeneracy; i.e., there must be pairs of levels that can be separated only by a magnetic field. These pairs are known as Kramer's doublets. Provided no other factor intervenes, they always yield an EPR spectrum, which is why EPR is obtained mostly for odd-electron systems.

10.1.2.9 Magnetic Dilution. If one calculates the magnetic field produced by a paramagnetic center, acting as a point dipole, at another such center at a distance of 7 Å, such as might well occur in a crystal of a transition metal complex, one deduces a value that depends on the direction of the dipole but can be as high as

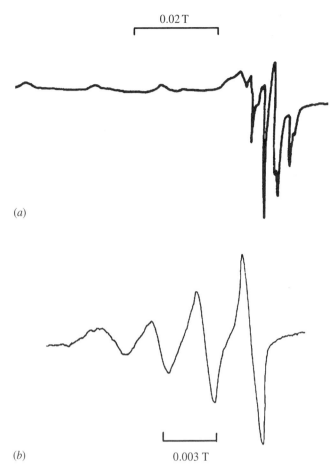

0.02 T

(a)

(b)   0.003 T

Figure 10.4. (a) EPR spectrum of a frozen solution of Cu(acac)$_2$ in CHCl$_3$. [From "Electron Spin Resonance of Transition-Metal Complexes" by B. R. McGarvey in volume 3 of *Transition Metal Chemistry*, R. L. Carlin, Ed., Marcel Dekker, New York, 1966. Used with permission.] (b) EPR spectrum of a solution of Cu(3-ethylacac)$_2$ in toluene at room temperature. [From Germann, H. R. and Swalen, J. D. *J. Chem. Phys.* 1962, **36**, 3221. Used with permission.]

about 10 mT. This is some 3% of the value of the field for resonance in the X band. In some cases, with few and simply oriented paramagnetic neighbors, it is possible to study such interactions in detail. Generally, however, a range of distances and directions occurs, producing a general line broadening. Thus, in crystals of transition-metal compounds, EPR lines are usually quite broad, and much of the potential accuracy of the technique is lost. g value anisotropy, for instance, may not be resolved. On the other hand, over such distances, magnetic exchange effects may be appreciable, and when the frequency of electron exchange greatly exceeds the microwave frequency, this will *average* the

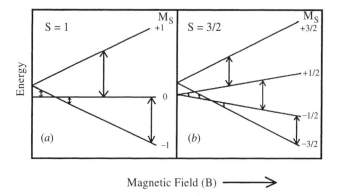

Figure 10.5. Allowed EPR transitions between the Zeeman levels of an $S = 1$ state (a) and an $S = \frac{3}{2}$ state (b) for a complex of tetragonal symmetry. The zero-field splittings are assumed to be small, and the applied magnetic field is parallel to the symmetry axis of the complex.

properties of the coupled sites. Sharp EPR lines again occur, but the information on g anisotropy is lost if the sites have a different orientation in the crystal lattice. This is the phenomenon known as *exchange narrowing*. $CuSO_4 \cdot 5H_2O$ illustrates these effects; it contains two distinct Cu^{2+} $(S = 1/2)$ sites, with some small magnetic exchange between them, probably mediated through the hydrogen bonding. At the X band, line widths vary strongly with the orientation of the crystal to the magnetic field; there is dipole broadening in some directions and exchange narrowing in others. At the higher frequency of the Q band, the two sites are always distinguished.

To minimize these magnetic effects in complexes, EPR on solids is often carried out on solid solutions, in which the complex of interest is incorporated (doped) to the extent of perhaps 1%, into a diamagnetic isomorphous host. For example, trace amounts of many transition metal ions have been studied in crystals such as MgO (M^{2+}) and Al_2O_3 (M^{3+}). The sensitivity of the technique allows good signals to be recorded despite the dilution. Then the lines observed are those of the isolated paramagnetic molecule, unbroadened and unnarrowed by the effects of neighbors. Of course, the site symmetry details in the host may not be exactly the same as those of the "native" crystal, and that must be borne in mind when comparing EPR results with those for other physical techniques performed on "concentrated" crystals. Liquid and frozen solutions, of course, also introduce suitable magnetic dilution.

10.1.2.10   Electron-Nuclear Spin Coupling. Atomic nuclei possess spin angular momentum. The nuclear magneton μ_N is some 1000 times smaller than the Bohr magneton μ_B, so nuclear magnetic phenomena are very weak. The wave functions of s electrons, however, have finite values at the nuclei of their atoms and serve to couple the nuclear and electronic spin angular momenta. Nuclear spin is specified by the quantum number I. The result of the coupling is that an

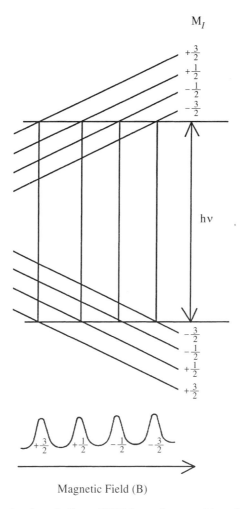

Figure 10.6. Energy levels and allowed EPR hyperfine transitions for a complex with a single unpaired electron and metal nuclear spin $I = \frac{3}{2}$. Peak positions are shown as normal absorption curves.

electronic spin state is perturbed so that each of its components becomes $2I + 1$ levels, equally spaced in energy and with the same transition moment. This is the *hyperfine* interaction. The selection rule $\Delta M_I = 0$ holds. The effect means that each EPR line of the earlier discussion is split into $2I + 1$ equally intense components, equally spaced in energy to first order, where I refers to the nucleus of the atom housing the unpaired electrons (Fig. 10.6). In marked contrast to the g based line, the separation of the hyperfine components does *not* depend on the value of the magnetic field used to obtain the resonance, so spectra taken at the X and at Q band can appear rather different. The line widths in the EPR experiment are such that hyperfine spectra are often well resolved.

It is even possible that the electrons involved in covalent bonding between ligands and the central paramagnetic atom may carry information from the ligand nuclei to the electronic spin system. This is the *superhyperfine* interaction. The effects are, of course, small, but if line widths are narrow and the coupling is sufficiently large, they may be seen clearly. The coupling rules must take into account the statistics of the contributions from different ligand M_I components. If there are n donor atoms with nuclear spin I each equally coupled to the central metal unpaired electrons, then there are $2nI + 1$ components to the parent EPR lines, equally spaced in energy to first order. A parent line itself may be a hyperfine line. The distribution of intensities among the components obeys a binomial law and is related to that found for rather similar coupling situations in NMR spectroscopy. Thus a single line owing to an EPR transition is split into components with an intensity distribution indicated below in the case of coupling with from one to four equivalent nuclear spins of 1/2. An example showing superhyperfine interaction effects is presented in Figure 10.7.

Electron				1			
1 nuclear spin			1		1		
2 nuclear spin		1		2		1	
3 nuclear spin	1		3		3		1
4 nuclear spin	1	4		6		4	1

The hyperfine splittings are defined in terms of a tensor **A**. In a similar fashion to the g values, the hyperfine and superhyperfine splittings differ depending upon the orientation of the molecule in the crystal and hence with respect to the applied magnetic field. Often one can measure components of **A**, $A_\parallel$ and $A_\perp$ in much the same way as for the g case. With the complexity of the spectra when

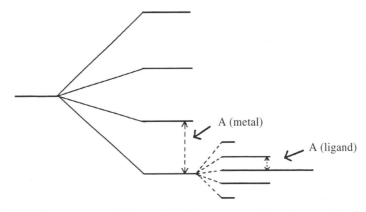

Figure 10.7. The pattern on the lowest line illustrates superhyperfine coupling with four equivalent ligand nuclei of spin $I = \frac{1}{2}$. The basic metal hyperfine interaction is for a nuclear spin of $I = \frac{3}{2}$.

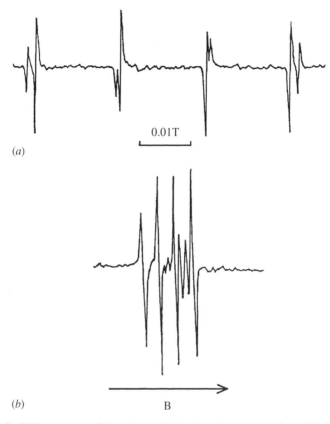

(a)

0.01T

(b)

B

Figure 10.8. EPR spectrum of the planar Cu(1-phenylacac)$_2$ complex with the magnetic field approximately parallel (a) and perpendicular (b) to the z molecular axis. [From "Electron Spin Resonance of Metal Chelates", T. F. Yen Ed., Plenum Press, New York, 1968, p. 102. Used with permission.]

such structure is present, single-crystal studies may be required to obtain an interpretation. This is presented in Figure 10.8, which shows results for a single crystal of a cupric complex present at high dilution in a diamagnetic host lattice. The ligand donor atoms (oxygen) have no nuclear spin, so that no superhyperfine coupling occurs. Copper consists of two isotopes, ^{63}Cu (69.1% abundance) and ^{65}Cu (30.9% abundance). Both have a nuclear spin $I = 3/2$ and hence give rise to a four-line pattern; however, the latter isotope has a slightly bigger nuclear magnetic moment, and hence slightly larger hyperfine parameters. This is apparent in the spectrum measured with the magnetic field approximately parallel to the symmetry axis (Fig. 10.8a), in which the hyperfine splitting is large and two sets of four lines are clearly resolved. The splitting when the magnetic field is approximately perpendicular to this direction is too small to allow the isotope effect to be resolved (Fig. 10.8b). The latter spectrum is complicated by the observation of additional formally 'forbidden' lines. These are owing to the neglect in the simple treatment of coupling with the nuclear quadrupole moment of the metal and the interaction of the nuclear magnetic moment with the applied magnetic field.

Although the resolution is generally lower, some features may be extracted from powder or solution spectra. In Figure 10.4, the spectrum of a frozen solution and of a room temperature liquid are given for another similar cupric complex. In Figure 10.4a, the spectrum is the superposition of all molecular orientations, and one sees again a large splitting of the $g_{\parallel}$ line into four components, corresponding to a high value of $A_{\parallel}$, and a similar but small splitting of the $g_{\perp}$ line, corresponding to a low value of $A_{\perp}$. In Figure 10.4b, there is fast tumbling of the molecules of the complex, and the g and A values are averaged. The single g line is split into the four components with medium separation. The line intensities are unequal because the tumbling rate is comparable to the EPR frequency. Information about the tumbling process may be obtained in such a case.

10.1.2.11 Orbital Degeneracy. As discussed in Chapter 9, orbitally degenerate ground terms of T symmetry are split into a series of energy levels by spin–orbit coupling and by small departures from cubic symmetry. For values of the spin–orbit coupling constants of the first-transition series, the resultant spread of energy levels is likely to be a few hundred cm^{-1} which is comparable to the thermal energy at ambient temperature, some $200\,cm^{-1}$. As a result, a number of the levels may be populated, and this provides a powerful mechanism for the relaxation of information about their spin properties. The EPR lines, therefore, become greatly broadened, and at room temperature it is rarely possible to carry out the experiment successfully in such systems.

When the temperature is reduced sufficiently, however, to say $10\,K$, only the lowest level may be thermally populated, and then the EPR spectrum becomes accessible. This is the case, for example, for octahedral complexes of Co^{2+} ($^4T_{2g}$), for which a number of low-temperature studies have been carried out. In such cases, the definition of the "spin" requires care, as is discussed in the next section.

10.2 THE SPIN HAMILTONIAN

As the last section emphasized, the EPR experiment involves application of a magnetic field to a ground state which may consist of several energy levels separated by up to a few cm^{-1}. The spectrum may be interpreted using a set of equations that describes the interactions present in each system. The resulting parameters provide a relatively simple way of comparing quantitatively the results of related but different EPR experiments. The expression involved is called the spin Hamiltonian (SH). It must be emphasized that the SH need not directly express real quantities. It is an empirical concept; its only constraints are that it conforms to the symmetry properties of the system under consideration and reproduces the EPR spectrum concerned succinctly. In particular, the angular momentum is generally represented by a fictitious spin operator $\mathbf{S}'$. The energy levels are the eigenvalues of the SH, and associated with each is a fictitious spin quantum number $\mathbf{S}' \cdot \mathbf{S}'$ and $\mathbf{S}'$ may or may not be the same as the real spin operator $\mathbf{S}$ and quantum number S for the parent ground term.

For the simple case in Figure 10.1, there are only two energy levels—at $g\mu_B\mathbf{B}$ and $-g\mu_B\mathbf{B}$ — so $S' = S = 1/2$. There, the fictitious and the real spin coincide. In Figure 10.5b, however, two cases occur. If the ZFS is small, all four levels of the $S = 3/2$, ground term can be reached by reasonable magnetic fields and need to be considered. Then $S' = S = 3/2$, and the real and fictitious spin coincide. On the other hand, if the ZFS is large, only the lower pair of levels enters into the EPR experiment at reasonable magnetic fields. Then S' is taken to be $1/2$, which is different from the real spin, S.

Suppose that g for the ground term in Figure 10.5b is 2.10. As given, the lower pair of levels is specified by $M_S = \pm 1/2$, and 2.10 is also the value for g that pertains to the SH. If the ZFS is of the opposite sign to that given in the Figure however, then the lower pair of levels is specified by $M_S = \pm 3/2$ of the ground term. For this pair, $S' = 1/2$ again, but now the energy splitting may be represented in the SH by an *effective* g of 6.30, rather than the "true" g value of 2.10.

A general form for the SH, to the level of the hyperfine and superhyperfine interactions, may be written as:

$$\mathbf{H} = \mu_B(\mathbf{B} \cdot \mathbf{g} \cdot \mathbf{S}') + \mathbf{S}' \cdot \mathbf{D} \cdot \mathbf{S}' + \mathbf{I} \cdot \mathbf{A} \cdot \mathbf{S}' \qquad (10.3)$$

This simple form hides much complexity, the details of which are beyond the scope of this book; for a fuller discussion of the meaning of the spin Hamiltonian, see Abragam and Bleaney (1) and Pilbrow (2). The magnetic field $\mathbf{B}$ is a vector; $\mathbf{I}$, the nuclear magnetic moment, and $\mathbf{S}'$ are operators, and $\mathbf{g}$, $\mathbf{D}$, and $\mathbf{A}$ are 3×3 matrices that specify the Zeeman effect, the spin–spin (ZFS) interaction, and the hyperfine interactions, respectively. These matrices each have six independent elements, which can be regarded as parameters specifying the lengths of the principal axes of an ellipsoid and the orientation (a set of three direction cosines) of these axes with respect to the molecular framework. Thus one may visualize and speak of, for instance, the g ellipsoid. The elements of $\mathbf{g}$ are dimensionless, whereas those of $\mathbf{D}$ and of $\mathbf{A}$ are energies, and are usually quoted in cm^{-1}.

The components of the above matrices provide the basis on which the EPR spectrum is interpreted. For technical reasons, there are some limitations of the completeness with which the relevant parameter set can be determined. The signs of the elements of $\mathbf{g}$ can rarely be obtained, and often it is experimentally difficult to determine the orientations of the components of $\mathbf{D}$. All the components of $\mathbf{A}$, however, can be deduced in favorable circumstances. The matrix representations of physical properties of this kind are conventionally called *tensors*, but see Abragam and Bleaney (1) for a detailed discussion of this terminology. It is, therefore, customary to refer to, for instance, the g tensor, which is the set of elements that describe the way g varies in a specified coordinate system.

For the present purposes, it is sufficient to use only some of the information potentially available in the tensors. If one assumes tetragonal (or sometimes trigonal) site symmetry at the paramagnetic metal center, with the tetragonal axis along a principal molecular direction, then the tensors become much simpler and can be specified, as noted, by the parameters $g_{\parallel}$ and $g_{\perp}$ for **g**, D for **D**, and $A_{\parallel}$ and $A_{\perp}$ for **A**. Then the SH can be put in the simplified form:

$$\mathbf{H}_{\text{tetrag}} = \mu_B\{g_{\parallel}\mathbf{B}_z\mathbf{S}'_z + g_{\perp}(\mathbf{B}_x\mathbf{S}'_x + \mathbf{B}_y\mathbf{S}'_y)\} + D\{\mathbf{S}'^2_z - S(S+1)/3\}$$
$$+ A_{\parallel}\mathbf{I}_z\mathbf{S}'_z + A_{\perp}(\mathbf{I}_x\mathbf{S}'_x + \mathbf{I}_y\mathbf{S}'_y) \qquad (10.4)$$

10.3   INTERPRETATION OF THE SPIN HAMILTONIAN PARAMETERS

10.3.1   The g Tensor

The **g** tensor includes the expectation values of the angular momenta of the ground state of a system, and it is defined by the first terms in the SH of Eqs. 10.3 and 10.4. The interpretation of the **g** tensor of a complex may be illustrated by considering some examples (1–5). The complex $VO(H_2O)_5^{2+}$ has C_{4v} symmetry and has a $^2B_2$ ground state with a single unpaired electron in the d_{xy} orbital. The g values may be estimated by considering the energy levels that result when a magnetic field lies along the direction of the V=O bond, the z axis. Replacing the fictitious spin $\mathbf{S}'_z$ by separate operators representing the spin and orbital angular momenta, the first term in Eq. 10.4 becomes $\mu_B\mathbf{B}_z \cdot (\mathbf{L}_z + g_e\mathbf{S}_z)$. This must be applied to the ground state wavefunctions $^2B_2(xy)$ (+) and $^2B_2(xy)$ (−), where + and − represent the two possible spin wavefunctions of the electron. The effect of the components of the orbital angular momentum on the real d orbitals is shown in Table 10.1. Note that some of the d functions are complex (contain i, $\sqrt{-1}$). The eigenvalues of the spin wavefunctions are $+1/2$ and $-1/2$ $h/2\pi$, respectively, and as the real d orbitals are orthogonal, it follows from the relationships in Table 10.1 that:

$$\langle ^2B_2(xy)|\mathbf{L}_z|^2B_2(xy)\rangle = -2i\langle ^2B_2(xy)|^2B_1(d_{x^2y^2})\rangle = 0 \qquad (10.5)$$

The orbital angular momentum component of the operator, therefore, makes no contribution, and the two eigenvalues resulting from the magnetic field are $\pm\mu_B\mathbf{B}_zg_e$. The energy difference between these is $\mu_B\mathbf{B}_zg_e$ so that $g=g_e=2.00$. The same result is obtained if the magnetic field is applied along the x or y directions. The calculation, therefore, suggests that the **g** tensor should be isotropic, with the g value equal to g_e. Indeed, because the real d functions are not eigenfunctions of **L**, whenever an unpaired electron occupies a single d orbital, it gives rise to no orbital angular momentum, as was shown in Chapter 9.

TABLE 10.1. Effect of the Components of the Orbital Angular Momentum Operator L on the Real d Orbitals

Orbital	Operators		
	L_x	L_y	L_z
d_{z^2}	$-i\sqrt{3}\,d_{yz}$	$\sqrt{3}\,d_{xz}$	0
$d_{x^2-y^2}$	$-i\,d_{yz}$	$-i\,d_{xz}$	$2i\,d_{xy}$
d_{xy}	$i\,d_{xz}$	$-i\,d_{yz}$	$-2i\,d_{x^2-y^2}$
d_{xy}	$-i\,d_{xy}$	$-i\sqrt{3}\,d_{z^2}+i\,d_{x^2-y^2}$	$i\,d_{yz}$
d_{yz}	$i\sqrt{3}\,d_{z^2}+i\,d_{x^2-y^2}$	$i\,d_{xy}$	$-i\,d_{xz}$

The g values actually observed for the $VO(H_2O)_5^{2+}$ ion—$g_z = g_{\parallel} = 1.933$, $g_x = g_y = g_{\perp} = 1.981$—are indeed quite close to $g_e \approx 2.002$. The deviation may be explained by extending the model to include spin–orbit coupling, as is discussed in detail in Chapter 9. Following the procedure given there, the expressions for the g values become:

$$g_z = g_{\parallel} = 2.002 - 8\lambda/E(^2B_1) \tag{10.6}$$

$$g_x = g_y = g_{\perp} = 2.002 - 2\lambda/E(^2E) \tag{10.7}$$

Where, $E\,(^2B_1)$ and $E\,(^2E)$ are the energies of the two excited states relative to the $^2B_2$ ground state. The electronic spectrum of the $VO(H_2O)_5^{2+}$ ion in the visible region consists of two bands at 13,000 and 16,000 cm^{-1}, which have been assigned to transitions to the ^2E and $^2B_1$ excited states, respectively (Section 8.2.1). The spin–orbit coupling constant for the free V^{4+} ion is 250 cm^{-1}, and substitution of this and the appropriate excited state energies in Eqs. 10.6 and 10.7 yields the calculated values $g_{\parallel} = 1.877$ and $g_{\perp} = 1.964$. These calculated shifts from g_e are somewhat larger than the experimental values. The difference can be attributed to covalency effects, as set out in Chapter 9. These may be taken into account by including orbital reduction coefficients $k_{\parallel}$ and $k_{\perp}$ in the g expressions:

$$g_z = g_{\parallel} = 2.002 - 8\lambda k_{\parallel}^2/E(^2B_1) \tag{10.8}$$

$$g_x = g_y = g_{\perp} = 2.002 - 2\lambda k_{\perp}^2/E(^2E) \tag{10.9}$$

Here the higher accuracy of the EPR experiment allows k to be anisotropic, a feature that could not be supported by the magnetic susceptibility results in Chapter 9. Substitution of the observed g values into Eqs. 10.8 and 10.9 yields the estimate $k_{\parallel} = k_{\perp} = 0.74$ for the $VO(H_2O)_5^{2+}$ ion.

The orbital reduction parameters are closely analogous to the nephelauxetic ratio β used to describe the reduction of the interelectron repulsion parameters of a transition metal ion upon complexation (Section 8.4.2). Just as β derives from two effects—one caused by the expansion of the d orbitals owing to the lowering of the effective nuclear charge of the metal and one owing to transfer of the d electrons to the ligands—so the orbital reduction parameters are influenced by both the effective charge on the metal and the delocalization of the unpaired spin density onto the ligands. The effective charge on the metal influences the g values by reducing the spin–orbit interaction. It seems likely that in vanadyl complexes the effective metal charge is approximately half the formal oxidation state; i.e., the metal behaves more like a VO^{2+} group than a V^{4+} ion. For the V^{2+} ion, $\lambda = 170\,cm^{-1}$, and application of this in Eqs. 10.8 and 10.9 yields the orbital reduction parameters $k_{\parallel} = k_{\perp} = 0.90$. The delocalization of the unpaired electron onto the ligands that this implies, $\sim 10\%$, seems reasonable. This provides confirmation that the assignment of the electronic spectrum is valid. The alternative assignment of placing the ^2E state at $16,000\,cm^{-1}$ and $^2B_1$ at $13,000\,cm^{-1}$ yields the estimates $k_{\parallel} = 0.81$ and $k_{\perp} = 0.99$, which seem less plausible.

Generally, a similar procedure may be used to interpret the g values of other metal ions with a single unpaired electron. The main requirements for the approach to be valid are that the energy separations to the excited d states are much greater than the spin–orbit interaction and that the unpaired spin-density delocalized onto the ligands does not contribute significantly to the orbital angular momentum of the complex. The first requirement should almost always be met for vanadyl complexes and planar and tetragonally distorted 6-coordinate Cu^{2+} complexes. When the excited states are relatively close to the ground state, however, as may be the case, for instance, for many Ti^{3+} complexes, the perturbation approach used to derive Eqs. 10.8 and 10.9 is a rather poor approximation. The second requirement should be valid for the relatively ionic complexes formed by electronegative ligands with low spin–orbit coupling constants, such as the fluoride ion and oxygen donors, but may break down for more covalent complexes involving ligand atoms with high spin–orbit coupling parameters, such as iodide ion and those with selenium donor atoms.

When it can be measured, because of its sensitivity to the d orbital composition of the ground state, the EPR spectrum provides a powerful method of probing the ligand field and stereochemistry of a complex. For instance, six-coordinate Cu^{2+} complexes always undergo a Jahn–Teller distortion, generally to give a tetragonally elongated octahedral geometry. Rarely, the distortion occurs in the opposite direction, forming a tetragonally compressed octahedral complex (Section 7.1.2.1). Although both distortion types have D_{4h} symmetry and quite similar transition energies, the ground state wavefunctions, and hence the **g** tensors, are quite different. The ground states are $^2B_{1g}(x^2 - y^2)$ and $^2A_{1g}(z^2)$, respectively, for the elongated and compressed forms of the Jahn–Teller distortion, and when the effects of spin–orbit coupling are included to first order, the following g value expressions result. For $^2B_{1g}(x^2 - y^2)$:

$$g_{\parallel} = 2.002 - 8\lambda k_{\parallel}^2/E[^2B_{2g}(xy)] \qquad (10.10)$$

$$g_{\perp} = 2.002 - 2\lambda k_{\perp}^2/E[^2E_g(xz, yz)] \qquad (10.11)$$

For $^2A_{1g}(z^2)$:

$$g_\| = 2.002 \tag{10.12}$$

$$g_\perp = 2.002 - 6\lambda k_\perp^2 / E[^2E_g(xz, yz)] \tag{10.13}$$

The values $g_\| = 2.454$ and $g_\perp = 2.096$ are observed for the $Cu(H_2O)_6^{2+}$ ion, which has a tetragonally elongated octahedral geometry. Note that although the g expressions for this complex are similar to those associated with a $^2B_2(xy)$ ground state (Eqs. 10.8 and 10.9), much larger shifts from g_e are observed than for the VO^{2+} complex, and they are positive rather than negative. This is because the spin–orbit coupling constant of the Cu^{2+} ion is opposite in sign and considerably larger in magnitude than that of the V^{4+} ion (Section 5.5). Substitution of $\lambda = -830\,cm^{-1}$ and the observed excited state energies $^2B_{2g}(xy) \cong {}^2E_g(xy, yz) \cong 12,000\,cm^{-1}$ into Eqs. 10.10 and 10.11 yields the estimates $k_\| \cong 0.90$ and $k_\perp \cong 0.84$. Here the analysis is only approximate, because the individual excited state energies cannot be resolved in the electronic spectrum. For later transition metals such as copper, the spin–orbit coupling constant is rather insensitive to the effective charge on the metal (Table 5.5). It is of interest to compare this EPR result to that obtained from magnetic susceptibility measurements. Here the average value for k, $(k_\| + 2k_\perp)/3$, is 0.86, whereas, in Section 9.7, the value 0.8 was obtained for the same ion.

In the complex *trans*-$Cu(NH_3)_2Cl_4^{2-}$, the ammonia ligands produce a considerably stronger ligand field than do the chloride ions, and this acts to stabilize the unusual tetragonally compressed geometry with two short axial and four long in-plane bonds. The observed g values—$g_\| = 2.00$ and $g_\perp = 2.22$—confirm that the complex indeed has a $^2A_{1g}(z^2)$ ground state. The optical spectrum indicates an energy of $13,800\,cm^{-1}$ for the $^2E_g(xz, yz)$ excited state, and substitution of this in Eq. 10.13 yields the estimate $k_\perp = 0.78$, implying a somewhat greater spin delocalization than in the $Cu(H_2O)_6^{2+}$ ion. This is reasonable, although it should be noted that for this complex the present simple model can give only an approximate guide to the covalency.

Virtually all the complexes of transition ions with spin doublet ground states have a highly anisotropic orbital contribution to the **g** tensor. This is because Jahn–Teller coupling causes all such complexes to deviate from cubic symmetry. Even if this deviation is relatively small, as may be the case for Ti^{3+} complexes, the unpaired spin density is highly anisotropic, because the unpaired electron is largely localized in a single d orbital. Metal ions with a spin multiplicity higher than two, on the other hand, may form complexes with a regular octahedral or tetrahedral geometry, and the unpaired spin density is often close to isotropic. The **g** tensors then also are close to isotropic (although it must be stressed that quite small deviations from cubic symmetry can cause a zero-field splitting, which would make the EPR *spectrum* highly anisotropic). The g values are likely to exhibit a contribution from orbital angular momentum caused by admixture of excited states into the ground state by spin–orbit coupling, and expressions for the g values may be derived using perturbation theory in a manner similar to that outlined in Chapter 9.

For the $^3A_{2g}$ and $^4A_{2g}$ ground states of regular octahedral Ni^{2+} and Cr^{3+} complexes, the g anisotropy is rather small and may be neglected for many purposes. Then, if only the first excited term is considered, the g value is given by essentially the same expression as used for the magnetic susceptibility in Chapter 9:

$$g = 2.002 - 8\lambda k^2/\Delta E \qquad (10.14)$$

where, ΔE is the $A_{2g}-T_{2g}$ spin triplet (Ni) or quartet (Cr) term energy separation. For the $Ni(H_2O)_6^{2+}$ complex, $g = 2.25$, and $\lambda = -315\,cm^{-1}$ for the free ion. The first excited state $^3T_{2g}$ occurs at $8,700\,cm^{-1}$, which yields the estimate $k \cong 0.925$, implying that the bonding is quite ionic. This agrees well with the result from magnetic susceptibility studies described in Chapter 9, viz 0.92. For the $Cr(H_2O)_6^{3+}$ ion, $g = 1.977$, and the $^4T_{2g}$ excited state occurs at $17,000\,cm^{-1}$. Using the spin–orbit coupling constant for Cr^{3+}, $\lambda = 92\,cm^{-1}$, in Eq. 10.14 leads to a rather low value of the orbital reduction parameter, $k \cong 0.76$. However, as noted for the $VO(H_2O)_5^{2+}$ ion, the orbital reduction parameter describes both the nephelauxetic and the delocalization effects of covalency. If the effective nuclear charge in the $Cr(H_2O)_6^{3+}$ ion is $\sim 1.5+$ the effective spin–orbit constant λ is $70\,cm^{-1}$, and substitution of this into Eq. 10.14 yields the estimate $k \cong 0.87$, which suggests relatively little delocalization of the unpaired electrons onto the water molecules. As outlined in Chapter 9, magnetic susceptibility measurements are not sufficiently precise to provide a result for this ion. For the later transition-metal ions such as Ni^{2+}, the spin–orbit constant is much less sensitive to the metal charge—an effective value of $1+$ implies a constant $\lambda \cong -285\,cm^{-1}$, which yields the estimate $k \cong 0.97$ for the orbital reduction parameter of the $Ni(H_2O)_6^{2+}$ ion. The treatment presented here for these many-electron ions is only approximate. In particular, spin–orbit interaction with higher excited states will also influence the g values.

For high-spin octahedral Co^{2+} complexes, the **g** tensors cannot usually be interpreted satisfactorily using simple expressions, because the $^4T_{1g}$ ground term of such a complex is subject to a weak Jahn–Teller distortion. This produces low-symmetry ligand field components that, in conjunction with spin–orbit coupling, split the ground state into a set of Kramer's doublets, as outlined in Chapter 9. These have separations comparable to kT at 300 K, which means that at the very low temperature necessary to observe the EPR, there is a set of excited states lying close to the ground state. Perturbation expressions, therefore, cannot properly represent the g values. Moreover, even though tetrahedral complexes of this metal ion are not subject to Jahn–Teller effects, the small ligand field splitting means that the excited states lie close to the ground state, so that the perturbation expressions given in Eqs. 10.10 to 10.14 usually do not describe the orbital contribution to the **g** tensor at all accurately.

The complexes of almost all metal ions with the high-spin d^5 configuration have g values close to g_e. This is because all excited states are high in energy and differ in spin from the $^6A_1$ ground state, so that no admixture of orbital angular momentum occurs via first-order spin–orbit coupling. Thus $g = 2.003$ for the $Fe(H_2O)_6^{3+}$ ion, and $g = 2.0015$ for the Mn^{2+} ion doped into MgO. This latter system is often used as an internal standard when measuring EPR spectra. The g values of tetrahedral complexes of the Fe^{3+} and Mn^{2+} ions are similarly close to g_e.

10.3.2 The Hyperfine Tensor

10.3.2.1 The Metal Nucleus. Many of the transition metals have a nonzero nuclear spin, and this nuclear magnetic moment then couples with the spin and orbital angular momentum of the unpaired electron to produce a hyperfine splitting of the EPR signal. This is represented in the spin Hamiltonian by the operator $\mathbf{I} \cdot \mathbf{A} \cdot \mathbf{S}'$. A detailed analysis of the hyperfine tensor $\mathbf{A}$ is highly complicated (1,2), and we shall consider its interpretation only in general terms.

Expressions for the hyperfine parameters of a complex of tetragonal symmetry with a single unpaired electron in the d_{xy} or $d_{x^2-y^2}$ orbital are (1):

$$A_{\parallel} = P(-4\alpha^2/7 - K\alpha^2 + 3\delta g_{\perp}/7 + \delta g_{\parallel}) \tag{10.15}$$

$$A_{\perp} = P(2\alpha^2/7 - K\alpha^2 + 11\delta g_{\perp}/14) \tag{10.16}$$

The parameter $P = g_e g_N \mu_B \mu_N \overline{r^{-3}}$ represents the basic interaction between the electron spin and the nuclear magnetic moments, scaled by the average of the inverse cube distance of the unpaired electron from the nucleus. Here g_N is the nuclear g value of the metal and μ_N is the nuclear magneton. The first factor inside each bracket in Eqs. 10.15 and 10.16 represents the interaction between the unpaired spin in the d orbital and the nuclear magnetic moment. Because of the anisotropic shape of the d orbital, this gives rise to a highly anisotropic contribution to $\mathbf{A}$. The factor K describes the isotropic hyperfine contribution owing to unpaired electron spin density in s orbitals, both induced by polarization of core shells and, when allowed by symmetry, direct participation in the ground state. These two effects influence the isotropic hyperfine interaction in an opposing manner; polarization makes a positive, and direct participation a negative, contribution to K. Although the unpaired spin density in the s orbitals is small, the contribution to $\mathbf{A}$ is quite large. This is because, unlike d orbitals, s wavefunctions have a significant amplitude close to the nucleus. These first two factors in Eqs. 10.15 and 10.16 are moderated by the fractional unpaired spin density on the metal α^2. The remaining contributions to $\mathbf{A}$ result from the interaction of the orbital angular momentum of the unpaired electron with the nuclear spin. Because the g values of a complex are shifted from that of the free electron by the orbital contribution to the electron magnetic moment, the observed g shifts—$\delta g_{\perp} = g_{\perp} - g_e$ and $\delta g_{\parallel} = g_{\parallel} - g_e$—are conventionally used to represent this effect in the expressions for the hyperfine constants.

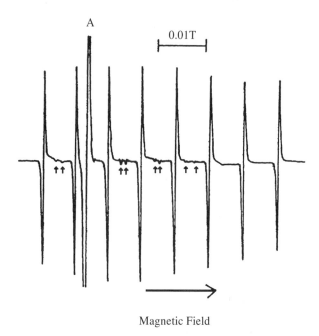

Figure 10.9. EPR spectrum of a typical vanadyl complex measured at the Q band frequency with the magnetic field perpendicular to the V=O direction. *Line A,* an organic free radical with g = 2.0036 used as a reference; *arrows,* weak forbidden transitions. [From Hitchman, M. A. and Belford, R. L., *Inorg. Chem.,* 1969, **8**, 958. Used with permission.]

The EPR spectrum of a typical vanadyl complex, which has a single unpaired electron localized largely in the d_{xy} orbital, is shown in Figure 10.9. The only stable isotope of vanadium ^{51}V has a nuclear spin of $I = 7/2$, so that eight equally intense lines arise from the hyperfine coupling. The weak features between the main lines are the result of forbidden transitions, which involve a change in both the electron and nuclear spin quantum numbers. The hyperfine parameters observed for the $VO(H_2O)_5^{2+}$ complex, $A_\perp = -70.10^{-4}$ cm^{-1} and $A_\parallel = -180 \cdot 10^{-4}$ cm^{-1}, must both be negative to give a positive value of K. Direct participation of metal s orbitals into the ground state is expected to be negligible here, because the $^2A_1(s)$ and $^2B_2(xy)$ states belong to different representations in the C_{4v} point group of the complex. As discussed for the spin–orbit coupling constant, the parameter P is rather sensitive to the effective nuclear charge on the metal. Assuming that this latter is $\sim 2+$ yields the value $P \approx 130 \cdot 10^{-4}$ cm^{-1}. For vanadyl complexes, the g shifts are quite small, so that the orbital angular momentum of the unpaired electron has little influence on the hyperfine parameters. Substitution of the observed values $\delta g_\perp = -0.021$ and $\delta g_\parallel = -0.069$ in Eqs. 10.15 and 10.16 yields the estimates $K \approx 0.84$ and $\alpha^2 \approx 0.94$. The small positive value of K is consistent with calculations of the polarization of the core s electrons, and the fractional unpaired spin density on the metal suggests that covalent bonding with the ligand water molecules has little influence on the ground state wavefunction. This is reasonable, because the d_{xy} orbital is involved in only a relatively weak π-interaction with the ligands. The covalency implied by the hyperfine parameters agrees well with that suggested by the orbital reduction parameters derived from the g values, $k \approx 0.90$, particularly when it is remembered that the latter also represents the spin delocalization in excited states, which involve either σ interaction or the π bonding within the VO^{2+} group.

The single unpaired electron in the complex $Cu(H_2O)_6^{2+}$ is in the $d_{x^2-y^2}$ orbital, and this has the hyperfine parameters $A_\perp = 17 \cdot 10^{-4}$ cm^{-1} and $A_\| = -110 \cdot 10^{-4}$ cm^{-1}. The EPR spectrum of the solid alone does not yield the sign of the parameters. Although $A_\|$ must be negative to give a reasonable value of α^2, the sign of $A_\perp$ is ambiguous. The average hyperfine splitting $\bar{A} = -27 \cdot 10^{-4}$ cm^{-1}, which is obtained from the solution spectrum measured at high temperature when the long axis of the complex switches direction more rapidly than the EPR time scale, may be used to determine this. The condition $\bar{A} = (A_x + A_y + A_z)/3$ reduces to $\bar{A} = (A_\| + 2A_\perp)/3$ for a complex of tetragonal symmetry. The value $\bar{A} = -25 \cdot 10^{-4}$ cm^{-1}, obtained if $A_\perp$ is positive, agrees with experiment. In contrast, that obtained if $A_\perp$ is negative, $-48 \cdot 10^{-4}$ cm^{-1}, does not agree as well, implying that $A_\perp$ is positive.

For this complex, the orbital angular momentum makes a significant positive contribution to the hyperfine parameters. Substituting the values $\delta g_\perp = 0.095$, and $\delta g_\| = 0.452$ in Eqs. 10.15 and 10.16, together with the value $P \approx 360 \cdot 10^{-4}$ cm^{-1} calculated for the Cu^{2+} ion, yields the estimates $K \approx 0.32$ and $\alpha^2 \approx 0.90$ for the $Cu(H_2O)_6^{2+}$ complex ion. The isotropic hyperfine parameter is quite similar to the value $K \approx 0.43$ calculated on the assumption that it is caused by the polarization of core s electrons, and the fractional unpaired spin density is close to that suggested by the orbital reduction parameters ($k_\| \approx 0.90$ and $k_\perp \approx 0.84$).

Somewhat different hyperfine expressions occur for a complex with a single unpaired electron localized largely in the d_{z^2} orbital:

$$A_\| = P(4\alpha^2/7 - K\alpha^2 - \delta g_\perp/7) \tag{10.17}$$

$$A_\perp = P(-2\alpha^2/7 - K\alpha^2 + 15\delta g_\perp/14) \tag{10.18}$$

In particular, the sign of the terms owing to the dipolar coupling is reversed compared to that when the unpaired electron is in a d_{xy} or $d_{x^2-y^2}$ orbital. The complex *trans*-$Cu(NH_3)_2Cl_4^{2-}$, which has a compressed tetragonal geometry with a $^2A_{1g}(z^2)$ ground state, has hyperfine parameters $A_\| = 240 \cdot 10^{-4}$ cm^{-1} and $A_\perp = 73 \cdot 10^{-4}$ cm^{-1}, and both must be positive to give a reasonable value for α^2. Substitution of the observed g shift, $\delta g_\perp = 0.218$, in Eqs. 10.17 and 10.18 yields the estimates $K \approx -0.25$ and $\alpha^2 \approx 0.85$. The fractional unpaired spin density agrees quite well with that suggested by the orbital reduction parameter, $k_\perp = 0.78$. The isotropic hyperfine constant, however, is negative, contrasting with the positive value expected if the coupling were owing solely to polarization of the core s electrons. The reversal in sign of K may be explained by a small admixture of the metal 4s orbital into the ground state. This may occur because, for this complex, the $^2A_{1g}(4s)$ excited state is of the same symmetry as the $^2A_{1g}(z^2)$ ground state. A similar conclusion is drawn from the transition energies observed for planar and linear complexes (Section 3.2.2).

The stable isotope of cobalt ^{59}Co has a nuclear spin of $I = 7/2$, so that hyperfine coupling produces an eight-line EPR spectrum. Low-spin Co^{2+} complexes usually have $^2A_{1g}(z^2)$ ground states, and when a complex has tetragonal symmetry, the hyperfine parameters may be interpreted satisfactorily using Eqs. 10.17 and 10.18. High-spin Co^{2+} complexes with both octahedral and tetrahedral geometries generally have excited states at quite low energy. The ground state wavefunctions often have a significant contribution from several d functions, and the simple expressions given above are inadequate to interpret the hyperfine parameters. Of the other first-row transition ions studied by EPR, only manganese has a common isotope with a significant nuclear magnetic moment, ^{55}Mn with $I = 5/2$. Most complexes are high spin, with a ground state of $^6A_1$ symmetry. Because the unpaired spin density is of cubic symmetry symmetric, there is no dipolar contribution to the hyperfine coupling. Moreover, the orbital angular momentum of the unpaired electrons is also vanishingly small. The hyperfine coupling is, therefore, dominated by the isotropic interaction and Eqs. 10.15 to and 10.18 reduce to:

$$\bar{A} = -PK\alpha^2 \qquad (10.19)$$

For complexes of near cubic symmetry, direct participation of the 4s orbital in the ground state does not occur, so that the hyperfine interaction is caused by polarization of the core s electrons. This makes a positive contribution to K, predicting that the hyperfine coupling constant should be negative. For the $Mn(H_2O)_6^{2+}$ ion $\bar{A} = -92 \cdot 10^{-4}$ cm^{-1}, and because $P \approx 180 \cdot 10^{-4}$ cm^{-1} for the Mn^{2+} ion, Eq. 10.19 yields the estimate $K\alpha^2 \approx 0.5$ in this case. Calculations suggest a value of $K \approx 0.64$ for the Mn^{2+} ion, which agrees well with experiment. Although the uncertainties are too large to allow the fractional unpaired spin density on the metal α^2 to be estimated with any confidence, it is noteworthy that the hyperfine parameter usually decreases in magnitude for complexes in which the bonding is expected to be more covalent. Thus, $\bar{A} = -81 \cdot 10^{-4}$ cm^{-1} when the Mn^{2+} ion is doped into MgO, $-74 \cdot 10^{-1}$ cm^{-1} in ZnO, and $-64 \cdot 10^{-4}$ cm^{-1} in ZnS.

In polymeric complexes, the unpaired electron may interact with the nuclear spins of several metal ions. A simple example is the dimer $Cu_2(CH_3COO)_4(H_2O)_2$, the spectrum of which is shown in Figure 10.2. Here each unpaired electron interacts with the two copper nuclei each of spin $I = 3/2$, to give seven lines in the ratio $1:2:3:4:3:2:1$. The average hyperfine splitting parameter $\bar{A} = 70$ cm^{-1} is about half that typically observed for monomeric copper(II) complexes, because each unpaired electron spends half of its time on each metal of the dimer.

10.3.2.2   The Ligand Nuclei. Many ligand donor atoms have a nonzero nuclear spin, and unpaired spin density transferred to the ligands then gives rise to the *superhyperfine* coupling. For a ligand, the principal axes x', y', and z' of the superhyperfine tensor generally lie along or perpendicular, to the metal–ligand bond. If the ligand atoms use only s and p orbitals in bonding to the metal, the superhyperfine parameters may usually be interpreted satisfactorily using simple expressions (1):

$$A_{z'} = 2 f_p A_p + f_s A_s \qquad (10.20)$$
$$A_{x'y'} = -f_p A_p + f_s A_s \qquad (10.21)$$

Here the tensor has axial symmetry about the metal–ligand bond z'. The parameters A_p and A_s represent the hyperfine coupling parameters expected for a single unpaired electron in ligand p and s orbitals, respectively, and f_p and f_s are the fractional occupancy of these orbitals in the metal-ligand bond. Note that although the s contribution is isotropic, the p contribution is highly anisotropic. In these expressions, the small contribution owing to the dipolar coupling with the unpaired spin density on the metal has been ignored.

As an example, we consider the EPR spectrum of the planar CuF_4^{2-} ion (Fig. 10.10). Fluorine has a nuclear spin of $I = 1/2$, and the interaction of the unpaired electron spin with four equivalent ligands is expected to produce a five-line pattern, with relative intensities $1:4:6:4:1$ (Fig. 10.7). When the direction of the magnetic field is parallel to the symmetry axis of the complex, it is normal to each metal–ligand bond. The ligands are equivalent, and the superhyperfine splitting $A_{x'y'}$ is much smaller than the relatively large copper hyperfine splitting $A_{\parallel}$ (Fig. 10.10a) (6). When the magnetic field is midway between two pairs of Cu–F bonds, so that all four ligands are again equivalent, the copper hyperfine splitting $A_{\perp}$ is small and occurs as a minor splitting to the basic superhyperfine pattern (Fig. 10.10b), which now has a significant contribution from $A_{z'}$. The parameters A_p and A_s are somewhat sensitive to the effective charge on the ligand. They have been estimated as 0.06 and 1.76 cm^{-1} for a fluorine atom, and 0.05 and 1.34 cm^{-1} for the fluoride ion. Substituting average values for each of these into Eqs. 10.20 and 10.21, together with the observed superhyperfine parameters $A_{z'} = 139 \cdot 10^{-4}$ and $A_{x'y'} = 38 \cdot 10^{-4}$ cm^{-1}, yields the estimates $f_p \approx 0.062$ and $f_s \approx 0.005$, corresponding to $f_p \approx 6.2\%$, and $f_s \approx 0.5\%$ for the fractional time the unpaired electron spends in the 2p and 2s orbitals of each fluorine, a total of $\sim 27\%$ in all.

It is interesting to compare these estimates of the covalency with those suggested by the copper hyperfine parameters, which are $A_{\parallel} = -148 \cdot 10^{-4}$ cm^{-1} and $A_{\perp} = -15 \cdot 10^{-4}$ cm^{-1}. Substituting these into Eqs. 10.15 and 10.16, together with the values $g_{\perp} = 2.076$ and $g_{\parallel} = 2.432$ observed for the complex,

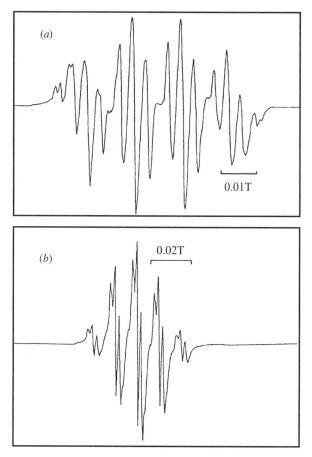

Figure 10.10. EPR spectrum of a planar CuF_4^{2-} complex when the magnetic field is approximately parallel to the z molecular axis (a) and midway between the Cu–F bonds (b) [From reference 6. Used with permission.]

yields the estimates $\alpha^2 \approx 0.94$ and $K \approx 0.39$. The isotropic hyperfine constant is fairly close to the calculated value $K = 0.43$, but the fractional time the unpaired electron spends on the metal, $\sim 94\%$, is considerably higher than the value implied by the analysis of the superhyperfine parameters. When comparing the two approaches, it must be remembered that the sum of the estimates of the percentage times spent in metal and ligand orbitals is in fact expected to be somewhat greater than 100%. This is because the unpaired electron couples with *both* nuclei when it is in the region of space where the metal and ligand atomic orbitals overlap. Even taking this into account, it seems likely that the estimate of α^2 is too high. It is quite sensitive to the value of K, and if the calculated value for this, 0.43, is substituted into Eqs. 10.20 and 10.21, it yields a best estimate of $\alpha^2 \approx 0.8$, which is more similar to that implied by the superhyperfine parameters.

TABLE 10.2 Fractional Percentage Unpaired Spin Density in s and p Orbitals [a]

Complex	Shape	f_s (%)	f_p (%)
$Cu(H_2^{17}O)_6^{2+}$	octahedral	0.85	4.7
CuF_4^{2-}	planar	0.5	6.2
$CuCl_4^{2-}$	planar	0.6	10
$CuBr_4^{2-}$	planar	0.5	12
VF_6^{4-}	octahedral	0.07	2.8
CrF_6^{3-}	octahedral	0.02	4.9
NiF_6^{4-}	octahedral	0.53	3.1

[a] f_s and f_p are estimated from the superhyperfine parameters of each ligand in some transition-metal complexes.

Superhyperfine coupling parameters provide probably the simplest and most direct method of obtaining experimental information of the delocalization of the d electrons onto the ligands, and the fractional unpaired spin densities obtained in this way for a range of complexes are shown in Table 10.2. In view of the various approximations involved in the interpretation, the data are to be considered only semiquantitative; however, other techniques give quite similar results. Scattering of polarized neutrons may also be used to estimate the delocalization of unpaired spin, and this leads to the estimates $f_p \approx 4.6\%$ for the Cr^{3+} ion and $f_s + f_p \approx 6\%$ for the Ni^{2+} ion in oxide lattices. Also, the values $f_s \approx 0.54\%$ and $f_p \approx 3.8\%$ have been derived from the NMR spectrum of $KNiF_3$. These estimates are quite similar to those obtained from the superhyperfine parameters of the fluoride complexes (Table 10.2). Moreover, the unpaired spin densities seem generally consistent with chemical expectations. Delocalization onto ligand s orbitals is relatively small, in line with the fact that these are considerably further in energy from the metal d orbitals than are the ligand p orbitals. The highest s character occurs with the hydrate complex, but even here the s:p ratio $\sim 1:5$ is considerably below the value of 1:3 required if the oxygen uses sp^3 hybrid orbitals.

Negligible unpaired spin density is transferred to the fluorine s orbitals for the complexes VF_6^{4-} and CrF_6^{3-}. This is expected because the t_{2g}^3 d configuration of the two metal ions means that the unpaired spin density is transferred to the ligands solely via π bonding. It is noteworthy, however, that the spin transferred by this π mechanism is quite large. For the CrF_6^{3-} ion, it is comparable to that transferred by σ bonding in the Cu^{2+} and Ni^{2+} fluoride complexes. This is consistent with the strong π donor character of fluoride ion toward the Cr^{3+} ion, as indicated by the position of this ligand in the two-dimensional spectrochemical series (Section 8.4.1). Considerably less spin density is transferred in the iso-electronic VF_6^{4-} complex, and this agrees with the smaller ligand field splitting observed for the divalent metal ion. Finally, it may be noted that, as expected, for the planar copper(II) tetrahalide complexes, the covalency increases progressively as the electronegativity of the halogen decreases.

10.3.3 The Zero-Field Splitting Tensor

The zero-field splitting of the ground states of metal ions with more than one unpaired electron is caused by admixture of excited states via spin–orbit coupling. For octahedral Ni^{2+} complexes with a slight tetragonal distortion, to a first approximation, the zero-field parameter D, which is equal to the energy separation between the $M_S = \pm 1$ and $M_S = 0$ spin levels in zero magnetic field as indicated in Figure 10.5(a), is related to the splitting d of the first excited state by:

$$D \approx 8\lambda^2\delta/E(^3T_{2g})^2 \tag{10.22}$$

Here $E(^3T_{2g})$ is the average energy of the components of first excited state $^3T_{2g}$, split by δ, and λ is the spin–orbit coupling constant. It is apparent that even a small asymmetry in the ligand field can cause a zero-field splitting that is large compared to the energy of the radiation used in EPR and so shift the allowed transitions outside the range available to this technique. Thus a splitting $\delta \approx 100\,cm^{-1}$ causes a zero-field splitting $D \approx 0.7\,cm^{-1}$ for a typical excited state energy $E(^3T_{2g}) \approx 10{,}000\,cm^{-1}$ and $\lambda = -300\,cm^{-1}$. This is larger than the energy of the radiation used in X band EPR. For mixed-ligand complexes, excited state splittings are usually much larger than this, and the zero-field splitting, therefore, usually cannot be measured by EPR for such complexes with currently available equipment. The technique is, however, a useful way of studying the effect of lattice perturbations on otherwise regular complexes. These often vary as a function of temperature and/or pressure. For instance, the D value observed for the compound $[Ni(H_2O)_6]SiF_6$ changes from $-0.5\,cm^{-1}$ at 295 K to $-0.12\,cm^{-1}$ on cooling to 20 K.

Similar arguments apply to the zero-field parameters of complexes formed by other metal ions with more than a single unpaired electron, although the problems are often not as pronounced as is the case for the Ni^{2+} ion. For the high-spin d^5 Mn^{2+} and Fe^{3+} ions, the high energies of the excited states and low values of the spin–orbit coupling constants mean that transitions between the ground state levels may be observed for mixed-ligand complexes, provided the ligand field asymmetry is not too large. The relationship between the observed EPR transitions and the ligand field parameters, however, is complicated and beyond the scope of the present treatment (2,3).

10.4 ELECTRON NUCLEAR DOUBLE RESONANCE

A sensitive method of measuring hyperfine interactions, electron nuclear double resonance (ENDOR), has been developed by combining NMR and EPR spectroscopy. The method uses the fact that in both forms of magnetic resonance the rate of decay from the excited state is on the same order of magnitude as the rate of absorption of photons. This means that by changing the intensity of the radiation source one can alter the relative populations of the levels involved in the transition. The technique is best illustrated by an example (only a greatly

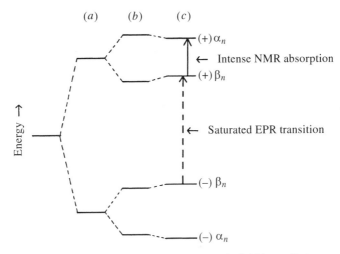

Figure 10.11. Energy levels that occur when a magnetic field is applied to a system with a single unpaired electron coupled to a nuclear spin of $I = \frac{1}{2}$. (*a*), The electronic Zeeman interaction. (*b*), The hyperfine interaction. (*c*), The nuclear Zeeman interaction. *Arrows,* important EPR and NMR transitions in a typical ENDOR experiment.

simplified consideration of the principles underlying the approach can be given). Consider the energy levels that result from the application of an applied field to a complex having a single unpaired electron coupled to a nuclear spin of $I = 1/2$. (Fig. 10.11a). The very weak interaction between the applied magnetic field and the nuclear magnetic moment is included in Figure 10.11. To a good approximation, this may be neglected in the interpretation of EPR spectra, except for its involvement in the mechanism by which formally forbidden transitions gain intensity. Because of its greater sensitivity to the nuclear Zeeman effect, these interactions may be studied directly by ENDOR spectroscopy, although it should be noted that ENDOR spectra are considerably weaker and, therefore, more difficult to observe than are EPR spectra.

In a normal CW-EPR experiment, the energies of the transitions— $(-)(\alpha_n) \rightarrow (+)(\alpha_n)$ and $(-)(\beta_n) \rightarrow (+)(\beta_n)$—are obtained by measuring the absorption of microwave radiation. The signal intensity depends on the population difference between the upper and lower states. It is, therefore, important that the power of the microwave radiation source is kept low, so that the rate of absorption of photons is less than the rate at which the complex decays from the upper to the lower state by transferring the absorbed energy to the lattice. Otherwise, as the sample becomes 'saturated', the population difference decreases and the signal fades away. In ENDOR, one of the EPR lines is deliberately just saturated, so that no signal is observed. This is shown in Figure 10.11c for the transition $(-)(\beta_n) \rightarrow (+)(\beta_n)$. The frequency region of the NMR transitions is then swept with an intense pulse of radiation. When the $(+)(\beta_n) \rightarrow (+)(\alpha_n)$ transition occurs (Fig. 10.11c), the population of the $(+)(\beta_n)$ state is depleted. This reestablishes a population difference with respect to the $(-)(\beta_n)$ state, so that net absorption occurs in the EPR region. The transition, $(-)(\alpha_n) \to (-)(\beta_n)$, also perturbs the population difference of the saturated EPR transition, so that the energy of the second NMR transition may be measured in the same way. The EPR absorption thus provides a highly sensitive method of measuring the NMR transition energies.

It might be asked why the NMR transition cannot simply be measured directly. Why go to the trouble of detecting it through its effect on the EPR transition? The point is that when a system is at equilibrium, the population difference between two energy levels depends on their energy difference. This means that because μ_B is $\sim 10^3$ times larger than μ_N, the population difference between the different electron spin wavefunctions is much greater than that between the nuclear spin wavefunctions. In effect, therefore, ENDOR makes use of the larger population difference associated with the EPR experiment to greatly enhance the sensitivity of the measurement of the NMR spectrum. In the above example, the two NMR transitions occur at energies of $\bar{A} \pm 1/2 g_N \mu_B \mathbf{B}$. Thus the average of the line positions gives the hyperfine (or superhyperfine) splitting, whereas their difference yields the magnetic moment of the particular nucleus under study. Although it is often a difficult and time-consuming experiment to perform, ENDOR provides a powerful way of unraveling the complicated EPR spectra which are sometimes observed when superhyperfine and hyperfine interactions are comparable. Moreover, its high resolution means that super-hyperfine coupling that is unobserved by EPR can often be detected. For instance, superhyperfine structure is not observed in the EPR spectrum of the $VO(H_2O)_5^{2+}$ ion, but the ENDOR spectrum of the complex reveals the interaction of the unpaired electron spin with the protons of the water molecules quite clearly.

REFERENCES

1. Abragam, A., and Bleaney, B., *Electron Paramagnetic Resonance of Transition Ions*, Clarendon Press, Oxford, UK, 1970.

2. Pilbrow, J. R., *Transition Ion Electron Paramagnetic Resonance*, Clarendon Press, Oxford, UK, 1990.

3. Mabbs, F. E., and Collison, D., *Electron Spin Resonance of Transition-Metal Compounds*, Elsevier, Amsterdam, 1992.

4. McGarvey, B. R., *Electron Spin Resonance of Transition-Metal Complexes*, Vol. 3 of *Transition Metal Chemistry*, Marcel Dekker, New York, 1966.

5. Bencini, A., and Gatteschi, D., *ESR Spectra of Metal Complexes*, Vol. 8 of *Transition Metal Chemistry*, Marcel Dekker, New York, 1982.

6. Steffen, G., Reinen, D., Stratemeier, H., Riley, M. J., Hitchman, M. A., Matthies, H. E., Recker, K., Wallrafen, F., and Niklas, J. R., *Inorg. Chem.* 1990, **29**, 2123.

CHAPTER 11

ACTINIDE ELEMENT COMPOUNDS

11.1 LIGAND FIELDS AND f ELECTRON SYSTEMS

The 4f electrons of the lanthanide element ions are shielded from the perturbing effects of ligands by outerlying s and p electrons (1). It is this shielding that accounts for the great similarity of the chemical properties of compounds of the lanthanides. It also explains the facts that the magnitudes of their ligand field splitting parameters are nearly two orders of magnitude smaller than those of the transition series ions and that superhyperfine splittings are not observed in their EPR spectra.

In the octahedral complexes of the transition ions, for example, the perturbations that act on a set of d electrons to remove their degeneracy are likely to be in the order:

$$e^2/r \sim \mathbf{V}_{\text{oct}} > \lambda \mathbf{L} \cdot \mathbf{S} \sim \mathbf{V}_{\text{tetrag}} > k\text{T} \tag{11.1}$$

That is, the interelectronic repulsions, which are responsible for the separation of the free ion energies into terms, are of the same order of magnitude as the primary ligand field splitting of d orbitals. These quantities are a good deal larger than the spin–orbit coupling and the comparable low-symmetry ligand field splittings, which in their turn are larger than the thermal energy available at ambient temperatures. In the lanthanide series ions, the 4f spin–orbit coupling coefficients ζ lie between about 600 and $2500\,\text{cm}^{-1}$, and the parameters for interelectronic repulsion are only about one-quarter of those for first-transition series ions of the same charge (1,2). The order of the effects of the perturbations then becomes:

$$e^2/r > \sim \lambda \mathbf{L} \cdot \mathbf{S} > \mathbf{V}_{lf} \sim kT \tag{11.2}$$

at ambient temperature. $\mathbf{V}_{lf}$ is $\mathbf{V}_{oct}$ for a site of octahedral symmetry, but in fact other stereochemistries with higher coordination numbers are more common.

Because the interelectronic repulsion and spin–orbit coupling parameters are so much larger than the ligand field splittings, there is little or no quenching of the orbital angular momentum associated with the ground state by the ligand field. It follows that there is no change of spin multiplicity even at the highest ligand-field strengths experienced. It was pointed out (Section 9.2.1) that Eq. 9.36, which pertains to a ground state only, holds quite well for the magnetic moments of compounds of the majority of the lanthanide element ions. The two special cases— Eu^{3+} (f^6) and Sm^{3+} (f^5)—were considered separately as examples of the application of the first- and second-order Zeeman effects, because the first excited state for those configurations lies much closer to the ground state than it does in the other configurations and is thermally accessible. In Table 11.1 the magnetic moments predicted on the basis of Eq. 9.36 are listed (3) along with the experimental magnetic moments for typical lanthanide ion complexes.

We observe in Table 11.1 that there is good, although not exact, agreement between the experimental magnetic moments and the predictions of Eq. 9.36. Of course, Eu^{3+} and Sm^{3+} are exceptions, and the moments calculated as described in Section 9.2 are given, rather than the result using Eq. 9.36. The differences between the experimental and calculated values can be attributed largely to the influence of ligand fields. Eq. 9.36 does not allow for variation of the magnetic moment with temperature nor from compound to compound of the same ion. In fact, there are small variations in the moments of compounds according to the nature of the attached ligands. In Table 11.1, the values of the Weiss constant θ in the Curie–Weiss law are listed. Except for Gd^{3+} and Eu^{2+} (both f^7, $^8S_{3\frac{1}{2}}$ states), they are generally nonzero, although fairly small. If Eq. 9.36 is obeyed, θ should be zero, and the departure is further evidence of the presence of ligand field effects. The $^8S_{3\frac{1}{2}}$ state is the only component of the ^8S term of f^7, and because there is no orbital degeneracy associated with this state, it leads to a magnetic moment very close to the spin-only value for seven unpaired electrons, 7.94 μ_B, and independent of ligand field effects and temperature. The position is equivalent to that of the $^6S_{2\frac{1}{2}}$ state and ^6S ground term of d^5 among the transition series ions, as for example in Mn^{2+}.

The electronic spectra of the free lanthanide ions consist of transitions between the states that arise from the appropriate f^n configuration with perturbation by interelectronic repulsions. A great many energy levels within the uv/vis range result, and the spectra are complicated (1). The peaks are usually quite sharp, and line widths of a few cm^{-1} are common, so that sufficient information is available for their interpretation in terms of spin–orbit and interelectronic repulsion parameters (1). The spectra of the ions in solution or

TABLE 11.1. The Magnetic Moment (μ_B) of Some Lanthanide Complexes

Number of f electrons	Ion	Ground State	ζ_{4f} (cm^{-1})	Compound	$\mu_{eff}^{300\,K}$ (Experimental)	$g_J[J(J+1)]^{1/2}$	θ (K)
1	Ce^{3+}	$^2F_{2.5}$	640	Ce$_2$Mg$_3$(NO$_3$)$_6\cdot$24H$_2$O	2.28	2.54	210
2	Pr^{3+}	$^3H_4$	710	Pr$_2$(SO$_4$)$_3\cdot$8H$_2$O	3.40	3.58	32
3	Nd^{3+}	$^4I_{4.5}$	900	Nd$_2$(SO$_4$)$_3\cdot$8H$_2$O	3.50	3.62	43
4	Pm^{3+}	$^5I_4$	—			2.68	[b]
5	Sm^{3+}	$^6H_{2.5}$	1180	Sm$_2$(SO$_4$)$_3\cdot$8H$_2$O	1.58	1.6[a]	[b]
6	Eu^{3+}	$^7F_0$	1360	Eu$_2$(SO$_4$)$_3\cdot$8H$_2$O	3.42	3.61[a]	[b]
	Sm^{2+}		800	SmBr$_2$	3.57	3.61[a]	
7	Gd^{3+}	$^8S_{3.5}$		Gd$_2$(SO$_4$)$_3\cdot$8H$_2$O	7.91	7.94	0
	Eu^{2+}			EuCl$_2$	7.91	7.94	0
8	Tb^{3+}	$^7F_6$	1720	Tb$_2$(SO$_4$)$_3\cdot$8H$_2$O	9.50	9.72	16
9	Dy^{3+}	$^6H_{7.5}$	1920	Dy$_2$(SO$_4$)$_3\cdot$8H$_2$O	10.4	10.63	5
10	Ho^{3+}	$^5I_8$	2080	Ho$_2$(SO$_4$)$_3\cdot$8H$_2$O	10.4	10.60	7
11	Er^{3+}	$^4I_7$	—	Er$_2$(SO$_4$)$_3\cdot$8H$_2$O	9.4	9.57	6
12	Tm^{3+}	$^3H_6$	—	Tm$_2$(SO$_4$)$_3\cdot$8H$_2$O	7.1	7.65	6
13	Yb^{3+}	$^2F_{3.5}$	—	Yb$_2$(SO$_4$)$_3\cdot$8H$_2$O	4.86	4.50	—

[a] Not $g_J[J(J+1)]^{1/2}$, but rather calculated as outlined in Section 9.2.2.
[b] Not relevant, because the behavior is not expected to approximate to the Curie law.

crystals are even more complicated; additional splittings are the result of ligand-field effects. The band intensities are low, as expected for parity-forbidden transitions essentially divorced from the chemical environment. The line widths are greater, up to a few 10's cm^{-1}, presumably because of vibrational modulation through ligand field effects. Nevertheless, because of the good resolution, there is sufficient information to permit an interpretation that includes the small ligand field parameters. These are even greater in number than is required for the transition-series complexes, as discussed below.

A great many analyses of lanthanide ion spectra, particularly in crystals, have been carried out, and ligand field parameters have been derived for most of the ions in various ligand environments, usually of lower than octahedral symmetry. Although the analyses are satisfying in the sense that they account well for the observed spectra, they have few chemical implications because the 4f electrons are so well shielded from the chemical environment. We do not pursue the subject further here, because it offers little reward for much labor. The preceding remarks do, however, provide a background for the discussion of actinide element complexes.

On the basis of Eq. 9.37, the EPR spectra of lanthanide free ions might be expected to be quite simple because this implies a ground state with an isotropic **g** tensor. In fact, interpretation of the EPR spectra of lanthanide complexes is relatively complicated because of interelectronic repulsion and ligand field perturbations (4,5). **g** is highly anisotropic, and spectra may usually be observed only at very low temperatures. Such results are expected when there are energy levels lying within a few hundred cm^{-1} of the ground state, which is likely to be the case for lanthanide complexes. It is possible to interpret the EPR spectra in terms of ligand field parameters, but for the same reasons as for the uv/vis spectra, the results convey little chemical information, and the topic is not followed further here.

11.2   ACTINIDE ELEMENT COMPOUNDS

For the actinide elements, the shielding of the 5f electrons by outer s and p electrons is not as effective as it is for the 4f lanthanide ions. Although it is not considered further here, there is a complication for the actinides in that it is not always clear that for these elements it is the 5f rather than the 6d electrons that are involved in valence phenomena (2). Ligand field effects for complexes of the actinide elements involve splittings of up to a few thousand cm^{-1} and are, therefore, intermediate in magnitude between those of the transition elements and lanthanides. Spin–orbit coupling constants are larger than those of the lanthanides, and ζ lies between 1400 and perhaps 4000 cm^{-1}. Interelectronic repulsion parameters are perhaps 30% smaller than for the lanthanides. Thus the order of perturbations of a 5f configuration is:

$$e^2/r > \sim \lambda \mathbf{L} \cdot \mathbf{S} > V_{lf} > kT \qquad (11.3)$$

TABLE 11.2. Calculated Spin–Orbit Coupling Constants (ζ, cm^{-1}) for Some Actinide Element Ions

	Element					
Charge	Pa	U	Np	Pu	Am	Cm
+1	—	1680	2215	2265	2605	2960
+2	1440	1725	2015	2295	2635	2985
+3	1705	1980	2270	2550	2880	3235
+4	1920	2210	2510	2810	3115	3483
+5	—	2240	2730	3065	3365	—
+6	—	—	2975	3295	3645	—

A list of some spin–orbit coupling constants for actinide ions is given in Table 11.2.

Octahedral complexes of the actinides are well known, but again stereochemistries involving higher coordination numbers are frequent. Because of the greater exposure of the 5f electrons to the chemical environment, the chemistry of the actinide elements is much more varied than that of the lanthanides; oxidation numbers range from 0 to VI, at least for the better known earlier members such as uranium. The greater involvement of the 5f electrons with the ligand environment, of course, makes their magnetic and spectral behavior of much higher chemical relevance. Also, ligand superhyperfine splittings are seen, as are effective reductions in the orbital angular momentum, reflecting the f orbital involvement in covalence in the M-L bonding.

11.3   f ELECTRONS AND V_{oct}

Although octahedral stereochemistry does not dominate for actinide complexes, the consideration of its ligand field effects serves to illustrate relationships to procedures for the transition metals and simultaneously problems peculiar to the actinides. As was pointed out in Chapter 1, it is sufficient for use in transition-metal chemistry to carry the expansion of the crystal-field potential $\mathbf{V}_{oct}$ to spherical harmonics of order four (Y_4^m) because integrals of the form $\langle(m)|Y_l^{m''}|(m')\rangle$, are zero if $l > 4$, because the d functions (m) are specified by spherical harmonics of order two, viz Y_2^m. We denote the f orbitals as $(3, m)$, $m = -3$ to 3, to specifically distinguish them from the d orbitals. Continuing the preceding argument, the f orbital functions are specified by spherical harmonics of order 3, viz Y_3^m, and integrals of the form $\langle(3, m)|Y_l^{m''}|(3, m')\rangle$ necessarily vanish only if $l > 6$. Thus for use with f orbital systems, it is necessary to continue the expansion of $\mathbf{V}_{oct}$ to order 6 in spherical harmonics (1,4–6). The result is:

$$\mathbf{V}_{oct} = (49/18)^{\frac{1}{2}}(2\pi)^{\frac{1}{2}}(zer^4/a^5)[Y_4^0 + (5/14)^{\frac{1}{2}}(Y_4^4 + Y_4^{-4})]$$
$$+ (9/104)^{\frac{1}{2}}(2\pi)^{\frac{1}{2}}(zer^6/a^7)[Y_6^0 + (7/2)^{\frac{1}{2}}(Y_6^4 + Y_6^{-4})] \quad (11.4)$$

Matrix elements of $\mathbf{V}_{\text{oct}}$ are evaluated between the f orbitals in the same way as described in Chapter 2 for d orbitals, but with more labor. A specific example is:

$$\langle(3,0)|\mathbf{V}_{\text{oct}}|(3,0)\rangle = (7/11)\text{ze }\overline{r^4}/a^5 + (25/143)\text{ze }\overline{r^6}/a^7 \qquad (11.5)$$

We can, for the present illustrative purposes only, put the results in the non standard form:

$$\langle(3,m)|\mathbf{V}_{\text{oct}}|(3,m')\rangle = \text{Dq}' + \text{Fr} \qquad (11.6)$$

Here Dq$'$ is essentially the same quantity as the Dq that we deduced for the equivalent quantities involving d orbitals, although there is a numerical difference between q and q$'$, because the coefficients in the f wavefunction expressions are different from those of d wavefunctions. F and r are quantities arising from the new Y_6^m terms in $\mathbf{V}_{\text{oct}}$ and are entirely equivalent to D and q, but involving a different coefficient from 1/6, and $\overline{r^6}/a^7$ rather than $\overline{r^4}/a^5$. F would introduce a term in x^6, y^6, and z^6 in the Cartesian form of $\mathbf{V}_{\text{oct}}$ (Section 2.3), but we do not examine the explicit details of Fr. In more conventional usage, linear combinations of Dq$'$ and Fr are taken such that Δ is the separation between the a_{1u} f$_{xyz}$ and the t$_{2u}$ f$_{z(x^2-y^2)}$, etc. orbitals, and Θ is the separation between those latter and the t$_{1u}$ f$_{z(5x^2-3r^2)}$, etc orbitals (Fig. 1.3), as shown in Figure 11.1.

The hierarchy of perturbations set out in Eq. 11.3 infers that the primary splitting of an f^n configuration is into terms, followed by *states* specified by the total angular momentum J. The states lie somewhere between the Russell–Saunders and the j-j coupling schemes in composition. The calculation of the splitting of these *states* by $\mathbf{V}_{\text{oct}}$ presents a similar problem to that of the splitting of free ion terms of the transition-series ions, but it is more complicated. It is not examined here, although some of the results are used below.

One important point is immediately obvious from the form of the matrix elements. *The cubic ligand field splittings of the states of f^n configurations cannot be described by means of a single parameter, as was possible for d^n* configurations. Accurate wavefunctions are not available for the 5f electrons of the actinide elements, but it can be estimated that for reasonable values of the metal–ligand distance a, the term in Fr may well be comparable in magnitude with that in Dq. Similar arguments apply for the lanthanides. Thus *the magnitudes of the interactions involving both the fourth- and the sixth-order components of the f radial functions must be known to describe the effects of an octahedral ligand field in f^n configurations.* It is obvious that *two* parameters are necessary to describe the effects of a cubic ligand field, and no simple generalizations such as hold for transition-metal complexes are available. Most f^n configurations must be considered independently.

Although the preceding deduction was obtained specifically for $\mathbf{V}_{\text{oct}}$, quite simple considerations show that it holds even more strongly for the lower symmetry stereochemistries that are common for lanthanide and acitinide complexes. More components in the crystal field potential occur for such

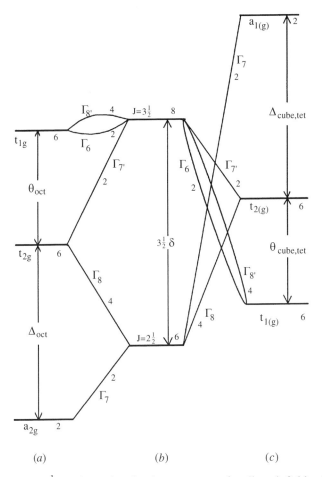

Figure 11.1. The f^1 configuration in the presence of a ligand field of octahedral stereochemistry, no spin–orbit coupling (*a*); spin–orbit coupling, no ligand field (*b*); and a ligand field of cube or tetrahedral stereochemistry, no spin–orbit coupling (*c*). The lines showing symmetry correlations (Γ_6, Γ_7, and $\Gamma_{8'}$) are only illustrative. The numbers 2, 4, 6, or 8 accompanying a level indicate its degeneracy.

stereochemistries than apply for V_{oct}, and each of these is repeated with higher order spherical harmonics when f orbitals are to be treated.

11.4   UV/VISIBLE SPECTRA OF ACTINIDE COMPLEXES

The uv/vis spectra of actinide complexes have many similarities in common with those of the lanthanides, and most features lie closer to those than to the spectra of transition-metal complexes. They are similarly complicated because many energy levels arise from the perturbation of a state by interelectronic repulsion

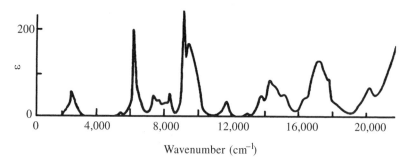

Figure 11.2. The vis/nir spectrum of U^{3+} in 1 M perchloric acid solution.

and ligand fields. The lines from transitions within the f orbital set are fairly sharp (1,6), at some tens to hundreds of cm^{-1}. They are much more intense than lanthanide spectra, as a result of the stronger interaction with the ligand environment, and on account of the sharpness, extinction coefficients may approach those of the transition-metal complexes. There may also be some very intense lines from parity-allowed $5f \rightarrow 6d$ transitions. An example, that of U^{3+} in a 1 M perchloric acid solution, is given in Figure 11.2. Again, a great deal of information can be obtained and, in principle, ligand field parameters can be extracted, including reductions in the orbital angular momentum and interelectronic repulsion parameters. The necessary analysis is, however, complicated and has not often been completed successfully. We do not deal in general with the spectra further here, except to note that splittings of the f orbitals by $1000 \, cm^{-1}$ or more are deduced from the spectra of the well-documented complexes of the M^{3+} ions; even larger splittings are observed for more highly charged ions (6). The case of f^1, however, with no interelectron repulsion effects, is sufficiently simple to be used for illustrative purposes.

Figure 11.1 shows the splitting of the f^1 configuration by high-symmetry ligand field environments and spin–orbit coupling. The left side indicates the effect of an octahedral field and the right side that of a tetrahedral or cube field, in each instance with no spin–orbit coupling. In the center is the case of spin–orbit coupling with no ligand field. The lines connecting levels of the same symmetry are illustrative only. For our purposes, the group theory symmetry symbols $\Gamma_{1-8'}$ in this and later figures are essentially arbitrary labels.

On the basis of Figure 11.1, it can be predicted that the spectra of an f^1 ion in an octahedral ligand arrangement should be similar to that in a cube or tetrahedral environment in that four transitions are available from the ground level. If the ligand field effects are substantial relative to the spin–orbit coupling, however, the relative positions of the resultant bands should be quite different for the two cases, because of the reversal in sign of the ligand field potential. These features have been observed for complexes of the Pa^{4+} ion. In $(NEt_4)_2PaCl_4$, the Pa stereochemistry is octahedral, and four bands are observed, originating from the Γ_7 ground level: 1730 $(\rightarrow \Gamma_8)$, 5330 $(\rightarrow \Gamma_{7'})$, 7022 $(\rightarrow \Gamma_{8'})$, and $8011 \, cm^{-1}$ $(\rightarrow \Gamma_6)$. These transitions have the ratios $0.22 : 0.67 : 0.88 : 1.00$. Pa^{4+} in $ThCl_4$

is in a site of distorted tetrahedral symmetry D_{2d}. The spectrum there consists of six bands, but two of the pairs result from a splitting owing to the distortion. Averaging these pairs, one obtains the spectrum of a tetrahedral complex consisting of four bands originating in a Γ_8 ground level: 1060 ($\rightarrow \Gamma_7$), 5130 ($\rightarrow \Gamma_6$), 5730 ($\rightarrow \Gamma_{8'}$), and 6510 cm^{-1} ($\rightarrow \Gamma_{7'}$), which give the ratios $0.16 : 0.79 : 0.88 : 1.00$. The first band, which involves the same two levels Γ_7 and Γ_8 in the two stereochemistries, exhibits a smaller splitting for the tetrahedral than for the octahedral stereochemistry, for the same reason that the splittings of the d orbitals show this trend. However, a splitting ratio as simple as $-4/9$ is not expected. These spectra do show the importance of ligand field effects in the actinide element complexes and that they are directly related to, and of magnitude approaching, those so commonly seen for transition metal compounds.

11.5   MAGNETIC PROPERTIES OF ACTINIDE COMPLEXES

Although magnetic data are available for actinide elements up to at least Cf, examples are taken only from the first few, for which radioactivity problems in the experimental work are not extreme (7).

The magnetic properties of actinide complexes can reflect the nature of the ground state rather strongly. In Figure 11.3, the effects of the relative magnitudes of the $\overline{r^4}/a^5$ and $\overline{r^6}/a^7$ components of $\mathbf{V}_{\text{oct}}$ in the octahedral ligand field splitting of the ground states of the configurations f^3 and f^4 are shown for likely values of the interelectronic repulsion and the spin–orbit coupling constants. It is seen that the multiplicity of the lowest level changes as we go from a ratio of much less, to one much greater than unity. The importance of the symmetry of the ligand field is shown in Figure 11.4. There the effects of changing from the field from an octahedron to a cube (or tetrahedron) are set out, again for differing ratios of the $\overline{r^4}/a^5$ and $\overline{r^6}/a^7$ components of the ligand field, for the $^3H_4$ ground state of f^2. It is seen that the multiplicity of the ground levels is changed by the reversal of the sign of the ligand field potential from $\mathbf{V}_{\text{oct}}$ to $\mathbf{V}_{\text{cube}}$, which was spelled out in Chapter 2.

11.5.1 f^1, $^2F_{2\frac{1}{2}}$

On the basis of Figure 11.1 it is seen that the $^2F_{2\frac{1}{2}}$ state is split by an octahedral ligand field to yield a Kramer's doublet ground level (Γ_7) with a degenerate set (Γ_8) above it. The g value for the free ion $J = 2\frac{1}{2}$ state is, using Eq. 9.37, $\frac{6}{7}$. This value is likely to be modified in a complex by mixing with the higher free ion state with $J = 3\frac{1}{2}$ caused by the ligand field and also by orbital-reduction effects, specified by the parameter k. The magnetic susceptibility is expected to be of the form of the Langevin–Debye expression (Eq. 9.26), arising from the first-order Zeeman effect of the ground level and its second-order effect with the higher lying Γ_8 levels.

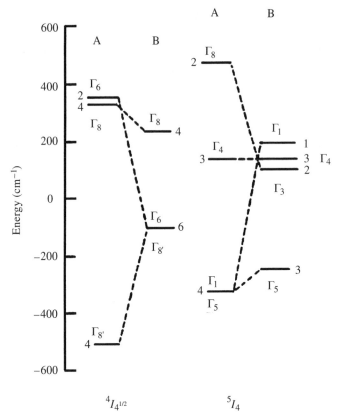

Figure 11.3. Ligand-field splitting of the ground states of f^3 (U^{3+}), $^5I_{4\frac{1}{2}}$ and f^4 (Np^{3+}), $^5I_4$ when the V_{oct} term in $\overline{r^6}$ is much smaller than that in $\overline{r^4}$ (*A*) and the term in $\overline{r^6}$ is much greater than that in $\overline{r^4}$ (*B*). The numbers at the end of the lines represent degeneracy.

Octahedral complexes with the f^1 configuration are well known, and some results are summarized in Table 11.3. The values of the ligand field parameters Δ and Θ and of ζ and k have been evaluated with the aid of optical spectral data. Indeed, the magnetic susceptibilities as a function of temperature follow the Langevin-Debye expression, with TIP terms on the order of $5 \cdot 10^{-9}\,m^3\,mol^{-1}$. The g values are seen to vary quite markedly, corresponding to substantial ligand field mixing and delocalization effects.

The Np^{6+} ion in the unit NpO_2^{2+}, which is linear, may be regarded as subject to a ligand field of axial symmetry. In contrast to the position for similar ions in d electron systems, however, this highly asymmetric field is not large enough to ensure that the magnetic properties approximate to the spin-only value. The unit occurs, for example, in $NaNpO_2(OOC \cdot CH_3)_3$, for which μ_{eff} is 2.0 μ_B, actually not too far from the spin-only value for one unpaired electron. This is to be contrasted with the free $^2F_{2\frac{1}{2}}$ state ion value of 2.54 μ_B, accompanied by g = 0.8. The EPR spectrum of the complex does not approximate to spin-only behavior;

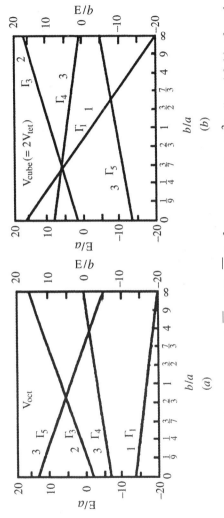

Figure 11.4. The effect of different ratios of the terms in $\overline{r^4}$ and for $\overline{r^6}$ the $^3H_4$ ground state of f^2 in ligand fields of octahedral (a) and cube or tetrahedral (b) stereochemistry. The numbers 1, 2, or 3 near a line indicate its degeneracy. The figures may also be used for the $^5I_4$ ground state of f^4 if the stereochemistries are reversed.

TABLE 11.3. Some Ligand Field and Spin–Orbit Coupling Parameters and g Values for Actinide f^1 Complexes

Compound	Stereochemistry	g	ζ	Δ (cm^{-1})	Θ (cm^{-1})	k
$(Et_4N)_2PaCl_6$	octahedron	1.18	1,690	600	3,530	0.83
$CsUF_6$	octahedron	0.71	2,210	3,340	8,050	0.84
$CsUBr_6$	octahedron	1.21	2,190	826	5,794	0.78
NpF_6	octahedron	0.61	2,700	4,780	16,900	0.85
$PaCl_4 / ThCl_4$	tetrahedron	1.6	1,525	2,800	1,145	—
$Pa(OOC \cdot H)_4$	cube	1.4	—	—	—	—
Na_3UF_8	cube	1.2	—	—	—	—

as $g_\parallel$ is 3.4 and $g_\perp$ is 0.2, both far from 2.00. Thus, owing to the presence of the large spin–orbit coupling constant, the axial ligand field has a substantial, but not overwhelming, effect on the magnetic behavior.

11.5.2 f^2, $^3H_4$

Octahedral and cube symmetry complexes that give rise to the f^2 configuration are available and present an interesting contrast. To simplify presentation, it is convenient to introduce two ligand field parameters—a proportional to the term in Dq' and b proportional to that in Fr—and to express the splitting of the $^3H_4$ state in terms of the ratio a/b. This is done in Figure 11.4. It is seen that the two types of stereochemistry lead to different ground terms. For the octahedral case, the ground term is nondegenerate (Γ_1) for any value of a/b and so can give no first-order Zeeman effect. For the cube/tetrahedron case, for most values of a/b the ground term is threefold degenerate (Γ_5) and so can show a first-order Zeeman effect. The position of a specific ion along the horizontal axis of Figure 11.4 depends on its values for $\overline{r^4}$ and, $\overline{r^6}$ which are not generally available except in so far that they may be inferred by interpretation of spectral and magnetic properties.

For octahedral complexes, provided the splitting $\Gamma_1 - \Gamma_4$ is $\gg kT$, the magnetic susceptibility is expected to be of the TIP type. The susceptibility of $(Me_4N)_2UCl_6$ is indeed independent of temperature, with χ_U $24 \cdot 10^{-9}$ m^3 mol^{-1}. For PuF_6, the TIP is only $1.9 \cdot 10^{-9}$ m^3 mol^{-1}; the difference reflects the larger spin–orbit coupling and larger $\Gamma_1 - \Gamma_4$ separation in this case.

For a complex with the stereochemistry of a cube, provided the Γ_5 level is lowest and the splitting between levels is much greater than kT, the susceptibility is expected to be of the Langevin–Debye form, with the Curie law term dominating. The free ion state g_J value is 0.8, so that the magnetic moment for the $^3H_4$ state, $g_J[J(J+1)]^{1/2}$, is 3.58 μ_B (Table 11.1). If the threefold degenerate Γ_4 level is treated as a state with fictitious spin $S' = 1$ and with the same value of g_J, the magnetic moment is calculated as $g_J[S'(S'+1)]^{1/2} = 1.13$ μ_B. If g has the spin-only value of 2.00, the calculated value is, of course, 2.83 μ_B. $Cs_4U(NCS)_8$

has $\mu_{\text{eff}}^{300\,\text{K}}$ 2.9 μ_{B}, which falls slowly on cooling to about 50 K and then rapidly below that. This behavior is explained if the $S' = 1$ level has g close to the spin-only value and a large zero-field splitting such that the singly degenerate component is lowest (Section 10.3.3). The change from the free ion state g_J to the spin-only value is caused by the ligand field, and the ZFS is caused by the large spin–orbit coupling constant of the U^{4+} ion. Spectral results can be combined with the magnetic ones (above) to give a rather complete ligand field analysis for this complex, which includes an orbital reduction effect.

$U(OOC \cdot H)_4$ also has cube stereochemistry, if somewhat distorted, and shows rather similar behavior, with $\mu_{\text{eff}}^{300\,\text{K}}$ 3.27 μ_{B}. In UCl_4, the stereochemistry is close to tetrahedral, and as expected, the magnetic behavior is again rather similar, with $\mu_{\text{eff}}^{300\,\text{K}}$ 3.1 μ_{B}. The splittings of the state are smaller than for the previous complex and are in agreement with the lower coordination number.

11.5.3 f^3, $^5I_{4\frac{1}{2}}$

The $J = 4\frac{1}{2}$ state is split into several components even by a high-symmetry ligand field. Detailed analysis of the spectrum of U^{3+} in solid solution in $LaCl_3$ has been made, although the magnetic behavior has not been analyzed in such detail. This state gives levels that cover 450 cm^{-1}. The free state magnetic moment is 3.62 μ_{B} (Table 11.1). A number of complexes giving the f^3 configuration are known, and they mostly have magnetic moments about 3.2 μ_{B} at ambient temperature, which do not change greatly at low temperatures. Examples of complexes exhibiting such behavior and for which the stereochemistry is well defined as octahedral are Cs_2NaUCl_6 and Cs_2NpCl_6.

11.5.4 f^4 to f^7

There are relatively few results available for the f^4 to f^7, configurations, and the interpretation can be complicated. A couple of points deserve mention.

The ground state of the f^4 configuration is $^5I_4$, and this has the same value of J as has $^3H_4$ from the f^2 configuration. This state is split by a ligand field in a similar manner to $^3H_4$ (Fig. 11.4), except that the sign is reversed. Thus, for the f^4 case, Figures 11.4a and 11.4b need to be interchanged, and so the magnetic behavior versus octahedral/cubic stereochemistry relationship should be reversed. This is indeed observed; $Cs_2NaNpCl_6$ shows approximate Curie law behavior, with $\mu_{\text{eff}}^{300\,\text{K}}$ 1.92 μ_{B}, in place of the TIP of Cs_2NaUCl_6. The magnetic moment is significantly different from the free ion state value of 2.68 μ_{B} (Table 11.1), again because of the effects of the ligand field. Conversely, for PuO_2, for which the stereochemistry is that of the cube, there is TIP, as expected from the same argument.

The half-filled shell f^7 configuration plays an equivalent role to the d^5 configuration of the transition series ions. The ground state is the orbitally nondegenerate $^8S_{3\frac{1}{2}}$. This is expected to be magnetically simple, with a g value and magnetic moment close to the spin-only values of 2.00 and 7.94 μ_{B}. This is generally observed. In the case of Cm^{3+} in $Cs_2NaLuCl_6$, for example, $\mu_{eff}^{300\,K} = 7.64$ μ_B and this does not vary until quite low temperature. The departures from spin-only behavior that do occur can be correlated accurately with the large value of ζ for Cm^{3+} (Table 11.2) which causes zero-field splitting.

REFERENCES

1. Wybourne, B. G., *Spectroscopic Properties of Rare Earths*, Interscience, New York, 1965, chapter 1.

2. Carnall, W. T., and Crosswhite, H. M., in Katz, J. J., Seaborg, G. T., and Morss, C. R., eds., *Chemistry of the Actinide Elements*, Vol. 2, Chapman & Hall, New York, 1986, p. 1235.

3. Van Vleck, J. H., *The Theory of Electric and Magnetic Susceptibilities*, Oxford University Press, Oxford, UK 1932.

4. Bleaney, B., and Stevens, K. W. H., *Rep. Prog. Phys.*, 1953, **16**, 108.

5. Bowers, K. D., and Owen, J., *Rep. Prog. Phys.*, 1955, **18**, 304.

6. Chan, S.-K., and Lam, D. J., in Freeman, A. J., and Darby, J. B., eds., *The Actinides: Electronic Structure and Related Properties*, Academic Press, New York, 1974, chap. 1.

7. Edelstein, N. M., and Goffart, J., in Ref. 2, p. 1361.

APPENDIX A1

A1.1 THE SPHERICAL HARMONICS Y_l^m

The normalized associated Legendre polynomials of order l are obtained as defining the well-behaved solutions of Legendre's differential equation (1). They are:

$$\Xi_l^m = [(2l+1) \cdot (l-|m|)!/2(l+|m|)!]^{1/2} P_l^{|m|}(\cos\theta) \qquad (A1.1)$$

with $|m| \leq l$ and θ the angle defined in Figure 1.1. Writing, for brevity, $x = \cos\theta$:

$$P_l^{|m|}(x) = [(-1)^l \cdot (1-x^2)^{|m|/2}/2^l l!] \cdot [\mathrm{d}^{l+|m|}(1-x^2)^l/\mathrm{d}x^{l+|m|}] \qquad (A1.2)$$

The spherical harmonics are then defined as:

$$Y_l^m(\theta,\phi) = (-1)^q \Xi_l^m \cdot (2\pi)^{-1/2} e^{-im\phi} \qquad (A1.3)$$

where q is 0 if m is negative and/or even and is 1 if m is positive *and* odd. (This is the Condon–Shortley phase convention and is most commonly but *not* always followed.) ϕ is the angle defined in Figure 1.1. It is not uncommon for the (θ,ϕ) specification to be presumed, and so omitted in the writing of the functions.

AI.2 INTEGRATION OF PRODUCTS OF SPHERICAL HARMONICS

In the application of quantum mechanics to atomic systems, the evaluation of matrix elements of the form

$$\langle \psi_i | \mathbf{H} | \psi_j \rangle = \int \psi_i^* \mathbf{H} \psi_j \, \mathrm{d}\tau \qquad (A1.4)$$

occurs frequently. The angular properties of ψ_i, and ψ_j, are usually defined in terms of spherical harmonics. Thus an important step in the evaluation of the matrix element reduces to the integration of products of spherical harmonics.

325

The general requirement is for integrals of the form:

$$\int_{\theta=0}^{\pi} \int_{\phi=0}^{2\pi} Y_{l_1}^{m_1} \cdot Y_{l_2}^{m_2} \cdot Y_{l_3}^{m_3} \cdot \sin\theta \cdot d\theta \cdot d\phi$$

$$= \int_{\theta=0}^{\pi} P_{l_1}^{m_1}(\cos\theta) \cdot P_{l_2}^{m_2}(\cos\theta) \cdot P_{l_3}^{m_3}(\cos\theta) \cdot \sin\theta \cdot d\theta$$

$$\cdot (2\pi)^{-\frac{1}{2}} \int_{\phi=0}^{2\pi} e^{-im_1\phi} \cdot e^{-im_2\phi} \cdot e^{-im_3\phi} \cdot d\phi \qquad (A1.5)$$

Simplifications occur; e.g., for matrix elements of the form $\langle \psi_i | \psi_j \rangle$ and $\langle \psi_i | \mathbf{H} | \psi_i \rangle$. Also, especially for the present purposes in which we deal only within d or f orbital sets, we have the restriction that two of the l values must be the same.

The examination of the integrations involving θ is facilitated by the fact that the Legendre polynomials are related by recursion formulae. In particular:

$$\cos\theta \cdot P_l^{|m|}(\cos\theta) = (2l+1)^{-1} \cdot [(l-|m|+1)P_{l+1}^{|m|}(\cos\theta)$$
$$+ (l+|m|)P_{l-1}^{|m|}(\cos\theta)] \qquad (A1.6)$$

and

$$\sin\theta \cdot P_l^{|m|}(\cos\theta) = (2l+1)^{-1} \cdot [P_{l+1}^{|m|+1} \cdot (\cos\theta) - P_{l-1}^{|m|+1}(\cos\theta)] \qquad (A1.7)$$

Employing these relationships and the standard integrals:

$$\int_{\theta=0}^{\pi} \sin^{2n}\theta \cdot \cos^{2m}\theta \cdot d\theta = [(2n)!(2m)!\pi]/[2^{2(n+m)}n!m!(n+m)!] \qquad (A1.8)$$

$$\int_{\theta=0}^{\pi} \sin^{2n+1}\theta \cdot \cos^{2m}\theta \cdot d\theta = \int_{\theta=0}^{\pi} \cos^{2n+1}\theta \cdot \sin^m\theta \cdot d\theta$$
$$= [2^{n+1}n!]/[(2m+1) \cdot (2m+3) \cdots$$
$$\cdot (2m+2n+1)] \qquad (A1.9)$$

$$\int_{\theta=0}^{\pi} \sin^{2n+1}\theta \cdot \cos^{2m+1}\theta \cdot d\theta = 0 \qquad (A1.10)$$

$$\int_{\phi=0}^{2\pi} e^{-im\phi}d\phi = 2\pi \qquad \text{if } m = 0$$

$$= 0 \qquad \text{otherwise} \qquad (A1.11)$$

Quantities in Eq. A1.5 have been extensively studied and tabulated. The integrals concerned appear widely in the coupling of angular momenta, but in, at least superficially, rather different guise (2). The Wigner coefficients, Clebsch–Gordan coefficients, Condon–Shortley coefficients, Racah symbols, and the 3-j and 6-j symbols all reflect aspects related to these integrals. Books have been devoted to the subject (3–5).

One of the simpler coupling requirements that is seen from these compilations is the *vector triangle rule* for the θ integrations. It states that for an integral to be nonzero l_1, l_2, and l_3 must be related vectorially as the edges of a triangle. From this, if $l_1 = l_3$, it is easy to deduce the requirement that $l_2 \leq 2l_1$.

REFERENCES

1. Eyring, H., Walter, J., and Kimball, W. E., *Quantum Chemistry*, Wiley, New York, 1944.

2. Gerloch, M., and Slade, R. C., *Ligand Field Parameters*, Cambridge University Press, Cambridge, UK 1973, chap. 2.

3. Judd, B. R., *Operator Techniques in Atomic Spectroscopy*, McGraw-Hill, New York, 1963.

4. Brink, D. M., and Satchler, G. R., *Angular Momentum*, Clarendon Press, Oxford, UK, 1968.

5. Rotenberg, M., Bivins, R., Metropolis, N., and Wooten, J. K., *3-j and 6-j Symbols*, Technology Press, Cambridge, MA, 1959.

APPENDIX A2

A2.1 THE ASSOCIATED LEGENDRE POLYNOMIALS Ξ_l^m TO ORDER 6

The spherical harmonics are defined, as in Appendix A1:

$$Y_l^m = (-1)^q \Xi_l^m \cdot (2\pi)^{-1/2} \cdot e^{im\phi} \tag{A2.1}$$

The Ξ_l^m functions are:

$$\Xi_0^0 = (1/2)^{1/2} \tag{A2.2}$$

$$\Xi_1^0 = (3/2)^{1/2} \cdot \cos\theta \tag{A2.3}$$

$$\Xi_1^1 = (3/4)^{1/2} \cdot \sin\theta \tag{A2.4}$$

$$\Xi_2^0 = (5/8)^{1/2} \cdot (3\cos^2\theta - 1) \tag{A2.5}$$

$$\Xi_2^1 = (15/4)^{1/2} \cdot \sin\theta \cos\theta \tag{A2.6}$$

$$\Xi_2^2 = (15/16)^{1/2} \cdot \sin^2\theta \tag{A2.7}$$

$$\Xi_3^0 = (7/8)^{1/2} \cdot (5\cos^3\theta - 3\cos\theta) \tag{A2.8}$$

$$\Xi_3^1 = (21/32)^{1/2} \cdot \sin\theta \cdot (5\cos^2\theta - 1) \tag{A2.9}$$

$$\Xi_3^2 = (105/16)^{1/2} \cdot \sin^2\theta \cdot \cos\theta \tag{A2.10}$$

$$\Xi_3^3 = (35/32)^{1/2} \cdot \sin^3\theta \tag{A2.11}$$

$$\Xi_4^0 = (9/128)^{1/2} \cdot (35\cos^4\theta - 30\cos^2\theta + 3) \tag{A2.12}$$

$$\Xi_4^1 = (45/32)^{1/2} \cdot \sin\theta \cdot (7\cos^3\theta - 3\cos\theta) \tag{A2.13}$$

$$\Xi_4^2 = (45/64)^{1/2} \cdot \sin^2\theta \cdot (7\cos^2\theta - 1) \tag{A2.14}$$

$$\Xi_4^3 = (315/32)^{1/2} \cdot \sin^3\theta \cdot \cos\theta \tag{A2.15}$$

$$\Xi_4^4 = (315/256)^{1/2} \cdot \sin^4\theta \tag{A2.16}$$

$$\Xi_5^0 = (11/128)^{1/2} \cdot (63\cos^5\theta - 70\cos^3\theta + 15\cos\theta) \tag{A2.17}$$

$$\Xi_5^1 = (165/256)^{1/2} \cdot \sin\theta \cdot (21\cos^4\theta - 14\cos^2\theta + 1) \tag{A2.18}$$

$$\Xi_5^2 = (1155/64)^{1/2} \cdot \sin^2\theta \cdot (3\cos^3\theta - \cos\theta) \tag{A2.19}$$

$$\Xi_5^3 = (385/512)^{1/2} \cdot \sin^3\theta \cdot (9\cos^2\theta - 1) \tag{A2.20}$$

$$\Xi_5^4 = (3465/256)^{1/2} \cdot \sin^4\theta \cdot \cos\theta \tag{A2.21}$$

$$\Xi_5^5 = (693/512)^{1/2} \cdot \sin^5 \theta \tag{A2.22}$$

$$\Xi_6^0 = (13/512)^{1/2} \cdot (231 \cos^6 \theta - 315 \cos^4 \theta + 105 \cos^2 \theta - 5) \tag{A2.23}$$

$$\Xi_6^1 = (273/256)^{1/2} \cdot \sin \theta \cdot (33 \cos^5 \theta - 30 \cos^3 \theta + 5 \cos \theta) \tag{A2.24}$$

$$\Xi_6^2 = (1365/2048)^{1/2} \cdot \sin^2 \theta \cdot (33 \cos^4 \theta - 18 \cos^2 \theta + 1) \tag{A2.25}$$

$$\Xi_6^3 = (1365/512)^{1/2} \cdot \sin^3 \theta \cdot (11 \cos^3 \theta - 3 \cos \theta) \tag{A2.26}$$

$$\Xi_6^4 = (819/1024)^{1/2} \cdot \sin^4 \theta \cdot (11 \cos^2 \theta - 1) \tag{A2.27}$$

$$\Xi_6^5 = (9009/512)^{1/2} \cdot \sin^5 \theta \cdot \cos \theta \tag{A2.28}$$

$$\Xi_6^6 = (3003/2048)^{1/2} \cdot \sin^6 \theta \tag{A2.29}$$

Appendix A3

A3.1   THE ENERGIES RESULTING FROM THE APPLICATION OF V_{trig}

The matrix elements $\langle (m)|\mathbf{V}_{trig}|(m')\rangle$ evaluate as (1):

$$\langle (0)|\mathbf{V}_{trig}|(0)\rangle = 3Cp \cdot (3\cos^2\theta - 1) + (9/7)Dq$$
$$\cdot (35\cos^4\theta - 30\cos^2\theta + 3) = a_{00} \qquad (A3.1)$$

$$\langle (\pm 1)|\mathbf{V}_{trig}|(\pm 1)\rangle = (3/2)Cp \cdot (3\cos^2\theta - 1) - (6/7)Dq$$
$$\cdot (35\cos^4\theta - 30\cos^2\theta + 3) = a_{11} \qquad (A3.2)$$

$$\langle (\pm 2)|\mathbf{V}_{trig}|(\pm 2)\rangle = -3Cp \cdot (3\cos^2\theta - 1) + (3/14)Dq$$
$$\cdot (35\cos^4\theta - 30\cos^2\theta + 3) = a_{22} \qquad (A3.3)$$

$$\langle (\mp 2)|\mathbf{V}_{trig}|(\pm 1)\rangle = \pm 15Dq \cdot \sin^3\theta \cdot \cos\theta = a_{1-2} \qquad (A3.4)$$

The energies are then $3Cp \cdot (3\cos^2\theta - 1) + (9/7)Dq \cdot (35\cos^4\theta - 30\cos^2\theta + 3)$ and the roots of the quadratic equation $E^2 - (a_{11} + a_{22})E + a_{11}a_{22} - a_{1-2}^2$, each two fold degenerate.

REFERENCE

1. Gerloch, M., and Slade, W., *Ligand Field Parameters*, Cambridge University Press, Cambridge, U.K. 1973, chap. 4.

Appendix A4

A4.1 RELATIONSHIPS BETWEEN SOME OF THE COEFFICIENTS IN THE OPERATORS DEFINED IN SECTION 2.8.1 (1)

$$A_2^0 = (2\pi)^{-1/2} \cdot (5/8)^{1/2} \cdot B_2^0 \tag{A4.1}$$

$$A_2^2 = (2\pi)^{-1/2} \cdot (15/4)^{1/2} \cdot B_2^2 \tag{A4.2}$$

$$A_4^0 = (2\pi)^{-1/2} \cdot (9/128)^{1/2} \cdot B_4^0 \tag{A4.3}$$

$$A_4^2 = (2\pi)^{-1/2} \cdot (45/16)^{1/2} \cdot B_4^2 \tag{A4.4}$$

$$A_4^3 = (2\pi)^{-1/2} \cdot (315/8)^{1/2} \cdot B_4^3 \tag{A4.5}$$

$$A_4^4 = (2\pi)^{-1/2} \cdot (315/64)^{1/2} \cdot B_4^4 \tag{A4.6}$$

$$A_6^0 = (2\pi)^{-1/2} \cdot (13/512)^{1/2} \cdot B_6^0 \tag{A4.7}$$

$$A_6^3 = (2\pi)^{-1/2} \cdot (1365/128)^{1/2} \cdot B_6^3 \tag{A4.8}$$

$$A_6^4 = (2\pi)^{-1/2} \cdot (819/256)^{1/2} \cdot B_6^4 \tag{A4.9}$$

$$A_6^6 = (2\pi)^{-1/2} \cdot (3003/512)^{1/2} \cdot B_6^6 \tag{A4.10}$$

$$Q_2^0 \equiv 3z^2 - r^2 \tag{A4.11}$$

$$Q_2^2 \equiv x^2 - y^2 \tag{A4.12}$$

$$Q_4^0 \equiv 35z^4 - 3z^2r^2 + 3r^4 \tag{A4.13}$$

$$Q_4^2 \equiv (7z^2 - r^2)(x^2 - y^2) \tag{A4.14}$$

$$Q_4^3 \equiv xz(x^2 - 3y^2) \tag{A4.15}$$

$$Q_4^4 \equiv x^4 - 6x^2y^2 + y^4 \tag{A4.16}$$

$$Q_6^0 \equiv 231z^6 - 315z^4r^2 + 105z^2r^4 - 5r^6 \tag{A4.17}$$

$$Q_6^3 \equiv (11z^2 - 3r^2)(x^2 - 3y^2)xz \tag{A4.18}$$

$$Q_6^4 \equiv (11z^2 - r^2)(x^4 - 6x^2y^2 + y^4) \tag{A4.19}$$

$$Q_6^6 \equiv x^6 - 15x^4y^2 + 15x^2y^4 - y^6 \tag{A4.20}$$

REFERENCE

1. Abragam, A., and Bleaney, B., *Electron Paramagnetic Resonance of Transition Ions*. Clarendon Press, Oxford, UK, 1970, chap. 16. Appendices Table 15.

Appendix A5

A5.1 MATRIX ELEMENTS OF THE CRYSTAL FIELD POTENTIAL V_{cf} FROM A GENERAL DISTRIBUTION OF EFFECTIVE POINT CHARGES

The charges are q_i at distances a_i, the real d orbitals are employed, and at least a twofold axis of symmetry is assumed (1,2). Put:

$$a_i^n = q_i r_2^n / a_i^{n+1} \tag{A5.1}$$

$$D^{lm} = \sum_{i=1}^{n} D_i^{lm} \tag{A5.2}$$

$$G^{lm} = \sum_{i=1}^{n} G_i^{lm} \tag{A5.3}$$

We use the abbreviation, for example,

$$H_{xy,xz} = \langle d_{xy} | V_{cf} | d_{xz} \rangle \tag{A5.4}$$

$$H_{z^2,z^2} = -D^{20}/7 + 3D^{40}/28 \tag{A5.5}$$

$$H_{x^2-y^2,x^2-y^2} = -D^{20}/7 + D^{40}/56 + 5D^{44}/24 \tag{A5.6}$$

$$H_{xy,xy} = -D^{20}/7 + D^{40}/56 - 5D^{44}/24 \tag{A5.7}$$

$$H_{xz,xz} = D^{20}/14 - D^{40}/14 + 3D^{22}/14 + 5D^{42}/42 \tag{A5.8}$$

$$H_{yz,yz} = D^{20}/14 - D^{40}/14 - 3D^{22}/14 - 5D^{42}/42 \tag{A5.9}$$

$$H_{z^2,x^2-y^2} = -3^{1/2}D^{22}/7 + 75^{1/2}D^{42}/84 \tag{A5.10}$$

$$H_{z^2,xy} = -3^{1/2}G^{22}/7 + 75^{1/2}G^{42}/84 \tag{A5.11}$$

$$H_{z^2,xz} = 3^{1/2}D^{21}/7 + 75^{1/2}D^{41}/14 \tag{A5.12}$$

$$H_{z^2,yz} = 3^{1/2}G^{21}/7 + 75^{1/2}G^{41}/14 \tag{A5.13}$$

$$H_{x^2-y^2,xy} = 5G^{44}/24 \tag{A5.14}$$

$$H_{x^2-y^2,xz} = 3D^{21}/7 - 5D^{41}/28 + 5D^{43}/12 \tag{A5.15}$$

$$H_{x^2-y^2,yz} = -3G^{21}/7 + 5G^{41}/28 + 5G^{43}/12 \tag{A5.16}$$

$$H_{xy,xz} = 3G^{21}/7 - 5G^{41}/28 + 5G^{43}/12 \tag{A5.17}$$

$$H_{xy,yz} = 3D^{21}/7 - 5D^{41}/28 - 5D^{43}/12 \tag{A5.18}$$

$$H_{xz,yz} = 3G^{22}/14 + 5G^{42}/42 \tag{A5.19}$$

Here:

$$D_i^{20} = (3\cos^2\theta_i - 1)a_i^2 \tag{A5.20}$$

$$D_i^{21} = \sin\theta_i \cdot \cos\theta_i \cdot \cos\phi_i \cdot a_i^2 \tag{A5.21}$$

$$D_i^{22} = \sin^2\theta_i \cdot \cos 2\phi_i \cdot a_i^2 \tag{A5.22}$$

$$D_i^{40} = (35\cos^4\theta_i - 30\cos^2\theta_i + 3) \cdot a_i^4/3 \tag{A5.23}$$

$$D_i^{41} = \sin\theta_i \cdot \cos\theta_i(7\cos^2\theta_i - 3) \cdot \cos\phi_i \cdot a_i^4/3 \tag{A5.24}$$

$$D_i^{42} = \sin^2\theta_i \cdot (7\cos^2\theta_i - 1) \cdot \cos 2\phi_i \cdot a_i^4 \tag{A5.25}$$

$$D_i^{43} = \sin^3\theta_i \cdot \cos\theta_i \cdot \cos 3\phi_i \cdot a_i^4 \tag{A5.26}$$

$$D_i^{44} = \sin^4\theta_i \cdot \cos 4\phi_i \cdot a_i^4 \tag{A5.27}$$

$$G_i^{21} = \sin\theta_i \cdot \cos\theta_i \cdot \sin 2\phi_i \cdot a_i^2 \tag{A5.28}$$

$$G_i^{22} = \sin^2\theta_i \cdot \sin 2\phi_i \cdot a_i^2 \tag{A5.29}$$

$$G_i^{41} = \sin\theta_i \cdot \cos\theta_i \cdot (7\cos^2\theta_i - 3) \cdot \sin\phi_i \cdot a_i^4/3 \tag{A5.30}$$

$$G_i^{42} = \sin^2\theta_i \cdot (7\cos^2\theta_i - 1) \cdot \sin 2\phi_i \cdot a_i^4 \tag{A5.31}$$

$$G_i^{43} = \sin^3\theta_i \cdot \cos\theta_i \cdot \sin 3\phi_i \cdot a_i^4 \tag{A5.32}$$

$$G_i^{44} = \sin^4\theta_i \cdot a_i^4 \tag{A5.33}$$

The energies of the d orbitals for the system are then obtained by diagonalizing the real symmetrical matrices H_{ij}, $i,j = d_{z^2} \cdots d_{yz}$. The corresponding linear combinations of the d orbitals, the eigenvectors, are produced in the diagonalization process.

REFERENCES

1. Companian, A. L., and Komarynsky, M. A., *J. Chem. Educ.*, 1964, **41**, 257.
2. Lever, A. B. P., *Inorganic Electronic Spectroscopy*, Elsevier, Amsterdam, The Netherlands, 1984.

Appendix A6

A6.1 ENERGIES OF THE TERMS OF d^n USING CONDON–SHORTLEY PARAMETERS

The contribution in F_0 is omitted. The results for d^{6-8} are equivalent, respectively, to those for d^{4-2} through the filled-shell-hole formalism (1,2) numbers.

Configuration	Term	Energy
d^2	^1S	$14F_2 + 126F_4$
	^1G	$4F_2 + F_4$
	^3P	$7F_2 - 84F_4$
	^1D	$-3F_2 + 36F_4$
	^3F	$-8F_2 - 9F_4$
d^3	^2P	$-6F_2 - 12F_4$
	$^2D_\pm$	$5F_2 + 3F_4 \pm (193F_2^2 - 1650F_2F_4 + 8325F_4^2)^{1/2}$
	^2F	$9F_2 - 87F_4$
	^2G	$-11F_2 + 13F_4$
	^2H	$-6F_2 - 12F_4$
	^4P	$-147F_4$
	^4F	$-15F_2 - 72F_4$
d^4	$^1S_\pm$	$10F_2 + 6F_4 \pm (3088F_2^2 - 26400F_2F_4 + 133200F_4^2)^{1/2}/2$
	$^1D_\pm$	$9F_2 - 153F_4/2 \pm (1296F_2^2 - 10440F_2F_4 + 30825F_4^2)^{1/2}/2$
	^1F	$84F_4$
	$^1G_\pm$	$-5F_2 - 13F_4/2 \pm (708F_2^2 - 7500F_2F_4 + 30825F_4^2)^{1/2}/2$
	^1I	$-15F_2 - 9F_4$
	$^3P_\pm$	$-5F_2 - 153F_4/2 \pm (912F_2^2 - 9960F_2F_4 + 38025F_4^2)^{1/2}/2$
	^3D	$-5F_2 - 129F_4$
	$^3F_\pm$	$-5F_2 - 153F_4/2 \pm (612F_2^2 - 4860F_2F_4 + 20025F_4^2)^{1/2}/2$
	^3G	$-12F_2 - 94F_4$
	^3H	$-17F_2 - 69F_4$
	^5D	$-21F_2 - 189F_4$

Configuration	Term	Energy
d^5	^2S	$-3F_2 - 195F_4$
	^2P	$20F_2 - 240F_4$
	$^2D_{\pm}$	$-3F_2 - 90F_4 \pm (513F_2^2 - 4500F_2F_4$ $+ 20700F_4^2)^{1/2}$
	^2D	$-4F_2 - 120F_4$
	^2F_a	$-9F_2 - 165F_4$
	^2F_b	$-25F_2 - 15F_4$
	^2G_a	$-3F_2 - 155F_4$
	^2G_b	$-13F_2 - 145F_4$
	^2H	$-22F_2 - 30F_4$
	^2I	$-24F_2 - 90F_4$
	^4P	$-28F_2 - 105F_4$
	^4D	$-18F_2 - 225F_4$
	^4F	$-13F_2 - 180F_4$
	^4G	$-25F_2 - 190F_4$
	^6S	$-35F_2 - 315F_4$

REFERENCES

1. Ballhausen, C. J, *Introduction to Ligand Field Theory*, McGraw-Hill, New York, 1962, chap. 2.
2. Condon, E. U., and Shortley, G. H., *The Theory of Atomic Spectra*, Cambridge University Press, Cambridge, U.K. 1957.

Appendix A7

A7.1 THE CURIE LAW FOR MAGNETIC BEHAVIOR

Consider a system of magnetic dipoles of magnitude μ distributed as n per unit volume and quantized to point either along or against an applied magnetic field H. The energy of such a moment, relative to the H $= 0$ condition, is, respectively, $\pm \mu$H. The energy that must be supplied to a set of the dipoles randomly oriented when H $= 0$ is, respectively, $\pm \mu H^2/2$. The relative population of the moments in the two orientation at a finite temperature is, from Boltzmann statistics:

$$
\begin{aligned}
(n_{\parallel}/n_{\perp}) &= \exp\left(\mu H/kT\right)/\exp\left(-\mu H/kT\right) \\
&= \exp\left(2\mu H/kT\right) \\
&\sim 1 + 2\mu H/kT \qquad \text{for } \mu H \ll kT
\end{aligned}
\tag{A7.1}
$$

The *excess* of moments aligned parallel:

$$
n_{\parallel} - n_{\perp} = n\mu H/kT \tag{A7.2}
$$

is the net magnetic moment per unit volume generated by the field, viz the intensity of magnetization I. Dividing by H, we have from Section 9.1:

$$
\chi_m = n\mu/\rho kT = C/T \tag{A7.3}
$$

which is the Curie law.

Alternatively, the change of the energy of the system, relative to H $= 0$, owing to the excess of moment aligned parallel to the field, is, per unit mass:

$$
W = -n\mu H^2/2\rho kT \tag{A7.4}
$$

Thus, again from Section 9.1, we repeat the result:

$$
-(1/H)\partial W/\partial H = n\mu/\rho kT \tag{A7.5}
$$

$$
\chi_m = C/T \tag{A7.6}
$$

Appendix A8

A8.1 THE OPERATORS L_x, L_y, S_x, AND S_y

The operators for angular momentum in directions other than z are required. They are most conveniently expressed using the "ladder operators" $\mathbf{L}_+$ and $\mathbf{L}_-$:

$$\mathbf{L}_+ = \mathbf{L}_x + i\mathbf{L}_y \tag{A8.1}$$

$$\mathbf{L}_- = \mathbf{L}_x - i\mathbf{L}_y \tag{A8.2}$$

$$\mathbf{S}_+ = \mathbf{S}_x + i\mathbf{S}_y \tag{A8.3}$$

$$\mathbf{S}_- = \mathbf{S}_x - i\mathbf{S}_y \tag{A8.4}$$

so that:

$$\mathbf{L}_x = \tfrac{1}{2}(\mathbf{L}_+ + \mathbf{L}_-) \tag{A8.5}$$

$$\mathbf{L}_y = \tfrac{-i}{2}(\mathbf{L}_+ - \mathbf{L}_-) \tag{A8.6}$$

$$\mathbf{S}_x = \tfrac{1}{2}(\mathbf{S}_+ + \mathbf{S}_-) \tag{A8.7}$$

$$\mathbf{S}_y = \tfrac{-i}{2}(\mathbf{S}_+ - \mathbf{S}_-) \tag{A8.8}$$

Equivalent expressions also hold for $\mathbf{l}_x$, $\mathbf{l}_y$, $\mathbf{s}_x$, and $\mathbf{s}_y$.

These ladder operators have the properties:

$$\langle\langle L, M_L + 1 | \mathbf{L}_+ | \langle\langle L, M_L \rangle = [(L - M_L)(L + M_L + 1)]^{1/2} \tag{A8.9}$$

$$\langle\langle L, M_L - 1 | \mathbf{L}_- | \langle\langle L, M_L \rangle = [(L + M_L)(L + M_L - 1)]^{1/2} \tag{A8.10}$$

$$\langle\langle S, M_S + 1 | \mathbf{S}_+ | \langle\langle S, M_S \rangle = [(S - M_S)(S + M_S + 1)]^{1/2} \tag{A8.11}$$

$$\langle\langle S, M_S - 1 | \mathbf{S}_- | \langle\langle S, M_S \rangle = [(S + M_S)(S + M_S - 1)]^{1/2} \tag{A8.12}$$

For example, for a ^3P term, in the notation of Section 5.4:

$$\langle\{1,1\}|\mathbf{L}_x|\{1,0\}\rangle = \tfrac{1}{2}(\langle\{1,1\}|\mathbf{L}_+ + \mathbf{L}_-|\{1,0\}\rangle$$
$$= \tfrac{1}{2}((1-0)(1+0+1))^{\frac{1}{2}}$$
$$= 2^{-\frac{1}{2}} \tag{A8.13}$$

$$\langle[1,0]|\mathbf{S}_y|[1,1]\rangle = -\tfrac{1}{2}(\langle[1,0]|\mathbf{S}_+ - \mathbf{S}_-|\{1,1\}\rangle$$
$$= \tfrac{1}{2}((1+1)(1+1-1))^{\frac{1}{2}}$$
$$= 2^{-\frac{1}{2}} \tag{A8.14}$$

Results relevant to the spin–orbit coupling operator are readily obtained by combining some of those above:

$$\mathbf{L} \cdot \mathbf{S} = \mathbf{L}_x \mathbf{S}_x + \mathbf{L}_y \mathbf{S}_y + \mathbf{L}_z \mathbf{S}_z \qquad (A8.15)$$

The component $\mathbf{L}_z \mathbf{S}_z$ is readily implemented because wavefunctions are referred to the z direction by definition. The components for the x and y directions can be rewritten:

$$
\begin{aligned}
\mathbf{L}_x \mathbf{S}_x + \mathbf{L}_y \mathbf{S}_y &= \frac{1}{4}[(\mathbf{L}_+ + \mathbf{L}_-)(\mathbf{S}_+ + \mathbf{S}_-) - (\mathbf{L}_+ - \mathbf{L}_-)(\mathbf{S}_+ - \mathbf{S}_-)] \\
&= \frac{1}{2}(\mathbf{L}_+ \mathbf{S}_- + \mathbf{L}_- \mathbf{S}_+)
\end{aligned} \qquad (A8.16)
$$

The matrix elements of $\mathbf{L}_x \mathbf{S}_x + \mathbf{L}_y \mathbf{S}_y$ take the form:

$$
\begin{aligned}
&\langle\langle L, M_L \pm 1\rangle[S, M_S \mp 1]|\mathbf{L}_x\mathbf{S}_x + \mathbf{L}_y\mathbf{S}_y|\langle L, M_L\rangle[S, M_S]\rangle \\
&= \frac{1}{2}[(L \mp M_L)(L \pm M_L + 1)(S \pm M_S)(S \mp M_S + 1)]^{1/2} \qquad (A8.17)
\end{aligned}
$$

Appendix A9

A9.1 EXPRESSIONS FOR THE MAGNETIC MOMENTS OF $^4T_{1(g)}$, $^2T_{2(g)}$, AND $^5T_{2(g)}$ TERMS IN CUBIC SYMMETRY

$x = \lambda/k\mathrm{T}$.

$^4T_{2(g)}$:

$$
\begin{aligned}
\mu_{\text{eff}} = 3\{ & 7(3-A)^2/5 + 1 + 12(2+A)^2/25\,Ax + [2(11-2A)^2/45 \\
& + 176(2+A)^2/675\,Ax]\exp\left(-5\,Ax/2\right) + [(5+A)^2/9 \\
& - 20(2+A)^2/27\,Ax]\exp\left(4\,Ax\right)/[3 + 2\exp\left(-5\,Ax/2\right) \\
& + \exp\left(-4\,Ax\right)]
\end{aligned}
\tag{A9.1}
$$

$^2T_{2(g)}$:

$$
\mu_{\text{eff}} = [8 + (3x-8)\exp\left(-3x/2\right)]/[x(2 + \exp\left(-3x/2\right))]
\tag{A9.2}
$$

$^5T_{2(g)}$:

$$
\begin{aligned}
\mu_{\text{eff}} = 3[& 28 + 28/3x + (45/2 + 25/6x)\exp\left(-3x\right) \\
& + (49/2 - 27/2x)\exp(-5x)]/[7 + 5\exp\left(-3x\right) + 3\exp(-5x)]
\end{aligned}
\tag{A9.3}
$$

LIST OF COMMONLY USED SYMBOLS AND WHERE THEY FIRST OCCUR

Symbol	Meaning
a	Distance of a point charge from the origin / 30
A	Coefficient of the **Q** in expansion of crystal field potential in operator form / 46
	Absorbance of radiation / 179
	Hyperfine tensor components / 292
A	Parameter defining contribution of orbital angular momentum / 251
A	Hyperfine tensor / 292
b	Distance of a point charge from the origin / 41
B	Coefficient of **O** in expansion of crystal field potential in operator form / 46
	Racah parameter for interelectronic repulsions / 102
B	Magnetic field (induction) strength / 228
c	Mixing coefficient / 66
C	Parameter for expansion of axial crystal field potential in Cartesian coordinates / 43
	Racah parameter for interelectronic repulsions / 102
	Normalization coefficient / 96
	Molar concentration / 179
	Vibronic coupling coefficient / 186
	Curie constant in Curie law / 232
d	Degeneracy factor / 180
D	Parameter for expansion of octahedral crystal field potential in Cartesian coordinates / 21
	Zero-field splitting parameter / 295
D	Zero-field splitting tensor / 295
f	Force constant for a bond / 146
	Oscillator strength / 180
	Parameter for ligand in empirical estimation of Δ / 218
	Fractional orbital occupancy / 305
F	Parameter in expansion of octahedral crystal field potential for f electrons / 50
	Matrix element of magnetic moment operator between states of a term / 239
F	Condon–Shortley parameter for interelectronic repulsion / 102
g	Spectroscopic splitting factor / 237
g	Parameter for ligand in empirical derivation of Δ / 218
g	Spectroscopic splitting tensor / 237
G	degeneracy Factor / 180
h	Parameter for ligand in empirical definition of nephelauxetic ratio / 219
H	Energy corresponding to Hamiltonian operator between wavefunctions / 28
	Magnetic field strength / 228

H	Bond Enthalpy / 168
$\mathbf{H}$	Hamiltonian operator / 5
i	Square root of minus 1 / 7
I	Intensity of radiation / 179
	Nuclear spin quantum number / 290
I	Intensity of magnetization / 228
$\mathbf{I}$	Nuclear spin operator / 295
j	Total angular momentum of an electron / 107
J	Total angular momentum of a term / 107
	Magnetic exchange integral / 272
k	Parameter for metal ion in empirical definition of nephelauxetic ratio / 219
k	Orbital angular momentum reduction factor / 256
K	Isotropic hyperfine spin density parameter / 303
l	Electron orbital angular momentum quantum number / 6
	Path length for radiation / 179
$\mathbf{l}$	Operator for electron orbital angular momentum / 11
L	Term orbital angular momentum quantum number / 95
$\mathbf{L}$	Operator for term orbital angular momentum / 97
m	Electron quantum number for z direction of orbital angular momentum / 6
M	Quantum number for z direction of orbital angular momentum for a set of electrons / 95
n	Electron principal quantum number / 6
$\mathbf{O}$	Operator in expansion of crystal field potential / 46
p	Coefficient of C in evaluation of axial crystal field potential / 43
P	Electron–nuclear spin interaction parameter / 301
q	Coefficient of D in evaluation of octahedral crystal-field potential / 21
Q	Normal coordinate in a complex / 146
	Transition moment / 180
$\mathbf{Q}$	Operator in expansion of crystal field potential / 46
r	Distance from the origin / 1
	Coefficient of F in evaluation of octahedral crystal field potential for f-electrons / 51
$\mathbf{r}$	Electric dipole operator / 180
R	Radial wavefunction / 6
s	Coefficient of D in parametrization of axial crystal field potential / 47
	Electron spin angular momentum quantum number / 98
$\mathbf{s}$	Spin angular momentum operator / 97
S	Overlap integral / 58
	Term spin angular momentum quantum number / 95
S	A contribution from interelectronic repulsions / 146

t	Coefficient of D in paramarization of axial crystal field potential / 47
T	Transmittance of radiation / 179
$\mathbf{v}$	Electrostatic potential from a point u energy. / 1
	Ratio of splitting of T term to 1 / 257
v	Vibrational quantum number / 197
$\mathbf{V}$	Crystal field potential / 27
V	Electronic coupling constant / 146
x	Ratio of temperature to spin orbit coupling constant / 252
y	Ratio of energy separation to kT / 269
Y	Spherical harmonic / 7
z	Charge on an ion / 27
α	Angle defining molecular geometry / 68
	Constant in temperature independent paramagnetism / 235
	Unpaired spin population / 301
	Symbol for an operator / 97
β	Stability constant / 171
	Nephelauxetic ratio / 219
δ	Axial crystal field splitting component / 257
Δ	Octahedral crystal field splitting / 22
ε	Molar extinction coefficient / 179
ζ	Electron spin–orbit coupling constant / 108
θ	Azimuthal angle / 6
	Crystal Field splitting for F electrons / 51
	Weiss constant in Curie–Weiss law / 233
λ	Term spin–orbit coupling constant / 108
	Site occupation ratio in spinels / 175
μ	Dipole moment / 39
	Magnetic moment / 232
μ	Magnetic moment operator / 32
ν	Energy in cm^{-1} / 180
ϕ	Basal angle in coordinate system / 6
	Wavefunction / 58
ρ	Density / 229
χ	Lowest unoccupied molecular orbital / 73
	Magnetic susceptibility / 229
ψ	Wavefunction / 9
	Highest occupied molecular orbital / 73
∇	The div-grad operator / 6

FUNDAMENTAL CONSTANTS

c	velocity of light	$2.9979248 \times 10^8 \, m \, s^{-1}$
e	electron charge	$1.6021773 \times 10^{-20} \, C$

m_e	electron mass	$9.109390 \times 10^{-31}\,\mathrm{kg}$
m_p	proton mass	$1.672623 \times 10^{-27}\,\mathrm{kg}$
h	Planck constant	$6.626076 \times 10^{-34}\,\mathrm{J\,s}$
k	Boltzmann constant	$1.38066 \times 10^{-23}\,\mathrm{J\,K^{-1}}$
a_0	Bohr radius	$5.2917725 \times 10^{-11}\,\mathrm{m}$
Å	Angstrom unit	$10^{-10}\,\mathrm{m}$
cm^{-1}	Wave number	$1.9864476 \times 10^{-23}\,\mathrm{J}$
		$0.0119266\,\mathrm{kJ\,mol^{-1}}$
eV	electron volt	$1.6021773 \times 10^{-19}\,\mathrm{J}$
		$8065.5402\,\mathrm{cm^{-1}}$
		$96.48531\,\mathrm{kJ\,mol^{-1}}$
R_∞	Rydberg constant	$2.269874 \times 10^{-18}\,\mathrm{J}$
		$109737.3153\,\mathrm{cm^{-1}}$
		$13.605698\,\mathrm{eV}$
N_A	Avagadro's number	6.022137×10^{23}
μ_0	Vaccum magnetic permeability	$4\pi \times 10^{-7}$
ε_0	Vaccum electric permittivity	$8.85418782\,\mathrm{F\,m^{-1}}$
F	Faraday	$96485.309\,\mathrm{C\,mol^{-1}}$
μ_B	Bohr magneton	$9.274015 \times 10^{-24}\,\mathrm{J\,T^{-1}}$
		$0.466864\,\mathrm{cm^{-1}\,T^{-1}}$
μ_N	Nuclear magneton	$5.050787 \times 10^{-27}\,\mathrm{J\,T^{-1}}$
		$2.542623 \times 10^{-4}\,\mathrm{cm^{-1}\,T^{-1}}$
g_e	electron gyromagnetic ratio	2.00231930439
$N_A\mu_B^2$	SI units: cm^{-1}	$3.27657 \times 10^{-5}\,\mathrm{m^3\,mol^{-1}\,cm^{-1}}$
	cgs units: cm^{-1}	$0.260741\,\mathrm{cm^3\,mol^{-1}\,cm^{-1}}$
$3k/N_A\mu_B^2$	SI units	$6.36357 \times 10^{5}\,\mathrm{m^{-3}\,mol^{-1}\,K^{-1}}$
	cgs units	$7.9967\,\mathrm{cm^{-3}\,mol^{-1}\,K^{-1}}$

INDEX

345